KNOTT'S

HANDBOOK FOR VEGETABLE GROWERS

FIFTH EDITION

KNOTT'S

HANDBOOK FOR VEGETABLE GROWERS

FIFTH EDITION

DONALD N. MAYNARD

University of Florida
Wimauma, Florida

GEORGE J. HOCHMUTH

University of Florida
Gainesville, Florida

JOHN WILEY & SONS, INC.

This book is printed on acid-free paper. ♾

Published by John Wiley & Sons, Inc., Hoboken, New Jersey
Published simultaneously in Canada

For general information on our other products and services or for technical support, please contact our Customer Care Department within the United States at (800) 762-2974, outside the United States at (317) 572-3993 or fax (317) 572-4002.

Wiley also publishes its books in a variety of electronic formats. Some content that appears in print may not be available in electronic books. For more information about Wiley products, visit our web site at www.wiley.com.

Library of Congress Cataloging-in-Publication Data:
Maynard, Donald N., 1932–
Knott's handbook for vegetable growers/Donald N. Maynard. George J. Hochmuth.—5th ed.
p. cm.
Includes bibliographical references.
ISBN-13: 978-0471-73828-2
ISBN-10: 0-471-73828-X
1. Truck farming—Handbooks, manuals, etc. 2. Vegetables—Handbooks, manuals, etc. 3. Vegetable gardening—Handbooks, manuals, etc. I. Title: Handbook for vegetable growers. II. Hochmuth, George J. (George Joseph) III. Knott, James Edward, 1897– Handbook for vegetable growers. IV. Title.
SB321.M392 2006
635—dc22

2006000893

Printed in the United States of America

10 9 8 7 6 5 4 3 2

CONTENTS

PREFACE

The pace of change in our personal and business lives continues to accelerate at an ever increasing rate. Accordingly, it is necessary to periodically update information in a long-running reference such as *Handbook for Vegetable Growers.* Our goal in this revision is to provide up-to-date information on vegetable crops for growers, students, extension personnel, crop consultants, and all those concerned with commercial production and marketing of vegetables.

Where possible, information in the Fourth Edition has been updated or replaced with current information. New technical information has been added on World Vegetable Production, Best Management Practices, Organic Crop Production, Food Safety, Pesticide Safety, Postharvest Diseases, and Minimally Processed Vegetables. The Internet has become a valuable source of information since 1997. Hundreds of websites relating to vegetables are included in this edition and are available online at www.wiley.com/college/Knotts.

We are grateful to our colleagues who have provided materials, reviewed portions of the manuscript, and encouraged us in this revision. We especially acknowledge the assistance of Brian Benson, California Asparagus Seed and Transplants, Inc.; George Boyhan, University of Georgia;

Wallace Chasson, Florida Department of Agriculture and Consumer Services; Steve Grattan, University of California; Tim Hartz, University of California; Richard Hassell, Clemson University; Larry Hollar, Hollar and Company; Adel Kader, University of California; Tom Moore, Harris-Moran Seed Co.; Stu Pettygrove, University of California; Steven Sargent, University of Florida; Pieter Vandenberg, Seminis Vegetable Seeds; and Jim Watkins, Nunhems USA.

We appreciate the outstanding assistance provided by Wiley editor Jim Harper, Senior Production Editor Millie Torres, and the attention to details and good humor in the preparation of this manuscript by Gail Maynard.

We hope that *Handbook for Vegetable Growers* will continue to be the timely and useful reference for those with interest in vegetable crops envisioned by Dr. J. E. Knott when it was first published in 1956. James E. Knott (1897–1977) was a Massachusetts native. He earned a B.S. degree at Rhode Island State College and an M.S. and Ph.D. at Cornell University. After distinguished faculty and administrative service at Pennsylvania State College and Cornell University he moved to the University of California, Davis, where he was administrator of the Vegetable Crops Department from 1940 to 1964. The department grew in numbers and stature to be one of the world's best vegetable centers. Dr. Knott was president of the American Society for Horticultural Science in 1948 and was made a Fellow in 1965.

Oscar A. Lorenz (1914–1994), senior author of the Second Edition (1980) and the Third Edition (1988) of *Handbook for Vegetable Growers,* was a native of Colorado. He earned a B.S. degree from Colorado State College and a Ph.D. from Cornell University before joining the University of California, Davis faculty in 1941. For the next 41 years he was an esteemed scientist and administrator at both the Riverside and Davis campuses. His research on vegetable crops nutrition was the first to establish the relationship between soil fertility, leaf nutrient composition, and yield. This concept has been used successfully by growers throughout the world. Oscar was recognized as a Fellow of the American Society for Horticultural Science and of the American Society of Agronomy and Soil Science Society of America, and received numerous industry awards. He was a friend to all and a personal mentor to me. (DNM)

DONALD N. MAYNARD
GEORGE J. HOCHMUTH

PART 1

VEGETABLES AND THE VEGETABLE INDUSTRY

01
BOTANICAL NAMES OF VEGETABLES

TABLE 1.1. BOTANICAL NAMES, COMMON NAMES, AND EDIBLE PARTS OF PLANTS USED AS VEGETABLES

Botanical Name	Common Name	Edible Plant Part
Division Sphendophyta		
Equisetaceae	HORSETAIL FAMILY	
Equisetum arvense L.	Horsetail	Young strobili
Division Pterophyta	FERN GROUP	
Dennstaedtiaceae		
Pteridium aquilinum (L.) Kuhn.	Bracken fern	Immature frond
Osmundaceae		
Osmunda cinnamomea L.	Cinnamon fern	Immature frond
Osmunda japonica Th.	Japanese flowering fern	Immature frond
Parkeriaceae		
Ceratopteris thalictroides (L.) Brongn.	Water fern	Young leaf
Polypodiaceae		
Diplazium esculentum (Retz.) Swartz.	Vegetable fern	Young leaf
Division Anthophyta		
Class Monocotyledons		
Alismataceae	WATER PLANTAIN FAMILY	
Sagittaria sagittifolia L.	Arrowhead	Corm
Sagittaria trifolia L. (Sieb.) Ohwi	Chinese arrowhead	Corm

Alliaceae	ONION FAMILY	
Allium ampeloprasum L. Ampeloprasum group	Great-headed garlic	Bulb and leaf
Allium ampeloprasum L. Kurrat group	Kurrat	Pseudostem
Allium ampeloprasum L. Porrum group	Leek	Pseudostem and leaf
Allium cepa L. Aggregatum group	Shallot	Pseudostem and leaf
Allium cepa L. Cepa group	Onion	Bulb
Allium cepa L. Proliferum group	Tree onion, Egyptian onion	Aerial bulb
Allium chinense G. Don.	Rakkyo	Bulb
Allium fistulosum L.	Welsh onion, Japanese bunching onion	Pseudostem and leaf
Allium grayi Regel	Japanese garlic	Leaf
Allium sativum L.	Garlic	Bulb and leaf
Allium schoenoprasum L.	Chive	Leaf
Allium scorodoprasum L.	Sand leek, giant garlic	Leaf and bulb
Allium tuberosum Rottler ex Sprengel	Chinese chive	Leaf, immature flower
Allium victorialis L. var. *platyphyllum*, Hult.	Longroot onion	Bulb, leaf
Allium × *wakegi* Araki	Turfed stone leek	Leaf
Araceae	ARUM FAMILY	
Alocasia macrorrhiza (L.) Schott	Giant taro, alocasia	Corm, immature leaf, petiole
Amorphophallus paeoniifolius (Dennst.) Nicolson	Elephant yam	Corm
Colocasia esculenta (L.) Schott	Taro, dasheen, cocoyam	Corm, immature leaf
Cyrtosperma chamissonis (Schott) Merr.	Giant swamp taro	Corm
Cyrtosperma merkusii (Hassk.) Schott.	Gallan	Corm
Xanthosoma brasiliense (Desf.) Engler	Tannier spinach, catalou	Immature leaf
Xanthosoma sagittifolium (L.) Schott	Tannia, yellow yautia	Corm and young leaf
Cannaceae	CANNA FAMILY	
Canna indica L.	Indian canna, arrowroot, edible canna	Rhizome

TABLE 1.1. BOTANICAL NAMES, COMMON NAMES, AND EDIBLE PARTS OF PLANTS USED AS VEGETABLES (*Continued*)

Botanical Name	Common Name	Edible Plant Part
Cyperaceae	SEDGE FAMILY	
Cyperus esculentus L.	Rushnut, chufa	Tuber
Eleocharis dulcis (Burm.f.) Trin. ex Henschel	Water chestnut, Chinese water chestnut	Corm
Eleocharis kuroguwai Ohwi	Wild water chestnut	Corm
Dioscoreaceae	YAM FAMILY	
Dioscorea alata L.	White yam, water yam	Tuber
Dioscorea batatas Decue.	Chinese yam	Tuber
Dioscorea bulbifera L.	Potato yam, aerial yam	Tuber
Dioscorea cayenensis Lam.	Yellow yam	Tuber
Dioscorea dumetorum (Kunth) Pax.	Bitter yam	Tuber
Dioscorea esculenta (Lour.) Burk.	Lesser yam	Tuber
Dioscorea rotundata Poir.	White Guinea yam	Tuber
Dioscorea trifida L. f.	Indian yam	Tuber
Iridaceae	IRIS FAMILY	
Tigridia pavonia Ker.-Gawl.	Common tiger flower	Bulb
Liliaceae	LILY FAMILY	
Asparagus acutifolius L.	Wild asparagus	Shoot
Asparagus officinalis L.	Asparagus	Shoot
Hemerocallis spp.	Daylily	Flower
Leopoldia comosa (L.) Parl.	Tuffed hyacinth	Bulb
Lilium spp.	Lily	Bulb

Limnocharitaceae	FLOWERING RUSH FAMILY	
Limnocharis flava (L.) Buchenau	Yellow velvetleaf	Young leaf, petiole, and floral shoot
Marantaceae	ARROWROOT FAMILY	
Calathea allouia (Aubl.) Lindl.	Sweet corn root	Tuber
Maranta arundinacea L.	West Indian arrowroot	Rhizome
Musaceae	BANANA FAMILY	
Musa × *paradisiaca* L. var. *paradisiaca*	Plaintain	Fruit, flower bud
Poaceae	GRASS FAMILY	
Bambusa spp.	Bamboo shoots	Young shoot
Dendrocalamus latiflorus Munro	Bamboo shoots	Young shoot
Pennisetum purpureum Schum.	Elephant grass, napier grass	Young spear
Phyllostachys spp.	Bamboo shoots	Young shoot
Saccharum edule Hassk.	Sugarcane inflorescence	Immature inflorescence
Setaria palmifolia Stapf.	Palm grass	Young plant
Zea mays L. subsp. *mays*	Sweet corn	Immature kernels and immature cob with kernel
Zizania latifolia (Griseb.) Turcz. ex Stapf.	Water bamboo, cobo	Swollen shoot/stem
Pontederiaceae	PICKERELWEED FAMILY	
Monochoria hastata (L.) Solms.	Hastate-leaved pondweed	Young leaf
Monochoria vaginalis (Brum.) Kunth	Oval-leaved pondweed	Young leaf
Taccaceae	TACCA FAMILY	
Tacca leontopetaloides (L.) Kuntze	East Indian arrowroot	Rhizome
Zingiberaceae	GINGER FAMILY	

TABLE 1.1. BOTANICAL NAMES, COMMON NAMES, AND EDIBLE PARTS OF PLANTS USED AS VEGETABLES (*Continued*)

Botanical Name	Common Name	Edible Plant Part
Alpinia galanga (L.) Sw.	Greater galangal	Floral sprout and flower, tender shoot, rhizome
Curcuma longa L.	Turmeric	Rhizome
Curcuma zedoaria (Christm.) Roscoe	Long zedoary	Rhizome
Zingiber mioga (Thunb.) Roscoe	Japanese wild ginger	Rhizome, tender shoot, leaf, flower
Zingiber officinale Roscoe	Ginger	Rhizome and tender shoot
Division Anthophyta		
Class Diocotyledons		
Acanthaceae	ACANTHUS FAMILY	
Justicia insularis T. And.	Tettu	Young shoot, leaf, root
Rungia klossii S. Moore	Rungia	Leaf
Aizoaceae	CARPETWEED FAMILY	
Mesembryanthemum crystallinum L.	Ice plant	Leaf
Tetragonia tetragoniodes (Pall.) O. Kuntze	New Zealand spinach	Tender shoot and leaf
Amaranthaceae	AMARANTH FAMILY	
Alternanthera philoxeroides (Martius) Griseb.	Alligator weed, Joseph's coat	Young top
Alternanthera sessilis (L.) R. Br.	Sessile alternanthera	Young top

Amaranthus spp.	Amaranthus, tampala	Tender shoot, leaf, sprouted seed
Celosia spp.	Cockscomb	Leaf and tender shoot
Apiaceae	CARROT FAMILY	
Angelica archangelica L.	Angelica	Tender shoot and leaf
Angelica keiskei (Miq.) Koidz.	Japanese angelica	Tender shoot and leaf
Anthriscus cerefolium (L.) Hoffm.	Chervil	Leaf
Apium graveolens L. var. *dulce* (Mill.) Pers.	Celery	Petiole, leaf
Apium graveolens L. var. *rapaceum* (Mill.) Gaud.	Celeriac, turnip-rooted celery	Root, leaf
Arracacia xanthorrhiza Bancroft	Arracacha, Peruvian carrot	Root
Centella asiatica (L.) Urban	Asiatic pennywort	Leaf and stolon
Chaerophyllum bulbosum L.	Tuberous chervil	Root
Coriandrum sativum L.	Coriander	Leaf and seed
Cryptotaenia japonica Hassk.	Japanese hornwort	Leaf
Daucus carota L. subsp. *sativus* (Hoffm.) Arcang.	Carrot	Root and leaf
Foeniculum vulgare var. *azoricum* (Miller) Thell.	Fennel	Leaf
Foeniculum vulgare var. *dulce* Fiori	Florence fennel	Leaf base
Glehnia littoralis F. Schm.	Coastal glehnia	Leaf, stem, root
Hydrocotyle sibthorpioides Lam.	Hydrocotyle	Young shoot and leaf
Myrrhis odorata (L.) Scop.	Garden myrrh	Leaf, root, and seed
Oenanthe javanica (*Blume*) DC. subsp. *javanica*	Oriental celery, water dropwort	Leaf and tender shoot
Pastinaca sativa L.	Parsnip	Root and leaf
Petroselinum crispum (Mill.) Nym. var. *crispum*	Parsley	Leaf
Petroselinum crispum (Mill.) Nym. *tuberosum*	Turnip-rooted parsley	Root and leaf
Petroselinum crispum (Mill.) Nym. var. *neapolitanum*	Italian parsley	Leaf
Sium sisarum L.	Skirret	Root
Araliaceae	ARALIA FAMILY	
Aralia cordata Thunb.	Spikenard	Tender shoot

TABLE 1.1. BOTANICAL NAMES, COMMON NAMES, AND EDIBLE PARTS OF PLANTS USED AS VEGETABLES (*Continued*)

Botanical Name	Common Name	Edible Plant Part
Aralia elata Seeman	Japanese aralia	Young leaf
Asteraceae	SUNFLOWER FAMILY	
Arctium lappa L.	Edible burdock	Root, petiole
Artemisia dracunculus L. var. *sativa* L.	French tarragon	Leaf
Artemisia indica Willd. var. *maximowiczii* (Nakai) Hara	Mugwort	Leaf
Aster scaber Thunb.	Aster	Leaf
Bidens pilosa L.	Bur marigold	Young shoot and leaf
Chrysanthemum spp.	Edible chrysanthemum	Leaf and tender shoot
Cichorium endivia L.	Endive, escarole	Leaf
Cichorium intybus L.	Chicory, witloof chicory	Leaf
Cirsium dipsacolepis (Maxim.) Matsum.	Gobouazami	Root
Cosmos caudatus Kunth	Cosmos	Leaf and young shoot
Crassocephalum biafrae (Oliv. et Hiern) S. Moore	Sierra Leone bologni	Young shoot and leaf
Crassocephalum crepidiodes (Benth.) S. Moore	Hawksbeard velvetplant	Young shoot and leaf
Cynara cardunculus L.	Cardoon	Petiole
Cynara scolymus L.	Globe artichoke	Immature flower bud
Emilia sonchifolia (L.) DC.	Emilia, false sow thistle	Young shoot and leaf
Enydra fluctuans Lour.	Buffalo spinach	Young shoot and leaf
Farfugium japonicum (L.) Kitamura	Japanese farfugium	Petiole
Fedia cornucopiae DC.	Horn of plenty, African valerian	Leaf

Galinsoga parviflora Cav.	Galinsoga	Young shoot
Gynura bicolor DC.	Gynura	Young leaf
Helianthus tuberosus L.	Jerusalem artichoke	Tuber
Lactuca indica L.	Indian lettuce	Leaf
Lactuca sativa L. var. *asparagina* Bailey	Asparagus lettuce, celtuce	Stem
Lactuca sativa L. var. *capitata* L.	Head lettuce, butterhead lettuce	Leaf
Lactuca sativa L. var. *longifolia* Lam.	Romaine lettuce, leaf lettuce	Leaf
Launaea taraxacifolia (Willd.) Amin ex C. Jeffrey	Wild lettuce	Leaf
Petasites japonicus (Sieb. & Zucc.) Maxim.	Butterbur	Petiole
Polymnia sonchifolia Poepp. & Endl.	Yacon strawberry	Root
Scolymus hispanicus L.	Golden thistle	Root and leaf
Scolymus maculatus L.	Spotted garden thistle	Leaf
Scorzonera hispanica L.	Black salsify	Root and leaf
Sonchus oleraceus L.	Milk thistle, sow thistle	Leaf
Spilanthes acmella (L.) Murr.	Brazil cress	Young leaf
Spilanthes ciliata HBK	Guasca	Young leaf
Spilanthes iabadicensus A.H. Moore	Getang	Young leaf and flower shoot
Spilanthes paniculata Wall ex DC.	Getang	Young leaf and flower shoot
Struchium sparganophora (L.) O. Ktze.	Bitter leaf	Young shoot
Taraxacum officinale Wiggers	Dandelion	Leaf, root
Tragopogon porrifolius L.	Salsify, vegetable oyster	Root and young leaf
Tragopogon pratensis L.	Goatsbeard, meadow salsify	Young root and shoot
Vernonia amygdalina Delile.	Bitter leaf	Young shoot
Basellaceae	BASELLA FAMILY	

TABLE 1.1. BOTANICAL NAMES, COMMON NAMES, AND EDIBLE PARTS OF PLANTS USED AS VEGETABLES (*Continued*)

Botanical Name	Common Name	Edible Plant Part
Basella alba L.	Indian spinach, Malabar spinach	Leaf and young shoot
Ullucus tuberosus Lozano	Ulluco	Tuber
Boraginaceae	BORAGE FAMILY	
Borago officinalis L.	Borage	Petiole
Symphytum officinale L.	Common comfrey	Leaf and tender shoot
Symphytum × *uplandicum* Nyman	Russian comfrey	Young leaf and shoot
Brassicaceae	MUSTARD FAMILY	
Armoracia rusticana Gaertn., Mey., Scherb.	Horseradish	Root, leaf, sprouted seed
Barbarea verna (Mill.) Aschers	Upland cress	Leaf
Brassica carinata A. Braun	Abyssinian mustard	Leaf
Brassica juncea (L.) Czernj. & Coss. var. *capitata* Hort.	Capitata mustard	Leaf
Brassica juncea (L.) Czernj. & Coss. var. *crassicaulus* Chen and Yang	Bamboo shoot mustard	Stem
Brassica juncea (L.) Czernj. & Coss. var. *crispifolia* Bailey	Curled mustard	Leaf
Brassica juncea (L.) Czernj. & Coss. var. *foliosa* Bailey	Small-leaf mustard	Leaf
Brassica juncea (L.) Czernj. & Coss. var. *gemmifera* Lee & Lin	Gemmiferous mustard	Stem and axillary bud

Brassica juncea (L.) Czernj. & Coss. var. *involuta* Yang & Chen	Involute mustard	Leaf
Brassica juncea (L.) Czernj. & Coss. var. *latipa* Li	Wide-petiole mustard	Leaf
Brassica juncea (L.) Czernj. & Coss. var. *leucanthus* Chen & Yang	White-flowered mustard	Leaf
Brassica juncea (L.) Czernj. & Coss. var. *linearifolia*	Line mustard	Leaf
Brassica juncea (L.) Czernj. & Coss. var. *longepetiolata* Yang & Chen	Long-petiole mustard	Leaf
Brassica juncea (L.) Czernj. & Coss. var. *megarrhiza* Tsen & Lee	Tuberous-rooted mustard	Root
Brassica juncea (L.) Czernj. & Coss. var. *multiceps* Tsen & Lee	Tillered mustard	Leaf
Brassica juncea (L.) Czernj. & Coss. var. *multisecta* Bailey	Flowerlike leaf mustard	Leaf
Brassica juncea (L.) Czernj. & Coss. var. *rugosa* Bailey	Brown mustard, mustard greens	Leaf
Brassica juncea (L.) Czernj. & Coss. var. *strumata* Tsen & Lee	Strumous mustard	Stem
Brassica juncea (L.) Czernj. & Coss. var. *tumida* Tsen & Lee	Swollen-stem mustard	Stem and leaf
Brassica juncea (L.) Czernj. & Coss. var. *utilis* Li	Penduncled mustard	Young flower stalk
Brassica napus L. var. *napobrassica* (L.) Reichb.	Rutabaga	Root and leaf
Brassica napus L. var. *napus*	Vegetable rape	Leaf and young flower stalk
Brassica napus L. var. *pabularia* (DC.) Reichb.	Siberian kale, Hanover salad	Leaf
Brassica nigra L. Koch.	Black mustard	Leaf
Brassica oleracea L. var. *acephala* DC.	Kale, collards	Leaf

TABLE 1.1. BOTANICAL NAMES, COMMON NAMES, AND EDIBLE PARTS OF PLANTS USED AS VEGETABLES (*Continued*)

Botanical Name	Common Name	Edible Plant Part
Brassica oleracea L. var. *alboglabra* Bailey	Chinese kale	Young flower stalk and leaf
Brassica oleracea L. var. *botrytis* L.	Cauliflower	Immature floral stalk
Brassica oleracea L. var. *capitata* L.	Cabbage	Leaf
Brassica oleracea L. var. *costata* DC.	Portuguese cabbage, tronchuda cabbage	Leaf and inflorescence
Brassica oleracea L. var. *gemmifera* Zenk.	Brussels sprouts	Axillary bud
Brassica oleracea L. var. *gongylodes* L.	Kohlrabi	Enlarged stem
Brassica oleracea L. var. *italica* Plenck.	Broccoli	Immature flower stalk
Brassica oleracea L. var. *medullosa* Thell. Marrow	Marrow stem kale	Leaf
Brassica oleracea L. var. *ramosa* Alef.	Thousand-headed kale	Leaf
Brassica oleracea L. var. *sabauda* L.	Savoy cabbage	Leaf
Brassica perviridis Bailey	Spinach mustard, tendergreen mustard	Leaf
Brassica rapa L. var. *chinensis* (Rupr.) Olsson	Pak choi, Chinese mustard	Leaf
Brassica rapa L. var. *narinosa* (Bailey) Olsson	Broad-beaked mustard	Leaf
Brassica rapa L. var. *parachinensis* (Bailey) Tsen & Lee	Mock pak choi, choy sum	Leaf
Brassica rapa L. var. *pekinensis* (Lour.) Olsson	Chinese cabbage, pe-tsai	Leaf
Brassica rapa L. var. (DC.) Metzg. *rapa*	Turnip	Enlarged root
Brassica rapa L. var. (DC.) Metzg. *utilis*	Turnip green	Leaf

Brassica rapa L. var. (DC.) Metzg. *septiceps*	Turnip broccoli, broccoli raab	Infloresence
Bunias orientalis L.	Hill mustard	Leaf
Capsella bursa-pastoris (L.) Medikus	Shepherd's purse	Young leaf
Cardamine pratensis L.	Cuckoo flower	Leaf
Crambe maritima L.	Sea kale	Petiole and young leaf
Crambe tatarica Jacq.	Tartar breadplant	Petiole and young leaf, root
Diplotaxis muralis (L.) DC.	Wallrocket	Leaf
Eruca sativa Miller	Rocket salad, arugula	Leaf
Lepidium meyenni Walp.	Maca	Root
Lepidium sativum L.	Garden cress	Leaf
Nasturtium officinale R. Br.	Watercress	Leaf
Raphanus sativus L. Caudatus group	Rat-tail radish	Immature seed pod
Raphanus sativus L. Radicula group	Radish	Root
Raphanus sativus L. Daikon group	Daikon	Root
Sinapis alba L.	White mustard	Leaf and young flower stalk
Wasabia japonica (Miq.) Matsum.	Wasabi, Japanese horseradish	Rhizome, young shoot
Cabombaceae	WATER LILY FAMILY	
Brasenia schreberi Gmelin	Watershield	Young leaf
Cactaceae	CACTUS FAMILY	
Opuntia ficus-indica (L.) Mill.	Prickly pear	Pad, fruit
Campanulaceae	BELLFLOWER FAMILY	
Campanula rapunculus L.	Rampion	Root and first leaf
Capparaceae	CAPER FAMILY	
Capparis spinosa L.	Capper	Flower bud
Cleome gynandra L.	Cat's whiskers	Leaf, young shoot, fruit

TABLE 1.1. BOTANICAL NAMES, COMMON NAMES, AND EDIBLE PARTS OF PLANTS USED AS VEGETABLES (*Continued*)

Botanical Name	Common Name	Edible Plant Part
Platycodon grandiflorum A. DC.	Chinese bellflower	Leaf
Chenopodiaceae	GOOSEFOOT FAMILY	
Atriplex hortensis L.	Orach	Leaf
Beta vulgaris L. Cicla group	Chard, Swiss chard	Leaf
Beta vulgaris L. Crassa group	Garden beet	Root and leaf
Chenopodium bonus-henricus L.	Good King Henry	Leaf
Chenopodium quinoa Willd.	Quinoa	Leaf
Kochia scoparia (L.) Schrader	Mock cypress	Tender shoot
Salsola komarovii Iljin.	Komarov Russian thistle	Leaf and young shoot
Salsola soda L.	Salsola	Leaf and young shoot
Spinacia oleracea L.	Spinach	Leaf
Suaeda asparagoides Mak.	Common seepweed	Young stem, leaf, plant
Convolvulaceae	BINDWEED FAMILY	
Convolvulus japonicus Thunb.	Rose glorybind	Root
Ipomoea aquatica Forssk.	Water spinach, kangkong	Tender shoot and leaf
Ipomoea batatas (L.) Lam.	Sweet potato	Root and leaf
Crassulaceae	ORPINE FAMILY	
Sedum sarmentosum Burge	Sedum	Leaf
Cucurbitaceae	GOURD FAMILY	
Benincasa hispida (Thunb.) Cogn.	Wax gourd	Immature/mature fruit

Citrullus lanatus (Thunb.) Matsum & Nakai	Watermelon	Ripe fruit and seed
Citrullus lanatus var. *citroides* (Bailey) Mansf.	Citron, preserving melon	Fruit
Coccinia grandis (L.) Voigt	Ivy gourd, tindora	Fruit, tender shoot, leaf
Cucumeropsis mannii Naudin	White-seeded melon	Fruit and seed
Cucumis anguria L.	West Indian gherkin	Immature fruit
Cucumis melo L. Cantaloupensis group	Cantaloupe	Fruit
Cucumis melo L. Chito group	Mango	Fruit
Cucumis melo L. Conomon group	Oriental pickling melon	Young fruit
Cucumis melo L. Flexuosus group	Japanese cucumber, snake melon	Immature fruit
Cucumis melo L. Inodorus group	Honeydew melon, casaba melon	Fruit
Cucumis melo L. Reticulatus group	Muskmelon (cantaloupe), Persian melon	Ripe fruit
Cucumis metuliferus E. Meyer ex Naudin	African horned cucumber	Fruit
Cucumis sativus L.	Cucumber	Immature fruit
Cucurbita argyrosperma Huber	Pumpkin	Young/mature fruit and seed
Cucurbita ficifolia Bouché	Fig-leaf gourd, Malabar gourd	Fruit
Cucurbita maxima Duchesne	Giant pumpkin, winter squash	Mature fruit and seed
Cucurbita moschata Duchesne	Butternut squash, tropical pumpkin	Young and mature fruit
Cucurbita pepo L.	Summer squash, zucchini	Young fruit
Cucurbita pepo L.	Common field pumpkin	Mature fruit and seed
Cyclanthera pedata (L.) Schrader var. *pedata*	Pepino	Immature fruit

TABLE 1.1. BOTANICAL NAMES, COMMON NAMES, AND EDIBLE PARTS OF PLANTS USED AS VEGETABLES (*Continued*)

Botanical Name	Common Name	Edible Plant Part
Lagenaria siceraria (Mol.) Standl.	Bottle gourd, calabash gourd	Immature fruit, tender shoot, and leaf
Luffa acutangula (L.) Roxb.	Angled loofah	Immature fruit
Luffa aegyptiaca Miller	Smooth loofah, sponge gourd	Immature fruit and leaf
Momordica charantia L.	Bitter gourd, balsam pear	Immature fruit and young leaf
Praecitrullus fistulosus (Stocks) Pang.	Squash melon	Fruit
Sechium edule (Jacq.) Swartz.	Chayote, mirliton, vegetable pear	Fruit, tender shoot, leaf
Sicana odorifera (Vell.) Naudin	Casabanana	Immature/mature fruit
Telfairia occidentalis Hook. f.	Fluted gourd, fluted pumpkin	Seed, leaf, tender shoot
Telfairia pedata (Smith ex Sims) Hook.	Oyster nut	Seed
Trichosanthes cucumerina L. var. *anguinea* (L.) Haines	Snake gourd	Immature fruit, leaf, and tender shoot
Trichosanthes cucumeroids (Ser.) Maxim.	Japanese snake gourd	Immature fruit
Trichosanthes dioica Roxb.	Pointed gourd	Immature fruit, tender shoot
Euphorbiaceae	SPURGE FAMILY	

Cnidoscolus aconitifolius (Miller) Johnston	Chaya	Leaf
Cnidoscolus chayamansa Mc Vaughn	Chaya	Leaf
Codiaeum variegatum (L.) Blume	Croton	Young leaf
Manihot esculenta Crantz	Yuca, cassava, manioc	Root and leaf
Sauropus androgynus (L.) Merr.	Common sauropus	Leaf
Fabaceae	PEA FAMILY	
Arachis hypogaea L.	Peanut, groundnut	Immature/mature seed
Bauhinia esculenta Burchell	Marama bean	Immature pod and root
Cajanus cajan (L.) Huth.	Cajan pea, pigeon pea	Immature pod/leaf
Canavalia ensiformis (L.) DC.	Jack bean, horse bean	Immature seed
Canavalia gladiata (Jacq.) DC.	Sword bean, horse bean	Immature seed
Cicer arietinum L.	Garbanzo, chickpea	Seed
Cyamposis tetragonoloba (L.) Taub.	Cluster bean, guar	Immature pod and seed
Flemingia vestita Benth. ex Bak.	Flemingia	Tuber
Glycine max (L.) Merr.	Soybean	Immature and sprouted seed
Lablab purpurus (L.) Sweet.	Hyacinth bean	Immature seed
Lathyrus sativus L.	Chickling pea	Immature pod/seed
Lathyrus tuberosus L.	Groundnut	Tuber
Lens culinaris Medikus	Lentil	Immature pod, sprouted seed
Lupinus spp.	Lupin	Seed
Macrotyloma geocarpum (Harms) Marechal and Baudet	Hausa groundnut	Seed
Macrotyloma uniflorum (Lam.) Verdc.	Horse gram	Seed

TABLE 1.1. BOTANICAL NAMES, COMMON NAMES, AND EDIBLE PARTS OF PLANTS USED AS VEGETABLES (*Continued*)

Botanical Name	Common Name	Edible Plant Part
Medicago sativa L.	Alfalfa, lucerne	Leaf, young shoot, sprouted seed
Mucuna pruriens (L.) DC.	Buffalo bean, velvet bean	Seed
Neptunia oleracea Lour.	Water mimosa	Leaf and tender shoot
Pachyrhizus ahipa (Wedd.) Parodi	Yam bean	Root
Pachyrhizus erosus (L.) Urban	Jicama, Mexican yam bean	Root, immature pod, and seed
Pachyrhizus tuberosus (Lam.) Sprengel	Potato bean	Root and immature pod
Phaseolus acutifolius A. Gray	Tepary bean	Seed, immature pod
Phaseolus coccineus L.	Scarlet runner bean	Immature pod and seed
Phaseolus lunatus L.	Lima bean	Immature seed, mature seed
Phaseolus vulgaris L.	Garden bean, snap bean	Immature pod and seed
Pisum sativum L. ssp. *sativum*	Pea, garden pea	Immature seed, tender shoot
Pisum sativum L. ssp. *sativum* f. *macrocarpon*	Snow pea, edible-podded pea	Immature pod
Psophocarpus tetragonolobus (L.) DC.	Goa bean, winged bean	Immature pod, seed, leaf, root

Pueraria lobata (Willd.) Ohwi	Kudzu	Root, leaf, tender shoot
Sphenostylis stenocarpa (Hochst. ex. A. Rich.) Harms.	African yam bean	Tuber and seed
Tetragonolobus purpureus Moench	Asparagus pea, winged pea	Immature pod
Trigonella foenum-graecum L.	Fenugreek	Leaf, tender shoot, immature pod
Vicia faba L.	Fava bean, broad bean, horse bean	Immature seed
Vigna aconitifolia (Jacq.) Maréchal	Moth bean	Immature pod and seed
Vigna angularis (Willd.) Ohwi & Ohashi	Adzuki bean	Seed
Vigna mungo (L.) Hepper	Black gram, urd	Immature pod and seed
Vigna radiata (L.) Wilcz.	Mung bean	Immature pod, sprouted seed, seed
Vigna subterranea (L.) Verdn.	Madagascar groundnut	Immature/mature seed
Vigna umbellata (Thunb.) Ohwi & Ohashi	Rice bean	Seed
Vigna unguiculata (L.) Walp. subsp. *cylindrica* (L.) Van Eselt. ex Verdn.	Catjang	Immature pod and seed
Vigna unguiculata (L.) Walp. subsp. *sesquipedalis* (L.)	Asparagus bean, yard-long bean	Immature pod and seed
Vigna unguiculata (L.) Walp. subsp. *unguiculata* (L.)	Southern pea, cowpea	Immature pod and seed
Gnetaceae	GNETUM FAMILY	
Gnetum gnemon L.	Bucko	Leaf, tender shoot and fruit
Haloragaceae	WATER MILFOIL FAMILY	

TABLE 1.1. BOTANICAL NAMES, COMMON NAMES, AND EDIBLE PARTS OF PLANTS USED AS VEGETABLES (*Continued*)

Botanical Name	Common Name	Edible Plant Part
Myriophyllum aquaticum (Vellozo) Verdc.	Parrot's feather	Shoot tip
Icacinaceae	ICACINA FAMILY	
Icacina senegalensis A. Juss.	False yam	Tuber
Lamiaceae	MINT FAMILY	
Lycopus lucidus Turcz.	Shiny bugleweed	Rhizome
Mentha pulegium L.	Pennyroyal mint	Leaf
Mentha spicata L. em. Harley	Spearmint	Leaf and inflorescence
Ocimum basilicum L.	Common basil, sweet basil	Leaf
Ocimum canum Sims.	Hoary basil	Young leaves
Origanum vulgare L.	Marjoram	Flowering plant and inflorescence
Perilla frutescens (L.) Britt. var. *crispa* (Thunb.) Deane	Perilla	Leaf and seed
Plectranthus esculentus N.E. Br.	Kaffir potato	Tuber
Satureja hortensis L.	Savory, summer savory	Leaf and young shoot
Solenostemon rotundifolius (Poir.) J. K. Morton	Hausa potato	Tuber
Stachys affinis Bunge	Japanese artichoke, Chinese artichoke	Tuber
Malvaceae	MALLOW FAMILY	
Abelmoschus esculentus (L.) Moench	Okra, gumbo	Immature fruit
Abelmoschus manihot (L.) Medikus	Hibiscus root	Leaf and tender shoot

Hibiscus acetosella Wel. ex Hiern	False roselle	Young leaf and shoot
Hibiscus sabdariffa L.	Jamaican sorrel	Calyx and leaf
Malva rotundifolia L.	Mallow	Leaf and young shoot
Moraceae	MULBERRY FAMILY	
Humulus lupulus L.	Hops	Tender shoot
Nelumbonaceae	LOTUS FAMILY	
Nelumbo nucifera Gaertn.	Lotus root	Rhizome, leaf, seed
		Seed
Nyctaginaceae	FOUR O'CLOCK FAMILY	
Mirabilis expansa (Ruiz & Paron) Standley	Mauka	Tuber
Nymphaeaceae	WATER LILY FAMILY	
Euryale ferox Salisb.	Foxnut	Seed, tender shoot, root
Nymphaea nouchali Burm. f.	Water lily	Rhizome, flower stalk, seed
Onagraceae	EVENING PRIMROSE FAMILY	
Oenothera biennis L.	Evening primrose	Leaf and tender shoot
Orobanchaceae	BROOMRAPE FAMILY	
Orobanche crenata Forsskal.	Broomrape	Shoot
Oxalidaceae	OXALIS FAMILY	
Oxalis tuberosa Molina	Oka, oca	Tuber
Passifloraceae	PASSION FLOWER FAMILY	
Passiflora biflora Lam.	Passion flower	Shoot, young leaf, flower
Pedaliaceae	PEDALIUM FAMILY	
Sesamum radiatum Schum. ex Thonn.	Gogoro	Young shoot
Phytolaccaceae	POKEWEED FAMILY	

TABLE 1.1. BOTANICAL NAMES, COMMON NAMES, AND EDIBLE PARTS OF PLANTS USED AS VEGETABLES (*Continued*)

Botanical Name	Common Name	Edible Plant Part
Phytolacca acinosa Roxb.	Indian poke	Leaf and young shoot
Phytolacca americana L.	Poke	Leaf and young shoot
Phytolacca esculenta Van Houtte	Pokeweed	Leaf and young shoot
Phytolacca octandra L.	Inkweed	Leaf and young shoot
Plantaginaceae	PLAINTAIN FAMILY	
Plantago coronopus L. var. *sativa* Fiori	Buckshorn plantain	Leaf
Polygonaceae	BUCKWHEAT FAMILY	
Rheum rhabarbarum L.	Rhubarb, pieplant	Petiole
Rumex acetosa L.	Sorrel	Leaf
Rumex patientia L.	Dock	Leaf
Rumex scutatus L.	French sorrel	Leaf
Portulacaceae	PURSLANE FAMILY	
Montia perfoliata (Donn. ex Willd.) Howell	Winter purslane, miner's lettuce	Leaf
Portulaca oleracea L.	Purslane	Leaf and young shoot
Talinum paniculatum (Jacq.) Gaertn.	Flameflower	Young shoot
Talinum triangulare (Jacq.) Willd.	Waterleaf, Suraim spinach	Leaf
Resedaceae	MIGNONETTE FAMILY	
Reseda odorata L.	Mignonette	Leaf and flower
Rosaceae	ROSE FAMILY	
Fragaria × *Ananassa* Duchesne.	Strawberry	Fruit
Saururaceae	LIZARD'S-TAIL FAMILY	
Houttuynia cordata Thunb.	Saururis, tsi	Leaf

Solanaceae	NIGHTSHADE FAMILY	
Capsicum annuum L. Grossum group	Bell pepper	Fruit
Capsicum annuum L. Longum group	Cayenne pepper, chile pepper	Mature fruit
Capsicum baccatum L. var. *baccatum*	Small pepper	Fruit
Capsicum chinense Jacq.	Scotch bonnet pepper, habanero pepper	Fruit
Capsicum frutescens L.	Tabasco pepper	Fruit
Capsicum pubescens Ruiz & Pavon	Rocoto	Fruit
Cyphomandra betacea (Cav.) Sendtner	Tamarillo, tree tomato	Ripe fruit
Lycium chinense Mill.	Boxthorn	Leaf
Lycopersicon esculentum Mill.	Tomato	Ripe fruit
Lycopersicon pimpinellifolium (L.) Mill.	Currant tomato	Ripe fruit
Physalis alkekengi L.	Chinese lantern plant	Ripe fruit
Physalis ixocarpa Brot. ex Hornem.	Tomatillo	Unripe fruit
Physalis peruviana L.	Cape gooseberry	Ripe fruit
Solanum aethiopicum L.	Golden apple	Fruit and leaf
Solanum americanum Mill.	American black nightshade	Tender shoot, leaf, unripe fruit
Solanum gilo Raddi	Gilo, jilo	Young shoot
Solanum incanum L.	Garden egg	Unripe fruit
Solanum integrifolium Poir.	Scarlet eggplant, tomato eggplant	Immature fruit
Solanum macrocarpon L.	African eggplant	Leaf and fruit
Solanum melongena L.	Eggplant, aubergine	Immature fruit
Solanum muricatum Ait.	Pepino, sweet pepino	Ripe fruit
Solanum nigrum L.	Black nightshade	Mature fruit, leaf, tender shoot
Solanum quitoense Lam.	Naranjillo	Ripe fruit
Solanum torvum Swartz	Pea eggplant	Tender shoot, immature fruit

TABLE 1.1. BOTANICAL NAMES, COMMON NAMES, AND EDIBLE PARTS OF PLANTS USED AS VEGETABLES (*Continued*)

Botanical Name	Common Name	Edible Plant Part
Solanum tuberosum L.	Potato	Tuber
Tiliaceae	BASSWOOD FAMILY	
Corchorus olitorius L.	Jew's marrow	Leaf and tender shoot
Trapaceae	WATER CHESTNUT FAMILY	
Trapa bicornis Osbeck	Water chestnut	Seed
Trapa natans L.	Water chestnut	Seed
Tropacolaceae	NASTURTIUM FAMILY	
Tropaeolum majus L.	Nasturtium	Leaf, flower
Tropaeolum tuberosum Ruiz & Pavon	Tuberous nasturtium	Tuber
Urticaceae	NETTLE FAMILY	
Pilea glaberrima (Blume) Blume	Pilea	Leaf
Pilea trinervia Wight	Pilea	Leaf
Urtica dioica L.	Stinging nettle	Leaf
Valerianaceae	VALERIAN FAMILY	
Valerianella eriocarpa Desv.	Italian corn-salad	Leaf
Valerianella locusta (L.) Laterrade em. Betcke	European corn-salad	Leaf
Violaceae	VIOLET FAMILY	
Viola tricolor L.	Violet, pansy	Flower, leaf
Vitaceae	GRAPE FAMILY	
Cissus javana DC.	Kangaroo vine	Leaf, young shoot

Adapted from S.J. Kays and J.C. Silva Dias, *Cultivated Vegetables of the World* (Athens, Ga.: Exon, 1996). Used with permission.

TABLE 1.2. NAMES OF COMMON VEGETABLES IN NINE LANGUAGES

English	Danish	Dutch	French	German	Italian	Portuguese	Spanish	Swedish
Artichoke	artiskok	artisjok	artischaut	Artischocke	carciofo	alcachofra	alcachofa	kronärtskocka
Asparagus	asparges	asperge	asperge	Spargel	asparago	espargo	espárrago	sparris
Broad bean	hestebønne	tuinboon	feve	Puffbohne	fava maggiore	fava	haba	bondböna
Snap bean	bønne	boon	haricot	Bohne	fagiolino	feijão	judia	böna
Beet	rødbede	kroot	betterave rouge	Rote Rübe	bietola da orta	beterraba de mesa	betabol	rödbeta
Broccoli	broccoli	broccoli	chou-brocoli	Brokkoli	cavolo broccolo	bróculo	bróculi	broccoli
Brussels sprouts	rosenkål	spruitkool	chou de Bruxelles	Rosenkohl	cavolo di Bruxelles	couve de Bruxelas	col de Bruselas	brysselkäl
Cabbage	kål	kool	chou	Kohl	cavolo	couve	col	käl
Carrot	karotte	peen	carotte	Karotte	carota	cenoura	zanahoria	morot
Cauliflower	blomkål	bloemkool	chou-fleur	Blumenkohl	cavolfiore	courve-flor	coliflor	blomkål
Celery	selleri	selderij	céleri	Schnittselleri	sedano da erbucci	aipo	apio	selleri
Celeriac	knoldselleri	knolselderij	céleri-rave	Knollensellerie	sedano rapa	aipo de cabega	apio nabo	rotselleri
Chicory	cikorie	cichorei	chicoree	Zichorienwurzel	cicoria	chicoria do café	achicoria de raiz	cikoria
Chinese cabbage	kinesisk kål	Chinese kool	chou de Chine	Chinesischer Kohl	cavolo cinese	couve da China	col de China	Salladkål
Sweet corn	sukkermajs	suikermais	mais sucré	Zuckermais	mais dolce	milho doce	maiz dulce	sockermajs

TABLE 1.2. NAMES OF COMMON VEGETABLES IN NINE LANGUAGES (*Continued*)

English	Danish	Dutch	French	German	Italian	Portuguese	Spanish	Swedish
Cucumber	agurk	komkommer	concombre	Gurke	cetriolino	pepino	pepino	gurka
Eggplant	aegplante	aubergine	aubergine	Aubergine	melanzana	beringela	berenjena	äggplanta
Endive	endivie	andijvie	chicorée frisée	Endivie	indivia	chicoria	escarola	endiviesallat
Horseradish	peberrod	mierikswortel	raifort	Meerrettich	barbafonte	rabano rústico	rábana	pepparrot
Kale	grønkål	boerekool	chou vert	Grunkohl	cavolo a foglia riccia	couve galega frizada	berza enana	grönkål
Kohlrabi	knudekål	koolrabi	chou-rave	Kohlrabi	cavolo-rapa	couve-rabano	colirrábano	kålrabbi
Leek	porre	prei	poireau	Porree	porro	alho porro	puerro	purjolök
Lettuce	salat	sla	laitue	Salat	lattuga	alface	lechuga	sallad
Melon	melon	meloen	melon	Melone	popone	meläo	melón	melon
Onion	løg	ui	ognon	Zwiebel	cipolla	cebola	cebolla	lök
Parsley	persille	peterselie	persil	Petersilie	prezzemola comune	salsa frisada	perejil	persilja
Parsnip	pastinak	pastinaak	panais	Pastinake	pastinaca	pastinaca	chircivía	pasternacka
Pea	haveaert	erwt	pois	Erbse	pisello	ervilha	guisante	ärt
Pepper	peberfrugt	paprika	poivron	Paprika	peperone dolce	pimento	pimentus	paprika
Potato	kartoffel	aardappel	pomme de terre	Kartoffel	patata	batata	patata	potatis

Pumpkin	centnergraeskar	reuzenpompoen	potiron	Zentnerkürbis	zucca gigante	abóbora	calabaza grande	jättepumpa
Radish	radis	radijs	radis	Radies	ravanello	rabanete	rábano	rädisa
Rhubarb	rabarber	rabarber	rhubarbe	Rhabarber	rabarbaro	ruibarbo	ruibarbo	rabarber
Rutabaga	kålrabi	koolraap	chou-navet	Kohlrübe	navone	rutabaga	colinabo	kälrot
Spinach	spinat	spinazie	épinard	Spinat	spinaci	espinafre	espinaca	spenat
New Zealand spinach	nyzeelandsk spinat	Nieuwzeelandse spinazie	tetragone	Neuseelándsk Spinat	spinacio di Nuova Zelanda	espinafre da Nova Zelândia	espinaca Nueva Zelandia	nyzeeländsk spenat
Strawberry	jordbaer	aardbei	fraise	Erdbeere	fragola	morango	fresa	jordgubbe
Summer squash	mandelgraeskavr	pompoen	citrouelle	Garten	zucca	abóbora porquiera	cababaza	matpumpa
Swiss chard	bladbede	snijbiet	poirée	Mangold	bietola da costa	acelga	acelga	mangold
Tomato	tomat	tomaat	tomate	Tomate	pomodoro	tomate	tomate	tomat
Turnip	majroe	meiraap	navet	Mairübe	rapa bianca	nabo	nabo-colza	rova
Watermelon	vandmelon	watermeloen	melon d'eau	Wassermelone	melore d'acqua	melancia	sandia	vattenmelon

Adapted from P. J. Stadhouders (chief ed.), *Elsevier's Dictionary of Horticultural and Agricultural Plant Production* (New York: Elsevier Science, 1990). Used with permission.

02
EDIBLE FLOWERS

TABLE 1.3. BOTANICAL NAMES, COMMON NAMES, FLOWER COLOR, AND TASTE OF SOME EDIBLE FLOWERS

Cautions:

- Proper identification of edible flowers is necessary.
- Edible flowers should be pesticide free.
- Flowers of plants treated with fresh manure should not be used.
- Introduce new flowers into the diet slowly so possible allergic reactions can be identified.

Botanical Name	Common Name	Flower Color	Taste
Agavaceae	CENTURY PLANT FAMILY		
Yucca filamentosa L.	Yucca	Creamy white with purple tinge	Slightly bitter
Allicaceae	ONION FAMILY		
Allium schoenoprasum L.	Chive	Lavender	Onion, strong
Allium tuberosum Rottl. ex. Sprengel	Chinese chive	White	Onion, strong
Tulbaghia violacea Harv.	Society garlic	Lilac	Onion
Apiaceae	CARROT FAMILY		
Anethum graveolens L.	Dill	Yellow	Stronger than leaves
Anthriscus cerefolium (L.) Hoffm.	Chervil	White, pink, yellow, red, orange	Parsley
Coriandrum sativum L.	Coriander	White	Milder than leaf
Foeniculum vulgare Mill.	Fennel	Pale yellow	Licorice, milder than leaf

Asteraceae	SUNFLOWER FAMILY		
Bellis perennis L.	English daisy	White to purple petals	Mild to bitter
Calendula officinalis L.	Calendula	Yellow, gold, orange	Tangy and peppery
Carthamus tinctorius L.	Safflower	Yellow to deep red	Bitter
Chamaemelum nobilis Mill.	English chamomile	White petals, yellow center	Sweet apple
Chrysanthemum coronarium L.	Garland chrysanthemum	Yellow to white	Mild
Chicorium intybus L.	Chicory	Blue to lavender	Similar to endive
Dendranthema × *grandifolium* Kitam.	Chrysanthemum	Various	Strong to bitter
Leucanthemum vulgare Lam.	Oxeye daisy	White, yellow center	Mild
Tagetes erecta L.	African marigold	White, gold	Variable, mild to bitter
Tageto tenuifolia Cav.	Signet marigold	White, gold, yellow, red	Citrus, milder than *T. erecta*
Taraxicum officianle L.	Dandelion	Yellow	Bitter
Begonaiceae	BEGONIA FAMILY		
Begonia tuberhybrida	Tuberous begonia	Various	Citrus
Boraginaceae	BORAGE FAMILY		
Borago officinalis L.	Borage	Blue, purple, lavender	Cucumber
Brassicaceae	MUSTARD FAMILY		
Brassica spp.	Mustard	Yellow	Tangy to hot
Eruca vesicaria Mill.	Arugala	White	Nutty, smoky
Raphanus sativus L.	Radish	White, pink	Spicy
Caryophyllaceae	PINK FAMILY		
Dianthus spp.	Pinks	Pink, white, red	Spicy, cloves
Cucurbitaceae	GOURD FAMILY		
Cucurbita pepo L.	Summer squash, pumpkin	Yellow	Mild, raw squash

TABLE 1.3. BOTANICAL NAMES, COMMON NAMES, FLOWER COLOR, AND TASTE OF SOME EDIBLE FLOWERS (*Continued*)

Botanical Name	Common Name	Flower Color	Taste
Fabaceae	PEA FAMILY		
Cercis canadensis L.	Redbud	Pink	Bean-like to tart apple
Phaseolus coccineus L.	Scarlet runner bean	Bright orange to scarlet	Mild raw bean
Pisum sativum L.	Garden pea	White, tinged pink	Raw pea
Trifolium pretense L.	Red clover	Pink, lilac	Hay
Geraniaceae	GERANIUM FAMILY		
Pelargonium spp. L'Hérit	Scented geraniums	White, red, pink, purple	Various, e.g., apple, lemon, rose, spice, etc.
Iridaceae	IRIS FAMILY		
Gladiolus spp. L.	Gladiolus	Various	Mediocre
Lamiaceae	MINT FAMILY		
Hyssopus officinalis L.	Hyssop	Blue, pink, white	Bitter, similar to tonic
Lavandula augustifolia Mill.	Lavender	Lavender, purple, pink, white	Highly perfumed
Melissa officinalis L.	Lemon balm	Creamy white	Lemony, sweet
Mentha spp. L.	Mint	Lavender, pink, white	Minty
Monarda didyma L.	Bee balm	Red, pink, white, lavender	Tea-like
Ocimum basilicum L.	Basil	White to pale pink	Spicy
Origanum vulgare L.	Oregano	White	Spicy, pungent
Origanum majorana L.	Marjoram	Pale pink	Spicy, sweet

Rosmarinus officinalis L.	Rosemary	Blue, pink, white	Mild rosemary
Salvia rutilans Carr.	Pineapple sage	Scarlet	Pineapple/sage overtones
Salvia officinalis L.	Sage	Blue, purple, white, pink	Flowery sage
Satureja hortensis L.	Summer savory	Pink	Mildly peppery, spicy
Satureja montana L.	Winter savory	Pale blue to purple	Mildly peppery, spicy
Thymus spp. L.	Thyme	Pink, purple, white	Milder than leaves
Liliaceae	LILY FAMILY		
Hemerocallis fulva L.	Daylily	Tawny orange	Cooked asparagus/ zucchini
Muscari neglectum Guss. ex Ten	Grape hyacinth	Pink, blue	Grapey
Tulipa spp. L.	Tulip	Various	Slightly sweet or bitter
Malvaceae	MALLOW FAMILY		
Abelmochus esculentus (L.) Moench.	Okra	Yellow, red	Mild, sweet, slightly mucilaginous
Alcea rosa L.	Hollyhock	Various	Slightly bitter
Hibiscus rosa-sinensis L.	Hibiscus	Orange, red, purple	Citrus, cranberry
Hibiscus syriacus L.	Rose of Sharon	Red, white, purple, violet	Mild, nutty
Myrtaceae	MYRTLE FAMILY		
Acca sellowiara O. Berg	Pineapple guava	White to deep pink	Papaya or exotic melon
Oleaceae	OLIVE FAMILY		
Syringa vulgaris L.	Lilac	White, pink, purple, lilac	Perfume, slightly bitter
Papaveraceae	POPPY FAMILY		
Sanguisorba minor Soep.	Burnet	Red	Cucumber
Rosacese	ROSE FAMILY		

TABLE 1.3. BOTANICAL NAMES, COMMON NAMES, FLOWER COLOR, AND TASTE OF SOME EDIBLE FLOWERS (*Continued*)

Botanical Name	Common Name	Flower Color	Taste
Malus spp. Mill.	Apple, crabapple	White to pink	Slightly floral to sour
Rubiaceae	MADDER FAMILY		
Galium odoratum (L.) Scop.	Sweet woodruff	White	Sweet, grassy, vanilla
Rutaceae	RUE FAMILY		
Citrus limon (L.) Burm.	Lemon	White	Citrus, slightly bitter
Citrus sinensis (L.) Osbeck.	Orange	White	Citrus, sweet/strong
Tropaeolaceae	NASTURTIUM FAMILY		
Tropaeolum majus L.	Nasturtium	Variable	Watercress, peppery
Violaceae	VIOLET FAMILY		
Viola odorata L.	Violet	Violet, pink, white	Sweet
Viola × *wittrockiana* Gams.	Pansy	Various, multicolored	Stronger than violets
Viola tricolor L.	Johnny-jump-up	Violet, white, yellow	Stronger than violets

Adapted from K.B. Badertsher and S.E. Newman, *Edible Flowers* (Colorado Cooperative Extension), http://www.ext.colostate.edu/pubs/garden/07237.html.

Useful reference: Jeanne Mackin, *Cornell Book of Herbs and Edible Flowers* (Ithaca, N.Y.: Cornell Cooperative Extension).

TABLE 1.4. U.S. VEGETABLE PRODUCTION STATISTICS: LEADING FRESH MARKET VEGETABLE STATES, 2004[1]

	Harvested Acreage		Production		Value	
Rank	State	% of Total	State	% of Total	State	% of Total
1	California	43.4	California	48.8	California	52.9
2	Florida	9.5	Florida	9.1	Florida	11.8
3	Georgia	7.0	Arizona	8.4	Arizona	8.7
4	Arizona	6.7	Georgia	4.5	Georgia	3.9
5	New York	4.0	Texas	3.7	Texas	3.5

Adapted from *Vegetables, 2004 Summary* (USDA, National Agricultural Statistics Service Vg 1–2, 2005), http://jan.mannlib.cornell.edu/reports/nassr/fruit/pvg-bban/vgan0105.pdf.

[1] Includes data for artichoke, asparagus, *lima bean, snap bean, broccoli, *Brussels sprouts, cabbage, carrot, cauliflower, *celery, cantaloupe, cucumber, eggplant, escarole/endive, garlic, honeydew melon, lettuce, onion, bell pepper, spinach, sweet corn, tomato, and watermelon.

*Includes fresh market and processing.

TABLE 1.5. IMPORTANT STATES IN THE PRODUCTION OF U.S. FRESH MARKET VEGETABLES BY CROP VALUE, 2004

Crop	First	Second	Third
Artichoke[1]	California	—	—
Asparagus[1]	California	Washington	Michigan
Bean, snap	Florida	California	Georgia
Broccoli[1]	California	Arizona	—
Cabbage	California	Texas	New York
Cantaloupe	California	Arizona	Texas
Carrot	California	Texas	Michigan
Cauliflower[1]	California	Arizona	New York
Celery	California	Michigan	—
Cucumber	Florida	Georgia	California
Garlic	California	Oregon	Nevada
Honeydew melon	California	Arizona	Texas
Lettuce, head	California	Arizona	Colorado
Lettuce, leaf	California	Arizona	—
Lettuce, romaine	California	Arizona	—
Mushroom[1]	Pennsylvania	California	Florida
Onion	California	Texas	Oregon
Pepper, bell	California	Florida	New Jersey
Pepper, chile	California	New Mexico	Texas
Pumpkin	New York	Pennsylvania	California
Spinach	California	Arizona	Texas
Squash	California	Florida	New York
Strawberry	California	Florida	North Carolina
Sweet corn	California	Florida	New York
Tomato	California	Florida	Texas
Watermelon	California	Florida	Texas

Adapted from *Vegetables, 2004 Summary* (USDA, National Agricultural Statistics Service Vg 1–2, 2005), http://jan.mannlib.cornell.edu/reports/nassr/fruit/pvg-bban/vgan0105.pdf.

[1] Includes fresh market and processing.

TABLE 1.6. HARVESTED ACREAGE, PRODUCTION, AND VALUE OF U.S. FRESH MARKET VEGETABLES, 2002–2004 AVERAGE

Crop	Acres	Production (1000 cwt)	Value ($1000)
Artichoke[1]	7,633	925	71,716
Asparagus[1]	58,833	1,086	176,870
Bean, snap	94,733	5,840	277,141
Broccoli[1]	133,300	19,520	119,995
Cabbage	75,460	23,967	316,398
Cantaloupe	88,583	21,608	356,867
Carrot	85,400	26,577	518,266
Cauliflower[1]	40,533	6,612	218,110
Celery[1]	27,300	18,932	260,904
Cucumber	55,357	10,005	202,636
Garlic[1]	33,133	5,705	151,452
Honeydew melon	23,100	5,133	93,235
Lettuce, head	185,400	78,785	1,282,088
Lettuce, leaf	55,100	13,461	422,747
Lettuce, romaine	72,000	22,649	534,087
Onion	165,153	74,702	870,217
Pepper, bell[1]	54,167	16,196	511,813
Pepper, chile[1]	29,700	4,223	110,460
Pumpkin[1]	41,657	8,878	90,867
Spinach	36,393	5,543	203,619
Squash[1]	51,867	8,078	207,571
Strawberry[1]	49,200	20,847	1,336,008
Sweet corn	246,243	28,031	559,580
Tomato	125,707	37,094	1,309,213
Watermelon	147,733	38,207	328,342

Adapted from *Vegetables, 2004 Summary* (USDA, National Agricultural Statistics Service Vg 1–2, 2005), http://jan.mannlib.cornell.edu/reports/nassr/fruit/pvg-bban/vgan0105.pdf.

[1] Includes fresh market and processing.

TABLE 1.7. AVERAGE U.S. YIELDS OF FRESH MARKET VEGETABLES, 2002–2004

Crop	Yield (cwt/acre)
Artichoke[1]	122
Asparagus[1]	31
Bean, snap	62
Broccoli[1]	146
Cabbage	317
Cantaloupe	244
Carrot	311
Cauliflower[1]	163
Celery[1]	693
Cucumber	181
Garlic[1]	172
Honeydew melon	223
Lettuce, head	371
Lettuce, leaf	244
Lettuce, romaine	315
Onion	452
Pepper, bell[1]	299
Pepper, chile[1]	142
Pumpkin[1]	211
Spinach	152
Squash[1]	156
Strawberry[1]	423
Sweet corn	114
Tomato	295
Watermelon	259

Adapted from *Vegetables, 2004 Summary* (USDA, National Agricultural Statistics Service Vg 1-2, 2005), http://jan.mannlib.cornell.edu/reports/nassr/fruit/pvg-bban/vgan0105.pdf.

[1] Includes fresh market and processing.

TABLE 1.8. LEADING U.S. PROCESSING VEGETABLE STATES, 2004[1]

	Harvested Acreage		Production		Value	
Rank	State	% of Total	State	% of Total	State	% of Total
1	California	24.1	California	67.8	California	51.2
2	Minnesota	16.0	Washington	6.3	Wisconsin	7.0
3	Wisconsin	15.0	Wisconsin	5.7	Minnesota	6.9
4	Washington	11.0	Minnesota	5.7	Washington	6.8
5	Oregon	5.0	Oregon	2.4	Michigan	4.1

Adapted from *Vegetables, 2004 Summary* (USDA, National Agricultural Statistics Service Vg 1–2, 2005), http://jan.mannlib.cornell.edu/reports/nassr/fruit/pvg-bban/vgan0105.pdf.

[1] Includes lima bean, snap bean, carrot, sweet corn, cucumber for pickles, pea, spinach, and tomato.

TABLE 1.9. HARVESTED ACREAGE, PRODUCTION, AND VALUE OF U.S. PROCESSING VEGETABLES, 2002–2004 AVERAGE

Crop	Acres	Production (tons)	Value ($1000)
Bean, lima	46,267	59,757	25,854
Bean, snap	196,600	781,630	122,141
Carrot	15,770	426,300	32,081
Cucumber	116,700	616,907	168,149
Pea, green	215,833	402,540	101,186
Spinach	12,640	118,140	13,354
Sweet corn	416,500	3,100,640	217,495
Tomato	302,247	11,252,313	658,516

Adapted from *Vegetables, 2004 Summary* (USDA, National Agricultural Statistics Service Vg 1–2, 2005), http://jan.mannlib.cornell.edu/reports/nassr/fruit/pvg-bban/vgan0105.pdf.

TABLE 1.10. IMPORTANT STATES IN THE PRODUCTION OF U.S. PROCESSING VEGETABLES BY CROP VALUE, 2004

Crop	First	Second	Third
Bean, snap	Wisconsin	Oregon	New York
Carrot	California	Washington	Wisconsin
Cucumber	Michigan	Florida	North Carolina
Pea, green	Minnesota	Washington	Wisconsin
Spinach	California	—	—
Sweet corn	Minnesota	Washington	Wisconsin
Tomato	California	Indiana	Ohio

Adapted from *Vegetables, 2004 Sumary* (USDA, National Agricultural Statistics Service Vg 1–2, 2005), http://jan.mannlib.cornell.edu/reports/nassr/fruit/pvg-bban/vgan0105.pdf.

TABLE 1.11. AVERAGE U.S. YIELDS OF PROCESSING VEGETABLES, 2002–2004

Crop	Yield (tons/acre)
Bean, lima	1.29
Bean, snap	3.97
Carrot	27.02
Cucumber	5.07
Pea, green	1.86
Spinach	9.44
Sweet corn	7.44
Tomato	37.20

Adapted from *Vegetables, 2004 Summary* (USDA, National Agricultural Statistics Service Vg 1–2, 2005), http://jan.mannlib.cornell.edu/reports/nassr/fruit/pvg-bban/vgan0105.pdf.

TABLE 1.12. U.S. POTATO AND SWEET POTATO PRODUCTION STATISTICS: HARVESTED ACREAGE, YIELD, PRODUCTION, AND VALUE, 2002–2004 AVERAGE

Crop	Acres	Yield (cwt/acre)	Production (1000 cwt)	Value ($1000)
Potato	1,227,533	373	457,449	2,765,300
Sweet Potato	89,400	168	15,029	269,176

Adapted from *Vegetables and Melons Outlook VGS-307* (USDA Economic Research Service, 2005), http://www.ers.usda.gov/publications/vgs/Feb05/vgs307.pdf.

TABLE 1.13. IMPORTANT U.S. STATES IN POTATO AND SWEET POTATO PRODUCTION BY CROP VALUE, 2003

Rank	Potato	Sweet Potato
1	Idaho	North Carolina
2	Washington	California
3	California	Louisiana
4	Wisconsin	Mississippi
5	Colorado	Alabama

Adapted from *Vegetables and Melons Outlook VGS-307* (USDA Economic Research Service, 2005), http://www.ers.usda.gov/publications/vgs/Feb05/vgs307.pdf.

TABLE 1.14. UTILIZATION OF THE U.S. POTATO CROP, 2001–2003 AVERAGE

Item	Amount	
	1000 cwt	% of Total
A. Sales	413,227	91
1. Table stock	129,936	29
2. Processing	256,808	57
a. Chips	52,825	12
b. Dehydration	46,845	10
c. Frozen french fries	126,033	28
d. Other frozen products	25,473	6
e. Canned potatoes	4,651	1
f. Starch and flour	981	<1
3. Other sales	26,483	6
a. Livestock feed	2,848	<1
b. Seed	23,634	5
B. Nonsales	37,992	8
1. Seed used on farms where grown	5,516	1
2. Shrinkage	32,476	7
Total production	451,219	

Adapted from *Vegetables and Melons Outlook VGS-307* (USDA Economic Research Service, 2005), http://www.ers.usda.gov/publications/vgs/Feb05/vgs307.pdf.

04
VEGETABLE CONSUMPTION

TABLE 1.15. TRENDS IN U.S. PER CAPITA CONSUMPTION OF VEGETABLES

	Amount (lb)[1]		
Year	Fresh	Processed	Total
1971	171	189	360
1975	180	187	367
1980	172	184	356
1985	187	198	385
1990	198	212	410
1995	210	223	433
2000	228	222	450
2004	227	219	446

Adapted from *Vegetables and Melons Outlook VGS-307* (USDA Economic Research Service, 2005), http://www.ers.usda.gov/publications/vgs/Feb05/vgs307.pdf.

[1] Fresh weight equivalent.

TABLE 1.16. U.S. PER CAPITA CONSUMPTION OF COMMERCIALLY PRODUCED VEGETABLES, 2004

	Amount (lb)			
Vegetable	Fresh	Canned	Frozen	Total
Artichoke, all	—	—	—	0.7
Asparagus	1.1	0.2	0.10	1.4
Bean, dry, all	—	—	—	6.7
Bean, snap	2.1	3.5	1.9	7.4
Broccoli	5.8	—	2.4	8.2
Cabbage	7.9	1.1	—	9.0
Cantaloupe	11.0	—	—	11.0
Carrot	8.4	1.5	1.7	11.5
Cauliflower	1.7	—	0.5	2.2
Celery	6.2	—	—	6.2
Cucumber	6.3	4.9	—	11.2
Eggplant, all	—	—	—	0.7
Escarole/endive	0.2	—	—	0.3
Garlic, all	—	—	—	2.8
Honeydew melon	2.2	—	—	2.2
Lettuce, head	21.3	—	—	21.3
Lettuce, leaf & romaine	10.0	—	—	10.0
Mushroom, all	2.6	1.6	—	4.2
Onion	19.3	—	—	20.8[1]
Pea, green	—	1.3	1.9	3.3
Pea and lentil, dry, all	—	—	—	0.6
Pepper, bell	7.2	—	—	7.2
Pepper, chile	—	5.7	—	5.7
Potato	45.6	33.8[2]	56.6	136.0
Spinach, all	—	—	—	1.8
Strawberry	5.4	—	1.7	7.1
Sweet corn[3]	9.7	8.8	9.5	27.8
Sweet potato, all	—	—	—	4.3
Tomato	19.1	69.8	—	88.9
Watermelon	14.0	—	—	14.0
Other vegetables, all	—	—	—	12.1

Adapted from *Vegetables and Melons Outlook VGS-307* (USDA Economic Research Service, 2005), http://www.ers.usda.gov/publications/vgs/Feb05/vgs307.pdf.

[1] Includes fresh and dehydrated onion.

[2] Other processed potato.

[3] On-cob basis.

TABLE 1.17. TRENDS IN U.S. PER CAPITA CONSUMPTION OF POTATO, SWEET POTATO, DRY BEAN, AND DRY PEA

Period	Amount (lb)			
	Potato[1]	Sweet Potato[2]	Dry Bean	Dry Pea
1947–1949 average	114	13	6.7	0.6
1957–1959 average	107	8	7.7	0.6
1965	108	6	6.6	0.4
1970	118	6	5.9	0.3
1975	122	5	6.5	0.4
1980	116	5	5.4	0.4
1985	122	5	7.1	0.5
1990	128	5	6.4	0.5
1995	139	4	7.4	0.5
2000	139	4	7.6	0.9
2004	136	4	6.7	0.6

Adapted from *Vegetable Outlook and Situation Report TVS-233* (1984); *Vegetables and Specialties TVS-260* (1993); *Vegetables and Specialties TVS-265* (1995); and *Vegetables and Melons Outlook VGS-307* (USDA Economic Research Service, 2005), http://www.ers.usda.gov/publications/vgs/Feb05/vgs307.pdf.

[1] Includes fresh and processed potato.

[2] Includes fresh and processed sweet potato.

05
WORLD VEGETABLE PRODUCTION

TABLE 1.18. IMPORTANT VEGETABLE-PRODUCING COUNTRIES, 2004

Crop	First	Second	Third
Artichoke	Italy	Spain	Argentina
Asparagus	China	Peru	United States
Bean, snap	United States	France	Mexico
Cabbage	China	India	Russian Federation
Cantaloupe	China	Turkey	United States
Carrot	China	United States	Russia
Cauliflower	China	India	Italy
Cucumber	China	Turkey	Iran
Eggplant	China	India	Turkey
Garlic	China	India	South Korea
Lettuce	China	United States	Spain
Mushroom	China	United States	Netherlands
Okra	India	Nigeria	Pakistan
Onion	China	India	United States
Pea, green	India	China	United States
Pepper	China	Mexico	Turkey
Potato	China	Russian Federation	India
Pumpkin	China	India	Ukraine
Spinach	China	United States	Japan
Strawberry[1]	United States	Spain	Japan
Sweet corn	United States	Nigeria	France
Sweet potato	China	Uganda	Nigeria
Tomato	China	United States	Turkey
Watermelon	China	Turkey	Iran
All	China	India	United States

Adapted from *Vegetables and Melons Situation and Outlook Yearbook VGS-2005* (USDA, Economic Research Service, 2005), http://www.ers.usda.gov/publications/vgs/JulyYearbook2005/VGS2005.pdf.

[1] http://usda.mannlib.cornell.edu/data-sets/specialty/95003/.

TABLE 1.19. WORLD VEGETABLE PRODUCTION, 2001–2003 AVERAGE

Country	Production (million cwt)	(%)
China	8,988.1	48.9
India	1,697.3	9.2
United States	823.6	4.5
Turkey	552.5	3.0
Russian Federation	326.0	1.7
Italy	325.5	1.7
Others	5,622.5	31.0
World	18,351.3	100.0

Adapted from *Vegetables and Melons Situation and Outlook Yearbook VGS-2005* (USDA, Economic Research Service, 2005), http://www.ers.usda.gov/publications/vgs/JulyYearbook2005/VGS2005.pdf.

TABLE 1.20. COMPOSITION OF THE EDIBLE PORTIONS OF FRESH, RAW VEGETABLES

	Amount/100 g Edible Portion										
Vegetable	Water (%)	Energy (kcal)	Protein (g)	Fat (g)	Carbohydrate (g)	Fiber (g)	Ca (mg)	P (mg)	Fe (mg)	Na (mg)	K (mg)
Artichoke	85	47	3.3	0.2	10.5	5.4	44	90	1.3	94	370
Asparagus	93	20	2.2	0.1	3.9	2.1	24	52	2.1	2	202
Bean, green	90	31	1.8	0.1	7.1	3.4	37	38	1.0	6	209
Bean, lima	70	113	6.8	0.9	20.2	4.9	34	136	3.1	8	467
Beet greens	91	22	2.2	0.1	4.3	3.7	117	41	2.6	226	762
Beet roots	88	43	1.6	0.2	9.6	2.8	16	40	0.8	78	325
Broccoli	89	34	2.8	0.4	6.6	2.6	47	66	0.7	33	316
Broccoli raab	93	22	3.2	0.5	2.9	2.7	108	73	2.1	33	196
Brussels sprouts	86	43	3.4	0.3	9.0	3.8	42	69	1.4	25	389
Cabbage, common	92	24	1.4	0.1	5.6	2.3	47	23	0.6	18	246
Cabbage, red	90	31	1.4	0.6	7.4	2.1	45	30	0.8	27	243
Cabbage, savoy	91	27	2.0	0.1	6.1	3.1	35	42	0.4	28	230
Carrot	88	41	0.9	0.2	9.6	2.8	33	35	0.3	69	320
Cauliflower	92	25	2.0	0.1	5.3	2.5	22	44	0.4	30	303
Celeriac	89	42	1.5	0.3	9.2	1.8	43	115	0.7	100	300
Celery	95	14	0.7	0.2	3.0	1.6	40	24	0.2	80	260

Chayote	95	17	0.8	0.1	3.9	1.7	17	18	0.3	2	125
Chicory, witloof	95	17	0.9	0.1	4.0	3.1	28	26	0.2	2	211
Chinese cabbage	95	13	1.5	0.2	2.2	1.0	105	37	0.8	65	252
Collards	91	30	2.5	0.4	5.7	3.6	145	10	0.2	20	169
Cucumber	95	15	0.7	0.1	3.6	0.5	16	24	0.3	2	147
Eggplant	92	24	1.0	0.2	5.7	3.4	9	25	0.2	2	230
Endive	94	17	1.3	0.2	3.4	3.1	52	28	0.8	22	314
Garlic	59	149	6.4	0.5	33.1	2.1	181	153	1.7	17	401
Kale	84	50	3.3	0.7	10.0	2.0	135	56	1.7	43	447
Kohlrabi	91	27	1.7	0.1	6.2	3.6	24	46	0.4	20	350
Leek	83	61	1.5	0.3	14.1	1.8	59	35	2.1	20	180
Lettuce, butterhead	96	13	1.4	0.2	2.3	1.1	35	33	1.2	5	238
Lettuce, crisphead	96	14	0.9	0.1	3.0	1.2	18	20	0.4	10	141
Lettuce, green leaf	94	18	1.3	0.3	3.5	0.7	68	25	1.4	9	264
Lettuce, red leaf	96	16	1.3	0.2	2.3	0.9	33	28	1.2	25	187
Lettuce, romaine	95	17	1.2	0.3	3.3	2.1	33	30	1.0	8	247
Melon, cantaloupe	90	34	0.8	0.2	8.2	0.9	9	15	0.2	16	267
Melon, casaba	92	28	1.1	0.1	6.6	0.9	11	5	0.2	9	182
Melon, honeydew	90	36	0.5	0.1	9.1	0.8	6	11	0.2	18	228
Mushroom	92	22	3.1	0.3	3.2	1.2	3	85	0.5	4	314
Mustard greens	91	26	2.7	0.2	4.9	3.3	103	43	1.5	25	354
Okra	90	31	2.0	0.1	7.0	3.2	81	63	0.8	8	303
Onion, bunching	90	32	1.8	0.2	7.3	2.6	72	37	1.5	16	276
Onion, dry	89	42	0.9	0.1	10.1	1.4	22	27	0.2	3	144
Parsley	88	36	3.0	0.8	6.3	3.3	138	58	6.2	56	554

TABLE 1.20. COMPOSITION OF THE EDIBLE PORTIONS OF FRESH, RAW VEGETABLES (*Continued*)

Vegetable	Amount/100 g Edible Portion										
	Water (%)	Energy (kcal)	Protein (g)	Fat (g)	Carbohydrate (g)	Fiber (g)	Ca (mg)	P (mg)	Fe (mg)	Na (mg)	K (mg)
Parsnip	80	75	1.2	0.3	18.0	4.9	36	71	0.6	10	375
Pea, edible-podded	89	42	2.8	0.2	7.6	2.6	43	53	2.1	4	200
Pea, green	79	81	5.4	0.4	14.5	5.1	25	108	1.5	5	244
Pepper, hot, chile	88	40	2.0	0.2	9.5	1.5	18	46	1.2	7	340
Pepper, sweet	94	20	0.9	0.2	4.6	1.7	10	20	0.3	3	175
Potato	79	77	2.0	0.1	17.5	2.2	12	57	0.8	6	421
Pumpkin	92	26	1.0	0.1	6.5	0.5	21	44	0.8	1	340
Radicchio	93	23	1.4	0.3	4.5	0.9	19	40	0.6	22	302
Radish	95	16	0.7	0.1	3.4	1.6	25	20	0.3	39	233
Rhubarb	94	21	0.9	0.2	4.5	1.8	86	14	0.2	4	288
Rutabaga	90	36	1.2	0.2	8.1	2.5	47	58	0.5	20	337
Salsify	77	82	3.3	0.2	18.6	3.3	60	75	0.7	20	380
Shallot	80	72	2.5	0.1	16.8	—	37	60	1.2	12	334
Southern pea	77	90	3.0	0.4	18.9	5.0	126	53	1.1	4	431
Spinach	91	23	2.9	0.4	3.6	2.2	99	49	2.7	79	558
Squash, acorn	88	40	0.8	0.1	10.4	1.5	33	36	0.7	3	347
Squash, butternut	86	45	1.0	0.1	11.7	2.0	48	33	0.7	4	352

Squash, Hubbard	88	40	2.0	0.5	8.7	—	14	21	0.4	7	320
Squash, scallop	94	18	1.2	0.2	3.8	—	19	36	0.4	1	182
Squash, summer	95	16	1.2	0.2	3.4	1.1	15	38	0.4	2	262
Squash, zucchini	97	16	1.2	0.2	3.4	1.1	15	38	0.4	10	262
Strawberry	91	32	0.7	0.3	7.7	2.0	16	24	0.4	1	153
Sweet corn	76	86	3.2	1.2	19.0	2.7	2	89	0.5	15	270
Sweet potato	77	86	1.6	0.1	20.1	3.0	30	47	0.6	55	337
Swiss chard	93	19	1.8	0.2	3.7	1.6	51	46	1.8	213	379
Taro	71	112	1.5	0.2	26.5	4.1	43	84	0.6	11	591
Tomato, green	93	23	1.2	0.2	5.1	1.1	13	28	0.5	13	204
Tomato, ripe	95	18	0.9	0.2	3.9	1.2	10	24	0.3	5	237
Turnip greens	90	32	1.5	0.3	7.1	3.2	190	42	1.1	40	296
Turnip roots	92	28	0.9	0.1	6.4	1.8	30	27	0.3	67	191
Watermelon	92	30	0.6	0.2	7.6	0.4	7	10	0.2	1	112
Waxgourd	96	13	0.4	0.2	3.0	2.9	19	19	0.4	111	6

Adapted from *USDA Nutrient Database for Standard Reference, Release 17* (2005), http://www.nal.usda.gov/fnic/foodcomp/Data/SR17/reports/sr17page.htm.

TABLE 1.21. VITAMIN CONTENT OF FRESH RAW, VEGETABLES

Vegetable	Amount/100 g Edible Portion					
	Vitamin A (IU)	Thiamine (mg)	Riboflavin (mg)	Niacin (mg)	Ascorbic Acid (mg)	Vitamin B_6 (mg)
Artichoke	0	0.07	0.07	1.05	11.7	0.12
Asparagus	756	0.14	0.14	0.98	5.6	0.09
Bean, green	690	0.08	0.11	0.75	16.3	0.07
Bean, lima	303	0.22	0.10	1.47	23.4	0.20
Beet greens	6,326	0.10	0.22	0.40	30.0	0.11
Beet roots	33	0.03	0.04	0.33	4.9	0.07
Broccoli	660	0.07	0.12	0.64	89.2	0.18
Broccoli raab	2,622	0.16	0.13	1.2	20.2	0.17
Brussels sprouts	754	0.14	0.09	0.75	85.0	0.22
Cabbage, common	171	0.05	0.04	0.30	32.2	0.10
Cabbage, red	1,116	0.06	0.07	0.42	57.0	0.21
Cabbage, savoy	1,000	0.07	0.03	0.30	31.0	0.19
Carrot	12,036	0.07	0.06	1.0	5.9	0.14
Cauliflower	13	0.06	0.06	0.53	46.4	0.22
Celeriac	0	0.05	0.06	0.70	8.0	0.17
Celery	449	0.02	0.06	0.32	3.1	0.07
Chayote	0	0.03	0.03	0.47	7.7	0.08
Chicory, witloof	29	0.6	0.03	0.16	2.8	0.04
Chinese cabbage	4,468	0.04	0.07	0.50	45.0	0.19

Collards	6,668	0.05	0.13	0.74	35.3	0.17
Cucumber	105	0.03	0.03	0.10	2.8	0.04
Eggplant	27	0.04	0.04	0.65	2.2	0.08
Endive	2,167	0.08	0.08	0.40	6.5	0.02
Garlic	0	0.20	0.11	0.70	31.2	1.20
Kale	15,376	0.11	0.13	1.00	120.0	0.27
Kohlrabi	36	0.05	0.02	0.40	62.0	0.15
Leek	1,667	0.06	0.03	0.40	12.0	0.23
Lettuce, butterhead	3,312	0.06	0.06	0.40	3.7	0.08
Lettuce, crisphead	502	0.04	0.03	0.12	2.8	0.04
Lettuce, green leaf	7,405	0.07	0.08	0.38	18.0	0.09
Lettuce, red leaf	7,492	0.06	0.08	0.32	3.7	0.10
Lettuce, romaine	5,807	0.10	0.10	0.31	24.0	0.07
Melon, cantaloupe	3,382	0.04	0.02	0.73	36.7	0.07
Melon, casaba	0	0.02	0.03	0.23	21.8	0.16
Melon, honeydew	40	0.08	0.02	0.60	24.8	0.06
Mushroom	0	0.09	0.42	3.85	2.4	0.12
Mustard greens	10,500	0.08	0.11	0.80	70.0	0.18
Okra	375	0.20	0.06	1.00	21.1	0.22
Onion, bunching	997	0.06	0.08	0.53	18.8	0.06
Onion, dry	2	0.05	0.03	0.08	6.4	0.15
Parsley	8,424	0.09	0.10	1.31	133.0	0.09
Parsnip	0	0.09	0.05	0.70	17.0	0.09
Pea, edible-podded	1,087	0.15	0.08	0.60	60.0	0.16
Pea, green	640	0.27	0.13	2.09	40.0	0.17
Pepper, hot, chile	1,179	0.09	0.09	0.95	242.5	0.28
Pepper, sweet	370	0.06	0.03	0.48	80.4	0.22
Potato	2	0.08	0.03	1.05	19.7	0.30

TABLE 1.21. VITAMIN CONTENT OF FRESH RAW VEGETABLES (*Continued*)

	Amount/100 g Edible Portion					
Vegetable	Vitamin A (IU)	Thiamine (mg)	Riboflavin (mg)	Niacin (mg)	Ascorbic Acid (mg)	Vitamin B_6 (mg)
Pumpkin	7,384	0.05	0.11	0.60	9.0	0.06
Radicchio	27	0.02	0.03	0.26	8.0	0.06
Radish	7	0.01	0.04	0.25	14.8	0.07
Rhubarb	102	0.02	0.03	0.30	8.0	0.02
Rutabaga	2	0.09	0.04	0.70	25.0	0.10
Salsify	0	0.08	0.22	0.50	8.0	0.28
Shallot	12	0.06	0.02	0.2	8.0	0.35
Southern pea	0	0.11	0.15	1.45	2.5	0.07
Spinach	9,377	0.08	0.19	0.72	28.1	0.20
Squash, acorn	367	0.14	0.01	0.70	11.0	0.15
Squash, butternut	10,630	0.10	0.02	1.20	21.0	0.15
Squash, Hubbard	1,367	0.07	0.04	0.50	11.0	0.15
Squash, scallop	110	0.07	0.03	0.60	18.0	0.11
Squash, summer	200	0.05	0.14	0.49	17.0	0.22
Squash, zucchini	200	0.05	0.14	0.49	17.0	0.22
Strawberry	12	0.02	0.02	0.39	58.8	0.05
Sweet corn	208	0.20	0.06	1.70	6.8	0.06
Sweet potato	14,187	0.08	0.06	0.56	2.4	0.80
Swiss chard	6,116	0.04	0.09	0.40	30.0	0.10

Taro	76	0.10	0.03	0.60	4.5	0.28
Tomato, green	642	0.06	0.04	0.50	23.4	0.08
Tomato, ripe	833	0.04	0.02	0.60	12.7	0.08
Turnip greens	0	0.07	0.10	0.60	60.0	0.26
Turnip roots	0	0.04	0.03	0.40	21.0	0.09
Watermelon	569	0.03	0.02	0.18	8.1	0.05
Waxgourd	0	0.04	0.11	0.40	13.0	0.04

Adapted from *USDA Nutrient Database for Standard Reference, Release 17* (2005), http://www.nal.usda.gov/fnic/foodcomp/Data/SR17/reports/sr17page.htm.
See also http://www.5aday.com.

PART 2

PLANT GROWING AND GREENHOUSE VEGETABLE PRODUCTION

TRANSPLANT PRODUCTION

Vegetable crops are established in the field by direct seeding or by use of vegetative propagules (see Part 3) or transplants. Transplants are produced in containers of various sorts in greenhouses, protected beds, and open fields. Either greenhouse-grown containerized or field-grown bare-root transplants can be used successfully. Generally, containerized transplants get off to a faster start but are more expensive. Containerized transplants, sometimes called "plug" transplants have become the norm for melons, pepper, tomato, and eggplant.

Transplant production is a specialized segment of the vegetable business that demands suitable facilities and careful attention to detail. For these reasons, many vegetable growers choose to purchase containerized or field-grown transplants from production specialists rather than grow them themselves.

TABLE 2.1. RELATIVE EASE OF TRANSPLANTING VEGETABLES (referring to bare-root transplants)[1]

Easy	Moderate	Require Special Care[2]
Beet	Celery	Sweet corn
Broccoli	Eggplant	Cantaloupe
Brussels sprouts	Onion	Cucumber
Cabbage	Pepper	Summer squash
Cauliflower		Watermelon
Chard		
Lettuce		
Tomato		

[1] Although containerized transplant production is the norm for most vegetables, information on bare-root transplants is available at http://pubs.caes.uga.edu/caespubs/pubs/PDF/B1144.pdf(2003).
[2] Containerized transplants are recommended.

Organic Vegetable Transplants

Organically grown vegetable transplants are not readily available from most commercial transplant producers. A good source of information on organic transplant production is at http://attra.ncat.org/attra-pub/plugs.html.

For information on organic seed production and seed handling, see J. Bonina and D. J. Cantliffe, *Seed Production and Seed Sources of Organic Vegetables* (University of Florida Cooperative Extension Service), http://edis.ifas.ufl.edu/hs227.

01
PLANT GROWING CONTAINERS

TABLE 2.2. ADVANTAGES AND DISADVANTAGES OF VARIOUS PLANT GROWING CONTAINERS

Container	Advantages	Disadvantages
Single peat pellet	No media preparation, low storage requirement	Requires individual handling in setup, limited sizes
Prespaced peat pellet	No media preparation, can be handled as a unit of 50	Limited to rather small sizes
Single peat pot	Good root penetration, easy to handle in field, available in large sizes	Difficult to separate, master container is required, dries out easily, may act as a wick in the field if not properly covered
Strip peat pots	Good root penetration, easy to handle in field, available in large sizes, saves setup and filling time	May be slow to separate in the field, dries out easily

Plastic flat with unit	Easily handled, reusable, good root penetration	Requires storage during off season, may be limited in sizes
Plastic pack	Easily handled	Roots may grow out of container causing handling problems, limited in sizes, requires some setup labor
Plastic pot	Reusable, good root penetration	Requires handling as single plant
Polyurethane foam flat	Easily handled, requires less media than similar sizes of other containers, comes in many sizes, reusable	Requires regular fertilization, plants grow slowly at first because cultural systems use low levels of nitrogen
Expanded polystyrene tray	Lightweight, easy to handle, various cell sizes and shapes, reusable, automation compatible	Need sterilization between uses, moderate investment, as trays age, roots can penetrate sidewalls of cells
Injection-molded trays	Various cell sizes, reusable, long life, compatible for automation	Large investment, need sterilization between uses
Vacuum-formed tray	Low capital investment	Short life span, needs sterilization between uses, automation incompatible due to damage to tray

Adapted in part from D. C. Sanders and G. R. Hughes (eds.), *Production of Commercial Vegetable Transplants* (North Carolina Agricultural Extension Service - 337, 1984).

02
SEEDS AND SEEDING

SEEDING SUGGESTIONS FOR GROWING TRANSPLANTS

1. *Media.* Field soil alone usually is not a desirable seeding medium because it may crust or drain poorly under greenhouse conditions. Adding sand or a sand and peat mix may produce a good seeding mixture. Many growers use artificial mixes (see page 65) because of the difficulty of obtaining field soil that is free from pests and contaminating chemicals.

   A desirable seeding mix provides good drainage but retains moisture well enough to prevent rapid fluctuations, has good aeration, is low in soluble salts, and is free from insects, diseases, and weed seeds.
2. *Seeding.* Adjust seeding rates to account for the stated germination percentages and variations in soil temperatures. Excessively thick stands result in spindly seedlings, and poor stands are wasteful of valuable bench or bed space. Seeding into containerized trays can be done mechanically using pelletized seeds. Pelletized seeds are seeds that have been coated with a clay material to facilitate planting by machine. Pelletized seeds also allow for easier singulation (one seed per cell in the tray).

   Carefully control seeding depth; most seeds should be planted at 1/4 to 1/2 in. deep. Exceptions are celery, which should only be 1/8 in. deep, and the vine crops, sweet corn, and beans, which can be seeded 1 in. or deeper.
3. *Moisture.* Maintain soil moisture in the desirable range by thorough watering after seeding and careful periodic watering as necessary. A combination of spot watering of dry areas and overall watering is usually necessary. Do not overwater.
4. *Temperature.* Be certain to maintain the desired temperature. Cooler than optimum temperatures may encourage disease, and warmer temperatures result in spindly seedlings. Seeded containerized trays can be placed in a germination room where temperature and humidity are controlled. Germination rate and germination uniformity are enhanced with this technique. Once germination has initiated, move the trays to the greenhouse.
5. *Disease control.* Use disease-free or treated seed to prevent early disease problems. Containers should be new or disease free. A disease-free seeding medium is essential. Maintain a strict sanitation program to prevent introduction of diseases. Carefully control

watering and relative humidity. Use approved fungicides as drenches or sprays when necessary. Keep greenhouse environment as dry as possible with air-circulation fans and anti-condensate plastic greenhouse covers.

6. *Transplanting.* Start transplanting when seedlings show the first true leaves so the process can be completed before the seedlings become large and overcrowded. Seedlings in containerized trays do not require transplanting to a final transplant growing container.
7. *Fertilization.* Developing transplants need light, water, and fertilization with nitrogen, phosphorus, and potassium to develop a stocky, vigorous transplant, ready for the field. Excessive fertilization, especially with nitrogen, leads to spindly, weak transplants that are difficult to establish in the field. Excessive fertilization of tomato transplants with nitrogen can lead to reduced fruit yield in the field. Only 40–60 ppm nitrogen is needed in the irrigation solution for tomato. Many commercial soilless transplant mixes have a starter nutrient charge, but this charge must be supplemented with a nutrient solution after seedlings emerge.

TABLE 2.3. APPROXIMATE SEED REQUIREMENTS FOR PLANT GROWING

Vegetable	Plants/oz Seed	Seed Required to Produce 10,000 Transplants
Asparagus	550	1¼ lb
Broccoli	5,000	2 oz
Brussels sprouts	5,000	2 oz
Cabbage	5,000	2 oz
Cantaloupe	500	1¼ lb
Cauliflower	5,000	2 oz
Celery	15,000	1 oz
Sweet corn	100	6¼ lb
Cucumber	500	1¼ lb
Eggplant	2,500	4 oz
Lettuce	10,000	1 oz
Onion	4,000	3 oz
Pepper	1,500	7 oz
Summer squash	200	3¼ lb
Tomato	4,000	3 oz
Watermelon	200	3¼ lb

To determine seed requirements per acre:

$$\frac{\text{Desired plant population}}{10{,}000} \times \text{seed required for 10,000 plants}$$

Example 1: To grow enough broccoli for a population of 20,000 plants/acre:

$$\frac{20{,}000}{10{,}000} \times 2 = 4 \text{ oz seed}$$

Example 2: To grow enough summer squash for a population of 3600 plants/acre:

$$\frac{3600}{10{,}000} \times 3\tfrac{1}{4} = 1\tfrac{1}{4} \text{ lb approximately}$$

03
TEMPERATURE AND TIME REQUIREMENTS

TABLE 2.4. RECOMMENDATIONS FOR TRANSPLANT PRODUCTION USING CONTAINERIZED TRAYS

Crop[1]	Cell Size (in.)	Seed Required for 10,000 Transplants	Seeding Depth (in.)	Optimum Germination Temperature (°F)	Germination (days)[2]	pH Tolerance[3]	Time Required (weeks)
Broccoli	0.8–1.0	2 oz	¼	85	4	6.0–6.8	5–7
Brussels sprouts	0.8–1.0	2 oz	¼	80	5	5.5–6.8	5–7
Cabbage	0.8–1.0	2 oz	¼	85	4	6.0–6.8	5–7
Cantaloupe	1.0	1¼ lb	½	90	3	6.0–6.8	4–5
Cauliflower	0.8–1.0	2 oz	¼	80	5	6.0–6.8	5–7
Celery	0.5–0.8	1 oz	⅛–¼	70	7	6.0–6.8	5–7
Collards	0.8–1.0	2 oz	¼	85	5	5.5–6.8	5–7
Cucumber	1.0	1¼ lb	½	90	3	5.5–6.8	2–3
Eggplant	1.0	4 oz	¼	85	5	6.0–6.8	5–7

TABLE 2.4. **RECOMMENDATIONS FOR TRANSPLANT PRODUCTION USING CONTAINERIZED TRAYS (*Continued*)**

Crop[1]	Cell Size (in.)	Seed Required for 10,000 Transplants	Seeding Depth (in.)	Optimum Germination Temperature (°F)	Germination (days)[2]	pH Tolerance[3]	Time Required (weeks)
Lettuce	0.5–0.8	1 oz	⅛	75	2	6.0–6.8	4
Onion	0.5–0.8	3 oz	¼	75	4	6.0–6.8	10–12
Pepper	0.5–0.8	7 oz	¼	85	8	5.5–6.8	5–7
Squash	0.5–0.8	3¼ lb	½	90	3	5.5–6.8	3–4
Tomato	1.0	3 oz	¼	85	5	5.5–6.8	5–7
Watermelon	1.0	3¼ lb	½	90	3	5.0–6.8	3–4

Adapted from C. S. Vavrina, *An Introduction to the Production of Containerized Transplants,* Florida Cooperative Extension Service Fact Sheet HS 849 (2002), http://edis.ifas.ufl.edu/HS126.

[1] Other crops can be grown as transplants by matching seed types and growing according to the above specifications (*example:* endive = lettuce). Sweet corn can be transplanted, but tap root is susceptible to breakage.

[2] Under optimum germination temperatures.

[3] Plug pH will increase over time with alkaline irrigation water.

04
PLANT GROWING MIXES

SOILLESS MIXES FOR TRANSPLANT PRODUCTION

Most commercial transplant producers use some type of soilless media for growing vegetable transplants. Most such media employ various mixtures of sphagnum peat and vermiculite or perlite, and growers may incorporate some fertilizer materials as the final media are blended. For small growers or on-farm use, similar types of media can be purchased premixed and bagged. Most of the currently used media mixes are based on variations of the Cornell mix recipe below:

TABLE 2.5. CORNELL PEAT-LITE MIXES

Component	Amount (cu yd)
Spagnum peat	0.5
Horticultural vermiculite	0.5

Additions for Specific Uses (amount/cu yd)

Addition	Seedling or Bedding Plants	Greenhouse Tomatoes Liquid Feed	Greenhouse Tomatoes Slow-release Feed
Ground limestone (lb)	5	10	10
20% superphosphate (lb)	1–2	2.5	2.5
Calcium or potassium nitrate (lb)	1	1.5	1.5
Trace element mix (oz)	2	2	2
Osmocote (lb)	0	0	10
Mag Amp (lb)	0	0	5
Wetting agent (oz)	3	3	3

Adapted from J. W. Boodley and R. Sheldrake Jr., *Cornell Peat-lite Mixes for Commercial Plant Growing*, New York State Agricultural Experiment Station, Station Agriculture Information Bulletin 43 (1982).

05
SOIL STERILIZATION

TABLE 2.6. STERILIZATION OF PLANT GROWING SOILS

Agent	Method	Recommendation
Heat	Steam	30 min at 180°F
	Aerated steam	30 min at 160°F
	Electric	30 min at 180°F
Chemical	Chloropicrin	3–5 cc/cu ft of soil. Cover for 1–3 days. Aerate for 14 days or until no odor is detected before using.
	Vapam	1 qt/100 sq ft. Allow 7–14 days before use.
	Methyl bromide	The phase-out of methyl bromide: *http://www.epa.gov/spdpublc/mbr/*

Caution: Chemical fumigants are highly toxic. Follow manufacturer's recommendations on the label.

Soluble salts, manganese, and ammonium usually increase after heat sterilization. Delay using heat-sterilized soil for at least 2 weeks to avoid problems with these toxic materials.

Adapted from K. F. Baker (ed.), *The UC System for Producing Healthy Container-grown Plants*, California Agricultural Experiment Station Manual 23 (1972).

TABLE 2.7. TEMPERATURES REQUIRED TO DESTROY PESTS IN COMPOSTS AND SOIL

Pests	30-min Temperature (°F)
Nematodes	120
Damping-off organisms	130
Most pathogenic bacteria and fungi	150
Soil insects and most viruses	160
Most weed seeds	175
Resistant weeds and resistant viruses	212

Adapted from K. F. Baker (ed.), *The UC System for Producing Healthy Container-grown Plants*, California Agricultural Experiment Station Manual 23 (1972).

TABLE 2.8. FERTILIZER FORMULATIONS FOR TRANSPLANT FERTILIZATION BASED ON NITROGEN AND POTASSIUM CONCENTRATIONS

	N and K_2O Concentrations (ppm)			
Fertilizer	50	100	200	400
	oz/100 gal[1]			
20-20-20	3.3	6.7	13.3	26.7
15-0-15	4.5	8.9	17.8	35.6
20-10-20	3.3	6.7	13.3	26.7
Ammonium nitrate	1.4	2.9	5.7	11.4
+ potassium nitrate	1.5	3.0	6.1	12.1
Calcium nitrate	3.0	6.0	12.0	24.0
+ potassium nitrate	1.5	3.0	6.0	12.0
Ammonium nitrate	1.2	2.5	4.9	9.9
+ potassium nitrate	1.5	3.0	6.0	12.0
+ monoammonium phosphate	0.5	1.1	2.2	4.3

Adapted from P. V. Nelson, "Fertilization," in E. J. Holcomb (ed.). *Bedding Plants IV: A Manual on the Culture of Bedding Plants as a Greenhouse Crop* (Batavia, Ill.: Ball, 1994), 151–176. Used with permission.

[1] 1.0 oz in 100 gal is equal to 7.5 g in 100 L.

TABLE 2.9. ELECTRICAL CONDUCTIVITY (EC) IN SOIL AND PEAT-LITE MIXES

Mineral Soils (mS)[1]	Peat-lite Mixes (mS)[1]		Interpretations
2.0+	3.5+	Excessive	Plants may be severely injured.
1.76–2.0	2.25–3.5	Very high	Plants may grow adequately, but range is near danger zone, especially if soil dries.
1.26–1.75	1.76–2.25	High	Satisfactory for established plants. May be too high for seedlings and cuttings.
0.51–1.25	1.0–1.76	Medium	Satisfactory for general plant growth. Excellent range for constant fertilization program.
0.0–0.50	0.0–1.0	Low	Low EC does no harm but may indicate low nutrient concentration.

Adapted from R. W. Langhans and E. T. Paparozzi, "Irrigation" in E. J. Holcomb (ed.), *Bedding Plant IV: A Manual on the Culture of Bedding Plants as a Greenhouse Crop* (Batavia, Ill.: Ball, 1994), 139–150. Used with permission.

[1] EC of soil determined from 1 part dry soil to 2 parts water. EC of mix determined from level tsp dry mix to 40 mL water.

TABLE 2.10. MAXIMUM ACCEPTABLE WATER QUALITY INDICES FOR BEDDING PLANTS

Variable	Plug Production	Finish Flats and Pots
pH[1] (acceptable range)	5.5–7.5	5.5–7.5
Alkalinity[2]	1.5 me/L (75 ppm)	2.0 me/L (100 ppm)
Hardness[3]	3.0 me/L (150 ppm)	3.0 me/L (150 ppm)
EC	1.0 mS	1.2 mS
Ammonium-N	20 ppm	40 ppm
Boron	0.5 ppm	0.5 ppm

Adapted from P. V. Nelson, "Fertilization," in E. J. Holcomb (ed.), *Bedding Plants IV: A Manual on the Culture of Bedding Plants as a Greenhouse Crop* (Batavia, Ill.: Ball, 1994), 151–176. Used with permission.

[1] pH not very important alone; alkalinity level more important.

[2] Moderately higher alkalinity levels are acceptable when lower amounts of limestone are incorporated into the substrate during its formulation. Very high alkalinity levels require acid injection into water source.

[3] High hardness values are not a problem if calcium and magnesium concentrations are adequate and soluble salt level is tolerable.

IRRIGATION OF TRANSPLANTS

There are two systems for application of water (and fertilizer solutions) to transplants produced in commercial operations: overhead sprinklers and subirrigation. Sprinkler systems apply water or nutrient solution by overhead water sprays from various types of sprinkler or emitter applicators. Advantages of sprinklers include the ability to apply chemicals to foliage and the ability to leach excessive salts from media. Disadvantages include high investment cost and maintenance requirements. Chemical and water application can be variable in poorly maintained systems, and nutrients can be leached if excess amounts of water are applied. One type of subirrigation uses a trough of nutrient solution in which the transplant trays are periodically floated, sometimes called *ebb and flow* or the *float system*. Water and soluble nutrients are absorbed by the media and move upward into the media. Advantages of this system include uniform application of water and nutrient solution to all flats in a trough or basin. Subirrigation with recirculation of the nutrient solution minimizes the potential for pollution because all nutrients are kept in an enclosed system.

Challenges with subirrigation include the need for care to avoid contamination of the entire trough with a disease organism. In addition, subirrigation systems restrict the potential to vary nutrient needs of different crops or developmental stages of transplants within a specific subirrigation trough.

With either production system, transplant growers must exercise care in application of water and nutrients to the crop. Excessive irrigation can leach nutrients. Irrigation and fertilization programs are linked. Changes in one program can affect the efficiency of the other program. Excessive fertilization can lead to soluble salt injury, and excessive nitrogen application can lead to overly vegetative transplants. More information on transplant irrigation and the float system is available from:

http://pubs.caes.uga.edu/caespubs/pubs/PDF/B1144.pdf (2003)
http://www.utextension.utk.edu/publications/pbfiles/PB819.pdf (1999)

07
PLANT GROWING PROBLEMS

TABLE 2.11. DIAGNOSIS AND CORRECTION OF TRANSPLANT DISORDERS

Symptoms	Possible Causes[1]	Corrective Measures
1. Spindly growth	Shade, cloudy weather, excessive watering, excessive temperature	Provide full sun, reduce temperature, restrict watering, ventilate or reduce night temperature, fertilize less frequently, provide adequate space.
2. Budless plants	Many possible causes; no conclusive cause	Maintain optimum temperature and fertilization programs.
3. Stunted plants	Low fertility	Apply fertilizer frequently in low concentration.
A. Purple leaves	Phosphorus deficiency	Apply a soluble, phosphorus-rich fertilizer at 50 ppm P every irrigation for up to 1 week.
B. Yellow leaves	Nitrogen deficiency	Apply N fertilizer solution at 50–75 ppm each irrigation for 1 week. Wash the foliage with water after application.
C. Wilted shoots	*Pythium* root rot, flooding damage, soluble salt damage to roots	Check for *Pythium* or other disease organism. Reduce irrigation amounts and reduce fertilization.
D. Discolored roots	High soluble salts from overfertilization; high soluble salts from poor soil sterilization	Leach the soil by excess watering. Do not sterilize at temperatures above 160°F. Leach soils before planting when soil tests indicate high amounts of soluble salts.
E. Normal roots	Low temperature	Maintain suitable day and night temperatures.

TABLE 2.11. DIAGNOSIS AND CORRECTION OF TRANSPLANT DISORDERS (*Continued*)

Symptoms	Possible Causes[1]	Corrective Measures
4. Tough, woody plants	Overhardening	Apply starter solution (10-55-10 or 15-30-15 at 1 oz/gal to each 6–12 sq ft bench area) 3–4 days before transplanting.
5. Water-soaked and decayed stems near the soil surface	Damping off	Use a sterile, well-drained medium. Adjust watering and ventilation practices to provide a less moist environment. Use approved fungicidal drenches.
6. Poor root growth	Poor soil aeration; poor soil drainage; low soil fertility; excess soluble salts; low temperature; residue from chemical sterilization; herbicide residue	Determine the cause and take corrective measures.
7. Green algae or mosses growing on soil surface	High soil moisture, especially in shade or during cloudy periods	Adjust watering and ventilation practices to provide a less moist environment. Use a better-drained medium.

[1] Possible causes are listed here; however, more than one factor may lead to the same symptom. Therefore, plant producers should thoroughly evaluate all possible causes of a specific disorder.

SUGGESTIONS FOR MINIMIZING DISEASES IN VEGETABLE TRANSPLANTS

Successful vegetable transplant production depends on attention to disease control. With the lack of labeled chemical pesticides, growers must focus on cultural and greenhouse management strategies to minimize opportunities for disease organisms to attack the transplant crop.

Greenhouse environment: Transplant production houses should be located at least several miles from any vegetable production field to avoid the entry of disease-causing agents in the house. Weeds around the greenhouse should be removed and the area outside the greenhouse maintained free of weeds, volunteer vegetable plants, and discarded transplants.

Media and water: All media and irrigation water should be pathogen free. If media are to be blended on site, all mixing equipment and surfaces must be routinely sanitized. Irrigation water should be drawn from pathogen-free sources. Water from ponds or recycling reservoirs should be avoided.

Planting material: Only pathogen-free seed or plant plugs should be brought into the greenhouse to initiate new transplant crops. Transplant producers should not accept seeds of unknown quality for use in transplant production. This can be a problem, especially when producing small batches of transplants from small packages of seed, e.g., for a variety trial.

Cultural practices: Attention must be given to transplant production practices such as fertilization, irrigation, and temperature so that plant vigor is optimum. Free moisture, from sprinkler irrigation or condensation, on plants should be avoided. Ventilation of houses by exhaust fans and horizontal airflow fans helps reduce free moisture on plants. Growers should follow a strict sanitation program to prevent introduction of disease organisms into the house. Weeds under benches must be removed. Outside visitors to the greenhouse should be strictly minimized, and all visitors and workers should walk through a disinfecting foot bath. All plant material and soil mix remaining between transplant crops should be removed from the house.

CONTROLLING TRANSPLANT HEIGHT

One aspect of transplant quality involves transplants of size and height that are optimum for efficient handling in the field during transplantation and for rapid establishment. Traditional means for controlling plant height included withholding water and nutrients and/or application of growth regulator chemicals. Today, growth regulator chemicals are not labeled for vegetable transplant production. Plant height control research focuses on nutrient management, temperature manipulation, light quality, and mechanical conditioning of plants.

Nutrient management: Nitrogen applied in excess often causes transplants to grow tall rapidly. Using low-N solutions with 30–50 parts per million (ppm) nitrogen helps control plant height when frequent (daily) irrigations are needed. Higher concentrations of N may be needed when irrigations are infrequent (every 3 to 4 days). Often, an intermediate N concentration (e.g., 80 ppm) is chosen for the entire transplant life cycle, and an excessive growth rate often results. Irrigation frequency should guide the N concentration. Research has shown that excessive N applied to the transplant can lead to reduced fruit yield in the field.

Moisture management: Withholding water is a time-tested method of reducing plant height, but transplants can be damaged by drought. Sometimes transplants growing in Styrofoam trays along the edge of a greenhouse walkway dry out faster than the rest of the transplants in the greenhouse. These dry plants are always shorter compared to the other transplants. Overwatering transplants should therefore be avoided, and careful attention should be given to irrigation timing.

Light intensity: Transplants grown under reduced light intensity stretch; therefore, growers must give attention to maximizing light intensity in the greenhouse. Aged polyethylene greenhouse covers should be replaced and greenhouse roofs and sides should be cleaned periodically, especially in winter. Supplementing light intensity for some transplant crops with lights may be justified.

Temperature management: Transplants grown under cooler temperatures (e.g., 50°F) are shorter than plants grown under warmer temperatures. Where possible, greenhouse temperatures can be reduced or plants moved outdoors. Under cool temperatures, the transplant production cycle is longer by several days and increased crop turnaround time may be unacceptable. For some crops, such as tomato, growing transplants under cool temperatures may lead to fruit quality problems, e.g., catfacing of fruits.

Mechanical conditioning: Shaking or brushing transplants frequently results in shorter transplants. Transplants can be brushed by several physical methods—for example, by brushing a plastic rod over the tops of the plants. This technique obviously should be practiced on dry plants only to avoid spreading disease organisms.

Day/night temperature management: The difference between the day and night temperatures (DIF) can be employed to help control plant height. With a negative DIF, day temperature is cooler than night temperature. Plants grown with a positive DIF are taller than plants grown with a zero or negative DIF. This system is not used during germination but rather is initiated when the first true leaves appear. The DIF system requires the capability to control the greenhouse temperature and is most applicable to temperate regions in winter and spring, when day temperatures are cool and greenhouses can be heated.

TABLE 2.12. VEGETABLE TRANSPLANT RESPONSE TO THE DIFFERENCE IN DAY AND NIGHT TEMPERATURE (DIF)

Common Name	Response to DIF[1]
Broccoli	3
Brussels sprouts	3
Cabbage	3
Cantaloupe	3
Cucumber	1–2
Eggplant	3
Pepper	0–1
Squash	2
Tomato	2
Watermelon	3

From E. J. Holcomb (ed.), *Bedding Plants IV* (Batavia, Ill.: Ball, 1994). Original source: J. E. Erwin and R. D. Heins, "Temperature Effects on Bedding Plant Growth," Bulletin 42:1–18, Minnesota Commercial Flower Growers Association (1993). Used with permission.

[1] 0 = no response; 3 = strong response

08
CONDITIONING TRANSPLANTS

Objective: To prepare plants to withstand stress conditions in the field. These may be low temperatures, high temperatures, drying winds, low soil moisture, or injury to the roots in transplanting. Growth rates decrease during conditioning, and the energy otherwise used in growth is stored in the plant to aid in resumption of growth after transplanting. *Conditioning* is used as an alternative to the older term, *hardening*.

Methods: Any treatment that restricts growth increases hardiness. Cool-season crops generally develop hardiness in proportion to the severity of the treatment and length of exposure and when well-conditioned withstand subfreezing temperatures. Warm-season crops, even when highly conditioned, do not withstand temperatures much below freezing.

1. *Water supply*. Gradually reduce water by watering lightly at less frequent intervals. Do not allow the plants to dry out suddenly, with severe wilting.
2. *Temperature*. Expose plants to lower temperatures (5–10°F) than those used for optimum growth. High day temperatures may reverse the effects of cool nights, making temperature management difficult. Do not expose biennials to prolonged cool temperatures, which induces bolting.
3. *Fertility*. Do not fertilize, particularly with nitrogen, immediately before or during the initial stages of conditioning. Apply a starter solution or liquid fertilizer 1 or 2 days before field setting and/or with the transplanting water (see page 78).
4. *Combinations*. Restricting water and lowering temperatures and fertility, used in combination, are perhaps more effective than any single approach.

Duration: Seven to ten days are usually sufficient to complete the conditioning process. Do not impose conditions so severe that plants are overconditioned in case of delayed planting because of poor weather. Overconditioned plants require too much time to resume growth, and early yields may be lower.

PRETRANSPLANT HANDLING OF CONTAINERIZED TRANSPLANTS

Field performance of transplants is related not only to production techniques in the greenhouse but also to handling techniques before field planting. In the containerized tray production system, plants can be delivered to the field in the trays if the transplant house is near the production fields. For long-distance transport, the plants are usually pulled from the trays and packed in boxes. Tomato plants left in trays until field planting tend to have more rapid growth rates and larger fruit yields than when transplants were pulled from the trays and packed in boxes. Storage of pulled and packed tomato plants also reduces yields of large fruits compared to plants kept in the trays. If pulled plants must be stored prior to planting, storage temperatures should be selected to avoid chilling or overheating the transplants. Transplants that must be stored for short periods can be kept successfully at 50–55°F.

TABLE 2.13. STARTER SOLUTIONS FOR FIELD TRANSPLANTING[1]

Materials	Quantity to Use in Transplanter Tank
Readily Soluble Commercial Mixtures	
8-24-8, 11-48-0 23-21-17, 13-26-13 6-25-15, 10-52-17	(Follow manufacturer's directions.) Usually 3 lb/50 gal water
Straight Nitrogen Chemicals	
Ammonium sulfate, calcium nitrate, or sodium nitrate	2½ lb/50 gal water
Ammonium nitrate	1½ lb/50 gal water
Commercial Solutions	
30% nitrogen solution	1½ pt/50 gal water
8-24-0 solution (N and P_2O_5)	2 qt/50 gal water
Regular Commercial Fertilizer Grades	
4-8-12, 5-10-5, 5-10-10, etc.	
1 lb/gal for stock solution; stir well and let settle	5 gal stock solution with 45 gal water

[1] Apply at a rate of about ½ pt/plant.

09
ADDITIONAL INFORMATION SOURCES ON TRANSPLANT PRODUCTION

Charles W. Marr, *Vegetable Transplants* (Kansas State University, 1994), http://www.oznet.ksu.edu/library/hort2/MF1103.pdf.

W. Kelley et al., *Commercial Production of Vegetable Transplants* (University of Georgia, 2003), http://pubs.caes.uga.edu/caespubs/pubcd/b1144.htm.

J. Bodnar and R. Garton, *Growing Vegetable Transplants in Plug Trays* (Ontario Ministry of Agriculture, Food, and Rural Affairs, 1996), http://www.omafra.gov.on.ca/english/crops/facts/96-023.htm.

L. Greer and K. Adam, *Organic Plug and Transplant Production* (2002), http://attra.ncat.org/attra-pub/plugs.html.

D. Krauskopf, *Vegetable Transplant Production Tips* (Michigan State University), http://www.horticulture.wisc.edu/freshveg/Publications/WFFVGC%202005/Vegetable%20Transplant%20Production%20Tips.doc.

R. Styer and D. Koranski, *Plug and Transplant Production: A Grower's Guide* (Batavia, Ill.: Ball).

GREENHOUSE CROP PRODUCTION

10
CULTURAL MANAGEMENT

CULTURAL MANAGEMENT OF GREENHOUSE VEGETABLES

Although most vegetables can be grown successfully in greenhouses, only a few are grown there commercially. Tomato, cucumber, and lettuce are the three most commonly grown vegetables in commercial greenhouses. Some general cultural management principles are discussed here.

Greenhouse Design

Successful greenhouse vegetable production depends on careful greenhouse design and construction. Consideration must be provided for environmental controls, durability of components, and ease of operations, among other factors. The publications listed at the end of this chapter offer helpful advice for construction.

Sanitation

There is no substitute for good sanitation for preventing insect and disease outbreaks in greenhouse crops.

To keep greenhouses clean, remove and destroy all dead plants, unnecessary mulch material, flats, weeds, etc. Burn or bury all plant refuse. Do not contaminate streams or water supplies with plant refuse. Weeds growing in and near the greenhouse after the cropping period should be destroyed. Do not attempt to overwinter garden or house plants in the greenhouses. Pests can also be maintained and ready for an early invasion of vegetable crops. To prevent disease organisms from carrying over on the structure of the greenhouse and on the heating pipes and walks, spray with formaldehyde (3 gal 37% formalin in 100 gal water). Immediately after spraying, close the greenhouse for 4–5 days, then ventilate. *Caution:* Wear a respirator when spraying with formaldehyde.

A 15–20-ft strip of carefully maintained lawn or bare ground around the greenhouse helps decrease trouble from two-spotted mites and other pests. To reduce entry of whiteflies, leafhoppers, and aphids from weeds and other plants near the greenhouses, spray the area growth occasionally with a labeled insecticide and control weeds around the greenhouse. Some pests can be excluded with properly designed screens.

Monitoring Pests

Insects such as greenhouse and silverleaf whiteflies, thrips, and leaf miners are attracted to shades of yellow and fly toward that color. Thus, insect traps can be made by painting pieces of board with the correct shade of yellow pigment and then covering the paint with a sticky substance. Similar traps are available commercially from several greenhouse supply sources. By placing a number of traps within the greenhouse range, it is possible to check infestations daily and be aware of early infestations. Control programs can then be commenced while populations are low.

Two-spotted mites cannot be trapped in this way, but infestations usually begin in localized areas. Check leaves daily and begin control measures as soon as the first infested areas are noted.

Spacing

Good-quality container-grown transplants should be set in arrangements to allow about 4 sq ft/plant for tomato, 5 sq ft/plant for American-type cucumber, and 7–9 sq ft/plant for European-type cucumber. Lettuce requires 36–81 sq in./plant.

Temperature

Greenhouse tomato varieties may vary in their temperature requirements, but most varieties perform well at a day minimum temperature of 70–75°F and a night minimum temperature of 62–64°F. Temperatures for cucumber seedlings should be 72–76°F day and 68°F night. In a few weeks, night temperature can be gradually lowered to 62–64°F. Night temperatures for lettuce can be somewhat lower than for tomato and cucumber.

In northern areas, provisions should be made to heat water to be used in greenhouses to about 70°F.

Pruning and Tying

Greenhouse tomatoes and cucumbers are usually pruned to a single stem by frequent removal of axillary shoots or suckers. Other pruning systems are possible and sometimes used. Various tying methods are used; one common method is to train the pruned plant around a string suspended from an overhead wire.

Pollination

Greenhouse tomatoes must be pollinated by hand or with bumblebees to assure a good set of fruit. This involves tapping or vibrating each flower cluster to transfer the pollen grains from the anther to the stigma. This should be done daily as long as there are open blossoms on the flower cluster. The pollen is transferred most readily during sunny periods and with the most difficulty on dark, cloudy days. The electric or battery-operated hand vibrator is the most widely accepted tool for vibrating tomato flower clusters. Most red-fruited varieties pollinate more easily than pink-fruited varieties and can often be pollinated satisfactorily by tapping the overhead support wires or by shaking flowers in the airstream of a motor-driven backpack air-blower. Modern growers now use bumblebees for pollinating tomato. Specially reared hives of bumblebees are purchased by the grower for this purpose.

Pollination of European seedless cucumbers causes off-shape fruit, so bees must be excluded from the greenhouse. To help overcome this, gynoecious cultivars have been developed that bear almost 100% female flowers. Only completely gynoecious and parthenocarpic (set fruits without pollination) cultivars are now recommended for commercial production.

American-type cucumbers require bees for pollination. One colony of honeybees per house should be provided. It is advisable to shade colonies from the afternoon sun and to avoid excessively high temperatures and humidities. Honeybees fly well in glass and polyethylene plastic houses but fail to work under certain other types of plastic. Under these conditions, crop failures may occur through lack of pollination.

Adapted from Ontario Ministry of Agriculture Publication 356 (1985–1986) and from G. Hochmuth, "Production of Greenhouse Tomatoes," *Florida Greenhouse Vegetable Production Handbook,* vol. 3, (2001), http://edis.ifas.ufl.edu/CV266, and from G. Hochmuth and R. Hochmuth, *Keys to Successful Tomato and Cucumber Production in Perlite Media* (2003), http://edis.ifas.ufl.edu/HS169. See also:

D. Ross, Planning and Building a Greenhouse, www.wvu.edu/~agexten/hortcult/greenhou/building.htm.

G. Hochmuth and R. Hochmuth, *Design Suggestions and Greenhouse Management for Vegetable Production in Perlite and Rockwool Media in Florida* (2004), http://edis.ifas.ufl.edu/CV195.

Ray Bucklin, "Physical Greenhouse Design Considerations," *Florida Greenhouse Vegetable Production Handbook,* vol. 2 (2001), http://edis.ifas.ufl.edu/CV254.

11
CARBON DIOXIDE ENRICHMENT

CARBON DIOXIDE ENRICHMENT OF GREENHOUSE ATMOSPHERES

The beneficial effects of adding carbon dioxide (CO_2) to the northern greenhouse environment are well established. The crops that respond most consistently to supplemental CO_2 are cucumber, lettuce, and tomato, although almost all other greenhouse crops also benefit. CO_2 enrichment of southern greenhouses probably has little benefit due to frequent ventilation requirements under the warm temperatures.

Outside air contains about 340 parts per million (ppm) CO_2 by volume. Most plants grow well at this level, but if levels are higher, the plants respond by producing more sugars. During the day, in a closed greenhouse, the plants use the CO_2 in the air and reduce the level below the normal 340 ppm. This is the point at which CO_2 addition is most important. Most crops respond to CO_2 additions up to about 1300 ppm. Somewhat lower concentrations are adequate for seedlings or when growing conditions are less than ideal.

Carbon dioxide can be obtained by burning natural gas, propane, or kerosene and directly from containers of pure CO_2. Each source has potential advantages and disadvantages. When natural gas, propane, or kerosene is burned, not only is CO_2 produced but also heat, which can supplement the normal heating system. Incomplete combustion or contaminated fuels may cause plant damage. Most sources of natural gas and propane have sufficiently low levels of impurities, but you should notify your supplier of your intention to use the fuel for CO_2 addition. Sulfur levels in the fuel should not exceed 0.02% by weight.

A number of commercial companies have burners available for natural gas, propane, and liquid fuels. The most important feature of a burner is that it burns the fuel completely.

Because photosynthesis occurs only during daylight hours, CO_2 addition is not required at night, but supplementation is recommended on dull days. Supplementation should start approximately 1 hour before sunrise, and the system should be shut off 1 hour before sunset. If supplemental lighting is used at night, intermittent addition of CO_2 or the use of a CO_2 controller may be helpful.

When ventilators are opened, it is not possible to maintain high CO_2 levels. However, it is often during these hours (high light intensity and temperature) that CO_2 supplementation is beneficial. Because maintaining optimal levels is impossible, maintaining at least ambient levels is suggested. A CO_2 controller, whereby the CO_2 concentration can be maintained at any level above ambient, is therefore useful.

One important factor is an adequate distribution system. The distribution of CO_2 mainly depends on the air movement in the greenhouse(s), for CO_2 does not travel far by diffusion. This means that if a single source of CO_2 is used for a large surface area or several connecting greenhouses, a distribution system must be installed. Air circulation (horizontal fans or a fanjet system) that moves a large volume of air provides uniform distribution within the greenhouse.

Adapted from Ontario Ministry of Agriculture and Food AGDEX 290/27 (1984) and from G. Hochmuth and R. Hochmuth (eds.), *Florida Greenhouse Vegetable Production Handbook,* vol. 3, "Greenhouse Vegetable Crop Production Guide," Florida Cooperative Extension Fact Sheet HS784 (2001), http://edis.ifas.ufl.edu/pdffiles/CV/CV26200.pdf.

12
SOILLESS CULTURE

SOILLESS CULTURE OF GREENHOUSE VEGETABLES

Well-managed field soils supply crops with sufficient water and appropriate concentrations of the 13 essential inorganic elements. A combination of desirable soil chemical, physical, and biotic characteristics provides conditions for extensive rooting, which results in anchorage, the third general quality provided to crops by soil.

When field soils are used in the greenhouse for repeated intensive crop culture, desirable soil characteristics deteriorate rapidly. Diminishing concentrations of essential elements and impaired physical properties are restored as in the field by applications of lime, fertilizer, and organic matter. Deterioration of the biotic quality of the soil by increased pathogenic microorganism and nematode populations is restricted mostly by steam sterilization.

Even with the best management, soils may deteriorate in quality over time. In addition, the costs—particularly of steam sterilization—of maintaining greenhouse soils in good condition have escalated so that soilless culture methods are competitive with or perhaps more economically favorable than soil culture. Accordingly, recent years have seen a considerable shift from soil culture to soilless culture in greenhouses. Liquid and solid media systems are used.

Liquid Soilless Culture

The nutrient film technique (NFT) is the most commonly used liquid system.

NFT growing systems consist of a series of narrow channels through which nutrient solution is recirculated from a supply tank. A plumbing system of plastic tubing and a submersible pump in the tank are the basic components. The channels are generally constructed of opaque plastic film or plastic pipe; asphalt-coated wood and fiberglass are also used. The basic characteristic of all NFT systems is the shallow depth of solution maintained in the channels. Flow is usually continuous, but sometimes systems are operated intermittently by supplying solution a few minutes every hour. The purpose of intermittent flow is to assure adequate aeration of the root systems. This also reduces the energy required, but under rapid growth conditions, plants may experience water stress if the flow period is too short or infrequent. Therefore, intermittent flow management seems better adapted to mild temperature periods or to plantings during the early stages of development. Capillary matting is sometimes used in the bottom of NFT channels, principally to avoid the side-to-side meandering of the solution stream around young root systems; it also acts as a reservoir by retaining nutrients and water during periods when flow ceases.

NFT channels are frequently designed for a single row of plants with a channel width of 6–8 inches. Wider channels of 12–15 in. are used to accommodate two rows of plants, but meandering of the shallow solution stream becomes a greater problem with greater width. To minimize this problem, small dams can be created at intervals down the channel by placing thin wooden sticks across the stream, or the channel may be lined with capillary matting. The channels should be sloped 4–6 in. per 100 ft to maintain gravity flow of the solution. Flow rate into the channels should be in the range of 1–2 qt/min.

Channel length should be limited to a maximum of 100 feet in order to minimize increased solution temperature on bright days. The ideal solution temperature for tomato is 68–77°F. Temperatures below 59° or above 86°F decrease plant growth and tomato yield. Channels of black plastic film increase solution temperature on sunny days. During cloudy weather, it may be necessary to heat the solution to the recommended temperature. Solution temperatures in black plastic channels can be decreased by shading or painting the surfaces white or silver. As an alternative to channels lined with black polyethylene, 4–6-in. PVC pipe may be used. Plant holes are spaced appropriately along the pipe. The PVC system is permanent once it is constructed; polyethylene-lined channels must be replaced for each crop. Initial costs are higher for the PVC, and sanitation between crops may be more difficult. In addition, PVC pipe systems are subject to root flooding if root masses clog pipes.

Solid Soilless Culture

Lightweight media in containers or bags and rockwool mats are the most commonly used media culture systems.

Media Culture

Soilless culture in bags, pots, or troughs with a lightweight medium is the simplest, most economical, and easiest to manage of all soilless systems. The most common media used in containerized systems of soilless culture are perlite, peat-lite, or a mixture of bark and wood chips. Container types range from long wooden troughs in which one or two rows of plants are grown to polyethylene bags or rigid plastic pots containing one to three plants. Bag or pot systems using bark chips or peat-lite are in common use throughout the United States and offer major advantages over other types of soilless culture:

1. These materials have excellent retention qualities for nutrients and water.
2. Containers of medium are readily moved in or out of the greenhouse whenever necessary or desirable.
3. They are lightweight and easily handled.
4. The medium is useful for several successive crops.
5. The containers are significantly less expensive and less time-consuming to install.
6. In comparison with recirculated hydroponic systems, the nutrient-solution system is less complicated and less expensive to manage.

From a plant nutrition standpoint, the latter advantage is of significant importance. In a recirculated system, the solution is continuously changing in its nutrient concentration because of differential plant uptake. In the bag or pot system, the solution is not recirculated. Nutrient solution is supplied from a fertilizer proportioner or large supply tank to the surface of the medium in a sufficient quantity to wet the medium. Any excess is drained away from the system through drain holes in the base of the containers. Thus, the concentration of nutrients in solution supplied to the plants is the same at each application. This eliminates the need to sample and analyze the solution periodically to determine necessary adjustments and avoids the possibility of solution excess or deficiencies.

In the bag or pot system, the volume of medium per container varies from about 1/2 cu ft in vertical polyethylene bags or pots to 2 cu ft in lay-flat bags. In the vertical bag system, 4-mil black polyethylene bags with prepunched drain holes at the bottom are common. One, but sometimes two, tomato or cucumber plants are grown in each bag. Lay-flat bags accommodate two or three plants. In either case, the bags are aligned in rows with spacing appropriate to the type of crop. It is good practice to place vertical bags or pots on a narrow sheet of plastic film to prevent root contact or penetration into the underlying soil. Plants in lay-flat bags, which have drainage slits (or overflow ports) cut along the sides an inch or so above the base, also benefit from a protective plastic sheet beneath them.

Nutrient solution is delivered to the containers by supply lines of black polyethylene tubing, spaghetti tubing, spray sticks, or ring drippers in the containers. The choice of application system is important in order to provide proper wetting of the medium at each irrigation. Texture and porosity of the growing medium and the surface area to be wetted are important considerations in making the choice. Spaghetti tubing provides a point-source wetting pattern, which might be appropriate for fine-textured media that allow water to be conducted laterally with ease. In lay-flat bags, single spaghetti tubes at individual plant holes provide good wetting of peat-lite. In a vertical bag containing a porous medium, a spray stick with a 90-degree spray pattern does a good job of irrigation if it is located to wet the majority of the surface. Ring drippers are also a good choice for vertical bags, although somewhat more expensive. When choosing an application system for bag or container culture, remember that the objective of irrigation is to distribute nutrient solution uniformly so that all of the medium is wet.

Rockwool and Perlite Culture

Rockwool is made by melting various types of rocks at very high temperatures. The resulting fibrous particles are formed into growing blocks or mats that are sterile and free of organic matter. The growing mats have a high water-holding capacity, no buffering capacity, and an initial pH of 7–8.5, which is lowered quickly with application of slightly acidic nutrient solutions. Uncovered mats, which are covered with polyethylene during setup, or polyethylene enclosed mats can be purchased. The mats are 8–12 in. wide, 36 in. long, and 3 in. thick. Perlite, a volcanic mineral, is heated and expanded into small, granular particles. Perlite has a high water-holding capacity but provides good aeration.

The greenhouse floor should be carefully leveled and covered with 3-mil black/white polyethylene, which restricts weed growth and acts as a light reflector with the white side up. The mats are placed end to end to form a row; single or double rows are spaced for the crop and greenhouse configuration.

A complete nutrient solution made with good-quality water is used for initial soaking of the mats. Large volumes are necessary because of the high water-holding capacity of the mats. Drip irrigation tubing or spaghetti tubing arranged along the plant row are used for initial soaking and, later, for fertigation. After soaking, uncovered mats are covered with polyethylene and drainage holes made in the bagged mats.

Cross-slits, corresponding in size to the propagating blocks, are made in the polyethylene mat cover at desired in-row plant spacings; usually two plants are grown in each 30-in.-long mat. The propagating blocks containing the transplant are placed on the mat, and the excess polyethylene from the cross-slit is arranged around the block. Frequent irrigation is required until plant roots are established in the mat; thereafter, fertigation is applied 4–10 times a day depending on the growing conditions and stage of crop growth. The mats are leached with good-quality water when samples taken from the mats with a syringe have increased conductivity readings.

Adapted in part from H. Johnson Jr., G. J. Hochmuth, and D. N. Maynard, "Soilless Culture of Greenhouse Vegetables," Florida Cooperative Extension Bulletin 218 (1985), and from M. Sweat and G. Hochmuth, "Production Systems," *Florida Greenhouse Vegetable Production Handbook*, vol. 3, Fact Sheet HS785, http://edis.ifas.ufl.edu/pdffiles/CV/CV26300.pdf.

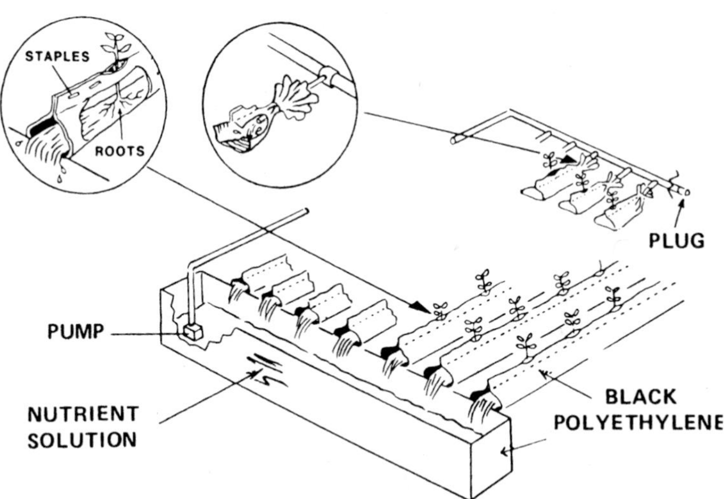

Figure 2.1. NFT culture system using polyethylene film to hold plants and supply nutrient solution through a recirculation system (From Florida Cooperative Extension Bulletin 218).

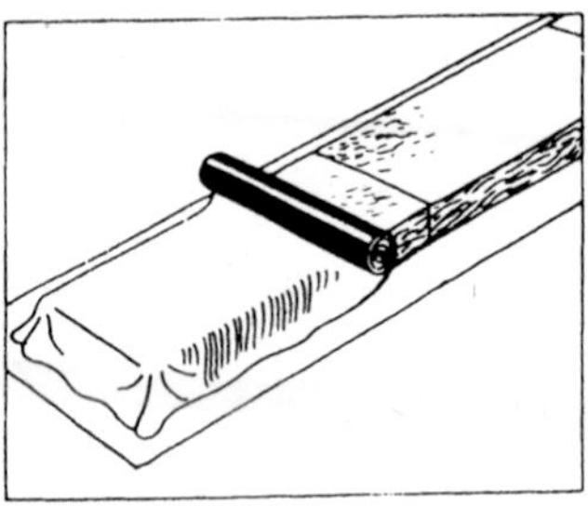

Figure 2.2. Arranged mats are covered with white/black polyethylene.

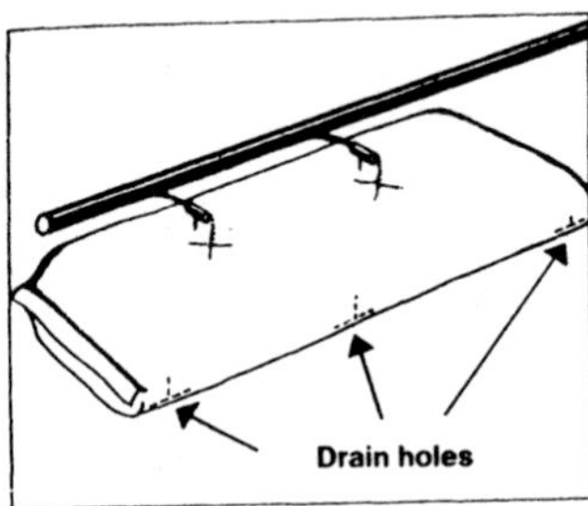

Figure 2.3. Irrigation system and drainage holes for rockwool mats enclosed in a polyethylene bag.

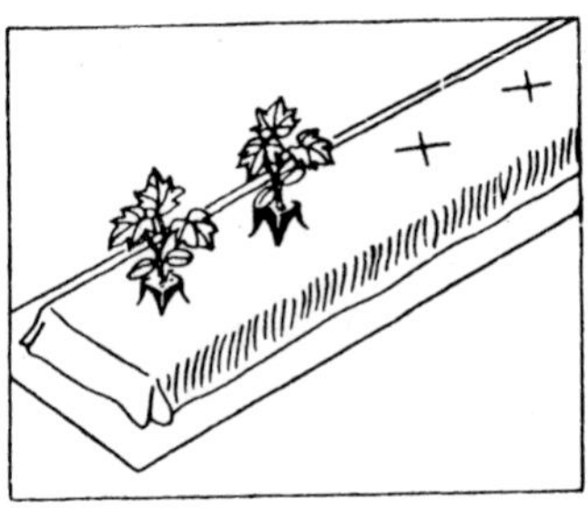

Figure 2.4. Cross-slits are made to accommodate transplants in propagation blocks.

Figure 2.5. Ordinarily, two plants are placed in each 30-in.-long mat.

Figure 2.6. Fertigation supplied by spaghetti tubing to each plant.

Figure 2.7. Fertigation supplied by drip irrigation tubing.

Figure 2.8. Removal of sample from rockwool mat with a syringe for conductivity determination.

Figures 2.2. through 2.8. Adapted from GRODAN® Instructions for cultivation—cucumbers, Grodania A/S, Denmark and used with permission.

13
NUTRIENT SOLUTIONS

NUTRIENT SOLUTIONS FOR SOILLESS CULTURE

Because the water and/or media used for soilless culture of greenhouse vegetables is devoid of essential elements, these must be supplied in a nutrient solution.

Commercially available fertilizer mixtures may be used, or nutrient solutions can be prepared from individual chemical salts. The most widely used and generally successful nutrient solution is one developed by D. R. Hoagland and D. I. Arnon at the University of California. Many commercial mixtures are based on their formula.

Detailed directions for preparation of Hoagland's nutrient solutions, which are suitable for experimental or commercial use, and the formulas for several nutrient solutions suitable for commercial use follow.

TABLE 2.14. HOAGLAND'S NUTRIENT SOLUTIONS

Salt	Stock Solution (g to make 1 L)	Final Solution (ml to make 1 L)
	Solution 1	
$Ca(NO_3)_2 \cdot 4H_2O$	236.2	5
KNO_3	101.1	5
KH_2PO_4	136.1	1
$MgSO_4 \cdot 7H_2O$	246.5	2
	Solution 2	
$Ca(NO_3)_2 \cdot 4H_2O$	236.2	4
KNO_3	101.1	6
$NH_4H_2PO_4$	115.0	1
$MgSO_4 \cdot 7H_2O$	246.5	2

Micronutrient Solution

Compound	Amount (g) Dissolved in 1 L Water
H_3BO_3	2.86
$MnCl_2 \cdot 4H_2O$	1.81
$ZnSO_4 \cdot 7H_2O$	0.22
$CuSO_4 \cdot 5H_2O$	0.08
$H_2MoO_4 \cdot H_2O$	0.02

Iron Solution

Iron chelate, such as Sequestrene 330, made to stock solution containing 1 g actual iron/L. Sequestrene 330 is 10% iron; thus, 10 g/L are required. The amounts of other chelates must be adjusted on the basis of their iron content.

Procedure: To make 1 L of Solution 1, add 5 ml $Ca(NO_3)_2 \cdot 4H_2O$ stock solution, 5 ml KNO_3, 1 ml KH_2PO_4, 2ml $MgSO_4 \cdot 7H_2O$, 1 ml micronutrient solution, and 1 ml iron solution to 800 ml distilled water. Make up to 1 L. Some plants grow better on Solution 2, which is prepared in the same way.

Adapted from D. R. Hoagland and D. I. Arnon, "The Water-culture Method for Growing Plants Without Soil," California Agricultural Experiment Station Circular 347 (1950).

SOME NUTRIENT SOLUTIONS FOR COMMERCIAL GREENHOUSE VEGETABLE PRODUCTION

These solutions are designed to be supplied directly to greenhouse vegetable crops.

TABLE 2.15. JOHNSON'S SOLUTION

Compound	Amount (g/100 gal water)
Potassium nitrate	95
Monopotassium phosphate	54
Magnesium sulfate	95
Calcium nitrate	173
Chelated iron (FeDTPA)	9
Boric acid	0.5
Manganese sulfate	0.3
Zinc sulfate	0.04
Copper sulfate	0.01
Molybdic acid	0.005

	N	P	K	Ca	Mg	S	Fe	B	Mn	Zn	Cu	Mo
ppm	105	33	138	85	25	33	2.3	0.23	0.26	0.024	0.01	0.007

TABLE 2.16. JENSEN'S SOLUTION

Compound	Amount (g/100 gal water)
Magnesium sulfate	187
Monopotassium phosphate	103
Potassium nitrate	77
Calcium nitrate	189
Chelated iron (FeDTPA)	9.6
Boric acid	1.0
Manganese chloride	0.9
Cupric chloride	0.05
Molybdic acid	0.02
Zinc sulfate	0.15

	N	P	K	Ca	Mg	S	Fe	B	Mn	Zn	Cu	Mo
ppm	106	62	156	93	48	64	3.8	0.46	0.81	0.09	0.05	0.03

Adapted from H. Johnson, Jr., G. J. Hochmuth, and D. N. Maynard, "Soilless Culture of Greenhouse Vegetables," Florida Cooperative Extension Service Bulletin 218 (1985).

TABLE 2.17. NUTRIENT SOLUTION FORMULATION FOR TOMATO GROWN IN PERLITE OR ROCKWOOL IN FLORIDA

	Stage of Growth				
	1 Transplant to First Cluster	2 First Cluster to Second	3 Second Cluster to Third	4 Third Cluster to Fifth	5 Fifth Cluster to Termination
Stock A	3.3 pt Phosphorus[1] 6 lb KCl 10 lb $MgSO_4$ 10 g $CuSO_4$ 35 g $MnSO_4$ 10 g $ZnSO_4$ 40 g Solubor 3 ml Molybdenum[2]	3.3 pt Phosphorus 6 lb KCl 10 lb $MgSO_4$ 10 g $CuSO_4$ 35 g $MnSO_4$ 10 g $ZnSO_4$ 40 g Solubor 3 ml Molybdenum	3.3 pt Phosphorus 6 lb KCl 10 lb $MgSO_4$ 2 lb KNO_3 10 g $CuSO_4$ 35 g $MnSO_4$ 10 g $ZnSO_4$ 40 g Solubor 3 ml Molybdenum	3.3 pt Phosphorus 6 lb KCl 12 lb $MgSO_4$ 2 lb KNO_3 10 g $CuSO_4$ 35 g $MnSO_4$ 10 g $ZnSO_4$ 40 g Solubor 3 ml Molybdenum	3.3 pt Phosphorus 6 lb KCl 12 lb $MgSO_4$ 6 lb KNO_3 1 lb NH_4NO_3 10 g $CuSO_4$ 35 g $MnSO_4$ 10 g $ZnSO_4$ 40 g Solubor 3 ml Molybdenum

TABLE 2.17. NUTRIENT SOLUTION FORMULATION FOR TOMATO GROWN IN PERLITE OR ROCKWOOL IN FLORIDA (*Continued*)

	Stage of Growth				
	1 Transplant to First Cluster	2 First Cluster to Second	3 Second Cluster to Third	4 Third Cluster to Fifth	5 Fifth Cluster to Termination
Stock B	2.1 gal $Ca(NO_3)_2$[3] or 11.5 lb dry $Ca(NO_3)_2$ 0.7 lb Fe 330[4]	2.4 gal $Ca(NO_3)_2$ or 13.1 lb dry $Ca(NO_3)_2$ 0.7 lb Fe 330	2.7 gal $Ca(NO_3)_2$ or 14.8 lb dry $Ca(NO_3)_2$ 0.7 lb Fe 330	3.3 gal $Ca(NO_3)_2$ or 18.0 lb dry $Ca(NO_3)_2$ 0.7 lb Fe 330	3.3 gal $Ca(NO_3)_2$ or 18.0 lb dry $Ca(NO_3)_2$ 0.7 lb Fe 330

Adapted from G. Hochmuth (ed.), *Florida Greenhouse Vegetable Production Handbook*, vol. 3, Florida Cooperative Extension Service SP-48 (1991).

[1] Phosphorus from phosphoric acid (13 lb/gal specific wt., 23% P)
[2] Molybdenum from liquid sodium molybdate (11.4 lb/gal specific wt., 17% Mo)
[3] Liquid $Ca(NO_3)_2$ from a 7-0-11 (N-P_2O_5-K_2O-Ca) solution
[4] Iron as Sequestrene 330 (10% Fe)

TABLE 2.18. RECOMMENDED NUTRIENT SOLUTION CONCENTRATIONS FOR TOMATO GROWN IN ROCKWOOL OR PERLITE IN FLORIDA

Nutrient	Stage of Growth				
	1 Transplant to First Cluster	2 First Cluster to Second	3 Second Cluster to Third	4 Third Cluster to Fifth	5 Fifth Cluster to Termination
	Final delivered nutrient solution concentration (ppm)				
N	70	80	100	120	150
P	50	50	50	50	50
K	120	120	150	150	200
Ca	150	150	150	150	150
Mg	40	40	40	50	50
S	50	50	50	60	60
Fe	2.8	2.8	2.8	2.8	2.8
Cu	0.2	0.2	0.2	0.2	0.2
Mn	0.8	0.8	0.8	0.8	0.8
Zn	0.3	0.3	0.3	0.3	0.3
B	0.7	0.7	0.7	0.7	0.7
Mo	0.05	0.05	0.05	0.05	0.05

Adapted from G. Hochmuth, "Fertilization Management for Greenhouse Vegetables," *Florida Greenhouse Vegetable Production Handbook*, vol. 3, Florida Cooperative Extension Service Fact Sheet HS787, http://edis.ifas.ufl.edu/pdffiles/CV/CV26500.pdf, and G. Hochmuth and R. Hochmuth, *Nutrient Solution Formulation for Hydroponic (Perlite, Rockwool, NFT) Tomatoes in Florida* (2001), http://edis.ifas.ufl.edu/CV216.

14
TISSUE COMPOSITION

TABLE 2.19. APPROXIMATE NORMAL TISSUE COMPOSITION OF HYDROPONICALLY GROWN GREENHOUSE VEGETABLES[1]

Element	Tomato	Cucumber
K	5–8%	8–15%
Ca	2–3%	1–3%
Mg	0.4–1.0%	0.3–0.7%
NO_3-N	14,000–20,000 ppm	10,000–20,000 ppm
PO_4-P	6,000–8,000 ppm	8,000–10,000 ppm
Fe	40–100 ppm	90–120 ppm
Zn	15–25 ppm	40–50 ppm
Cu	4–6 ppm	5–10 ppm
Mn	25–50 ppm	50–150 ppm
Mo	1–3 ppm	1–3 ppm
B	20–60 ppm	40–60 ppm

Adapted from H. Johnson, *Hydroponics: A Guide to Soilless Culture Systems,* University of California Division of Agricultural Science Leaflet 2947 (1977).

[1] Values are for recently expanded leaves, 5th or 6th from the growing tip, petiole analysis for macronutrients, leaf blade analysis for micronutrients. Expressed on a dry weight basis.

TABLE 2.20. SUFFICIENCY NUTRIENT RANGES FOR SELECTED GREENHOUSE VEGETABLE CROPS USING DRIED MOST RECENTLY MATURED WHOLE LEAVES

	Beginning of Harvest Season		Just Before Harvest
Element	Tomato	Cucumber	Lettuce
		percent	
N	3.5–4.0	2.5–5.0	2.1–5.6
P	0.4–0.6	0.5–1.0	0.5–0.9
K	2.8–4.0	3.0–6.0	4.0–8.0
Ca	0.5–2.0	0.8–6.0	0.9–2.0
Mg	0.4–1.0	0.4–0.8	0.4–0.8
S	0.4–0.8	0.4–0.8	0.2–0.5
		parts per million	
B	35–60	40–100	25–65
Cu	8–20	4–10	5–18
Fe	50–200	90–150	50–200
Mn	50–125	50–300	25–200
Mo	1–5	1–3	0.5–3.0
Zn	25–60	50–150	30–200

Adapted from G. Hochmuth, "Fertilizer Management for Greenhouse Vegetables," *Florida Greenhouse Vegetable Production Handbook,* vol. 3, Florida Cooperative Extension Fact Sheet HS787 (2001), http://edis.ifas.ufl.edu/pdffiles/cv/cv26500.pdf.

15
ADDITIONAL SOURCES OF INFORMATION ON GREENHOUSE VEGETABLES

University of Georgia, http://pubs.caes.uga.edu/caespubs/pubcd/B1182.htm.

Organic herbs, http://attra.ncat.org/attra-pub/gh-herbhold.html.

Mississippi State University, http://msucares.com/crops/comhort/index.html.

Mississippi State University, http://msucares.com/pubs/publications/p1828.htm.

North Carolina State University, http://www.ces.ncsu.edu/depts/hort/greenhouse_veg/.

University of Arizona, http://ag.arizona.edu/hydroponictomatoes/.

University of Florida, http://smallfarms.ifas.ufl.edu/greenhouse/.

List of greenhouse manufacturers/suppliers, http://nfrec-sv.ifas.ufl.edu/gh_suppliers.htm.

PART 3

FIELD PLANTING

01
TEMPERATURES FOR VEGETABLES

COOL-SEASON AND WARM-SEASON VEGETABLES

Vegetables generally can be divided into two broad groups. *Cool-season vegetables* develop edible vegetative parts, such as roots, stems, leaves, and buds or immature flower parts. Sweet potato and other tropical root crops (root used) and New Zealand spinach (leaf and stem used) are exceptions to this rule. *Warm-season vegetables* develop edible immature and mature fruits. Pea and broad bean are exceptions, being cool-season crops.

Cool-season crops generally differ from warm-season crops in the following respects:

1. They are hardy or frost tolerant.
2. Seeds germinate at cooler soil temperatures.
3. Root systems are shallower.
4. Plant size is smaller.
5. Some, the biennials, are susceptible to premature seed stalk development from exposure to prolonged cool weather.
6. They are stored near 32°F, except for the white potato. Sweet corn is the only warm-season crop held at 32°F after harvest.
7. The harvested product is not subject to chilling injury at temperatures between 32° and 50°F, as is the case with some of the warm-season vegetables.

TABLE 3.1. CLASSIFICATION OF VEGETABLE CROPS ACCORDING TO THEIR ADAPTATION TO FIELD TEMPERATURES

Cool-season Crops		
Hardy[1]		Half-hardy[1]
Asparagus	Kohlrabi	Beet
Broad bean	Leek	Carrot
Broccoli	Mustard	Cauliflower
Brussels sprouts	Onion	Celery
Cabbage	Parsley	Chard
Chive	Pea	Chicory
Collards	Radish	Chinese cabbage
Garlic	Rhubarb	Globe artichoke
Horseradish	Spinach	Endive
Kale	Turnip	Lettuce
		Parsnip
		Potato
		Salsify

Warm-season Crops	
Tender[1]	Very Tender[1]
Cowpea	Cantaloupe
New Zealand spinach	Cucumber
Snap bean	Eggplant
Soybean	Lima bean
Sweet corn	Okra
Tomato	Pepper, hot
	Pepper, sweet
	Pumpkin
	Squash
	Sweet potato
	Watermelon

Adapted from A. A. Kader, J. M. Lyons, and L. L. Morris, "Postharvest Responses of Vegetables to Preharvest Field Temperatures," *HortScience* 9:523–529 (1974).

[1] Relative resistance to frost and light freezes.

TABLE 3.2. GROWING DEGREE DAY BASE TEMPERATURES

Crop	Base Temperature (°F)[1]
Asparagus	40
Bean, snap	50
Beet	40
Broccoli	40
Cantaloupe	50
Carrot	38
Collards	40
Cucumber	55
Eggplant	60
Lettuce	40
Onion	35
Okra	60
Pea	40
Pepper	50
Potato	40
Squash	45
Strawberry	39
Sweet corn	48
Sweet potato	60
Tomato	51
Watermelon	55

Adapted from D. C. Sanders, H. J. Kirk, and C. Van Den Brink, "Growing Degree Days in North Carolina," North Carolina Agricultural Extension Service AG-236 (1980).

[1] Temperature below which growth is negligible.

TABLE 3.3. APPROXIMATE MONTHLY TEMPERATURES FOR BEST GROWTH AND QUALITY OF VEGETABLE CROPS

Some crops can be planted as temperatures approach the proper range. Cool-season crops grown in the spring must have time to mature before warm weather. Fall crops can be started in hot weather to ensure a sufficient period of cool temperature to reach maturity. Within a crop, varieties may differ in temperature requirements; hence this listing provides general rather than specific guidelines.

Temperatures (°F)			
Optimum	Minimum	Maximum	Vegetable
55–75	45	85	Chicory, chive, garlic, leek, onion, salsify, scolymus, scorzonera, shallot
60–65	40	75	Beet, broad bean, broccoli, Brussels sprouts, cabbage, chard, collards, horseradish, kale, kohlrabi, parsnip, radish, rutabaga, sorrel, spinach, turnip
60–65	45	75	Artichoke, cardoon, carrot, cauliflower, celeriac, celery, Chinese cabbage, endive, Florence fennel, lettuce, mustard, parsley, pea, potato
60–70	50	80	Lima bean, snap bean
60–75	50	95	Sweet corn, southern pea, New Zealand spinach
65–75	50	90	Chayote, pumpkin, squash
65–75	60	90	Cucumber, cantaloupe
70–75	65	80	Sweet pepper, tomato
70–85	65	95	Eggplant, hot pepper, martynia, okra, roselle, sweet potato, watermelon

TABLE 3.4. SOIL TEMPERATURE CONDITIONS FOR VEGETABLE SEED GERMINATION[1]

Vegetable	Minimum (°F)	Optimum Range (°F)	Optimum (°F)	Maximum (°F)
Asparagus	50	60–85	75	95
Bean	60	60–85	80	95
Bean, lima	60	65–85	85	85
Beet	40	50–85	85	95
Cabbage	40	45–95	85	100
Cantaloupe	60	75–95	90	100
Carrot	40	45–85	80	95
Cauliflower	40	45–85	80	100
Celery	40	60–70	70[2]	85[2]
Chard, Swiss	40	50–85	85	95
Corn	50	60–95	95	105
Cucumber	60	60–95	95	105
Eggplant	60	75–90	85	95
Lettuce	35	40–80	75	85
Okra	60	70–95	95	105
Onion	35	50–95	75	95
Parsley	40	50–85	75	90
Parsnip	35	50–70	65	85
Pea	40	40–75	75	85
Pepper	60	65–95	85	95
Pumpkin	60	70–90	90	100
Radish	40	45–90	85	95
Spinach	35	45–75	70	85
Squash	60	70–95	95	100
Tomato	50	60–85	85	95
Turnip	40	60–105	85	105
Watermelon	60	70–95	95	105

[1] Compiled by J. F. Harrington, Department of Vegetable Crops, University of California, Davis.
[2] Daily fluctuation to 60°F or lower at night is essential.

02
SCHEDULING SUCCESSIVE PLANTINGS

Successive plantings are necessary to ensure a continuous supply of produce. This seemingly easy goal is in fact extremely difficult to achieve because of interrupted planting schedules, poor stands, and variable weather conditions.

Maturity can be predicted in part by use of days to harvest or heat units. Additional flexibility is provided by using varieties that differ in time and heat units to reach maturity. Production for fresh market entails the use of days to harvest, while some processing crops may be scheduled using the heat unit concept.

Fresh Market Crops

Sweet corn is used as an example because it is an important fresh-market crop in many parts of the country and requires several plantings to obtain a season-long supply.

1. Select varieties suitable for your area that mature over time. We illustrate with five fictitious varieties maturing in 68–84 days from planting, with 4-day intervals between varieties.
2. Make the first planting as early as possible in your area.
3. Construct a table like the one following and calculate the time of the next planting, so that the earliest variety used matures 4 days after "Late" in the first planting. We chose to use "Mainseason" as the earliest variety in the second planting; thus, 88 days − 80 days = 8 days elapsed time before the second and subsequent plantings.
4. As sometimes happens, the third planting was delayed 4 days by rain. To compensate for this delay, "Midseason" is selected as the earliest variety in the third planting to provide corn 96 days after the first planting.

TABLE 3.5. EXAMPLES OF SWEET CORN PLANTINGS

Planting	Variety	Time (days) To Maturity	From First Planting	To Next Planting
First	Early	68	68	
	Second Early	72	72	
	Midseason	76	76	
	Mainseason	80	80	
	Late	84	84	
				8
Second	Mainseason	80	88	
	Late	84	92	
				12
Third	Midseason	76	96	
	Mainseason	80	100	
	Late	84	104	

Adapted from H. Tiessen, "Scheduled Planting of Vegetable Crops," Ontario Ministry of Agriculture and Food AGDEX 250/22 (1980).

Processing Crops

The heat unit system is used to schedule plantings and harvests for some processing crops, most notably pea and sweet corn. The use of this system implies that accumulated temperatures over a selected base temperature are a more accurate means of measuring growth than a time unit such as days.

In simplest form, heat units are calculated as follows:

$$\frac{\text{Maximum + minimum daily temperature}}{2} - \text{base temperature}$$

The base temperature is 40°F for pea and 50°F for sweet corn. A number of variations to this basic formula are proposed to further extend its usefulness.

Heat unit requirements to reach maturity are determined for most processing pea and sweet corn varieties and many snap bean varieties. Processors using the heat unit system assist growers in scheduling plantings to coincide with processing plant operating capacity.

TIME REQUIRED FOR SEEDLING EMERGENCE

TABLE 3.6. DAYS REQUIRED FOR SEEDLING EMERGENCE AT VARIOUS SOIL TEMPERATURES FROM SEED PLANTED ½ IN. DEEP

The days from planting to emergence constitute the time interval when a preemergence weed control treatment can be used safely and effectively. More days are required with deeper seeding because of cooler temperatures and the greater distance of growth.

	Soil Temperature (°F)								
Vegetable	32	41	50	59	68	77	86	95	104
Asparagus	NG	NG	53	24	15	10	12	20	28
Bean, lima	—	—	NG	31	18	7	7	NG	—
Bean, snap	NG	NG	NG	16	11	8	6	6	NG
Beet	—	42	17	10	6	5	5	5	—
Cabbage	—	—	15	9	6	5	4	—	—
Cantaloupe	—	—	—	—	8	4	3	—	—
Carrot	NG	51	17	10	7	6	6	9	NG
Cauliflower	—	—	20	10	6	5	5	—	—
Celery	NG	41	16	12	7	NG	NG	NG	—
Corn, sweet	NG	NG	22	12	7	4	4	3	NG
Cucumber	NG	NG	NG	13	6	4	3	3	—
Eggplant	—	—	—	—	13	8	5	—	—

TABLE 3.6. DAYS REQUIRED FOR SEEDLING EMERGENCE AT VARIOUS SOIL TEMPERATURES FROM SEED PLANTED ½ IN. DEEP (*Continued*)

	Soil Temperature (°F)								
Vegetable	32	41	50	59	68	77	86	95	104
Lettuce	49	15	7	4	3	2	3	NG	NG
Okra	NG	NG	NG	27	17	13	7	6	7
Onion	136	31	13	7	5	4	4	13	NG
Parsley	—	—	29	17	14	13	12	—	—
Parsnip	172	57	27	19	14	15	32	NG	NG
Pea	—	36	14	9	8	6	6	—	—
Pepper	NG	NG	NG	25	13	8	8	9	NG
Radish	NG	29	11	6	4	4	3	—	—
Spinach	63	23	12	7	6	5	6	NG	NG
Tomato	NG	NG	43	14	8	6	6	9	NG
Turnip	NG	NG	5	3	2	1	1	1	3
Watermelon	—	NG	—	—	12	5	4	3	—

Adapted from J. F. Harrington and P. A. Minges, "Vegetable Seed Germination," California Agricultural Extension Mimeo Leaflet (1954).

NG = No germination; — = not tested

TABLE 3.7. APPROXIMATE NUMBER OF SEEDS PER UNIT WEIGHT AND FIELD SEEDING RATES FOR TRADITIONAL PLANT DENSITIES

Vegetable	Seeds (no.)	Unit Weight	Field Seeding[1] (lb/acre)
Asparagus[2]	14,000–20,000	lb	2–3
Bean, baby lima	1,200–1,500	lb	60
Bean, Fordhook lima	400–600	lb	85
Bean, bush snap	1,600–2,000	lb	75–90
Bean, pole snap	1,600–2,000	lb	20–45
Beet	24,000–26,000	lb	10–15
Broad bean	300–800	lb	60–80
Broccoli[3]	9,000	oz	½–1½
Brussels sprouts[3]	9,000	oz	½–1½
Cabbage[3]	9,000	oz	½–1½
Cantaloupe[3]	16,000–20,000	lb	2
Cardoon	11,000	lb	4–5
Carrot	300,000–400,000	lb	2–4
Cauliflower[3]	9,000	oz	½–1½
Celeriac	72,000	oz	1–2
Celery[3]	72,000	oz	1–2
Chicory	27,000	oz	3–5
Chinese cabbage	9,000	oz	1–2
Collards	9,000	oz	2–4
Corn salad	13,000	oz	10
Cucumber	15,000–16,000	lb	3–5
Dandelion	35,000	oz	2
Eggplant[3]	6,500	oz	2
Endive	25,000	oz	3–4
Florence fennel	7,000	oz	3
Kale	9,000	oz	2–4
Kohlrabi	9,000	oz	3–5
Leek[3]	200,000	lb	4
Lettuce, head[3]	20,000–25,000	oz	1–3
Lettuce, leaf	25,000–30,000	oz	1–3

TABLE 3.7. APPROXIMATE NUMBER OF SEEDS PER UNIT WEIGHT AND FIELD SEEDING RATES FOR TRADITIONAL PLANT DENSITIES (*Continued*)

Vegetable	Seeds (no.)	Unit Weight	Field Seeding[1] (lb/acre)
Mustard	15,000	oz	3–5
New Zealand spinach	5,600	lb	15
Okra	8,000	lb	6–8
Onion, bulb[3]	130,000	lb	3–4
Onion, bunching	180,000–200,000	lb	3–4
Parsley	250,000	lb	20–40
Parsnip	192,000	lb	3–5
Pea	1,500–2,500	lb	80–250
Pepper[3]	4,200–4,600	oz	24
Pumpkin	1,500–4,000	lb	2–4
Radish	40,000–50,000	lb	10–20
Roselle	900–1,000	oz	3–5
Rutabaga	150,000–190,000	lb	1–2
Salsify	1,900	oz	8–10
Sorrel	30,000	oz	2–3
Southern pea	3,600	lb	20–40
Soybean	4,000	lb	20–40
Spinach	45,000	lb	10–15
Squash, summer	3,500–4,500	lb	4–6
Squash, winter	1,600–4,000	lb	2–4
Swiss chard	25,000	lb	6–8
Sweet corn, su, se	1,800–2,500	lb	12–15
Sweet corn, sh_2	3,000–5,000	lb	12–15
Tomato[3]	10,000–12,000	oz	½–1
Turnip	150,000–200,000	lb	1–2
Watermelon, small seed[3]	8,000–10,000	lb	1–3
Watermelon, large seed[3]	3,000–5,000	lb	2–4

[1] Actual seeding rates are adjusted to desired plant populations, germination percentage of the seed lot, and weather conditions that influence germination.

[2] 6–8 lbs/acre for crown production

[3] Transplants are used frequently instead of direct field seeding. See pages 62–64 for seeding rates for transplants.

05
PLANTING RATES FOR LARGE SEEDS

Weigh out a 1-oz sample of the seed lot and count the number of seeds.

The following table gives the approximate pounds of seed per acre for certain between-row and in-row spacings of lima bean, pea, snap bean, and sweet corn. These are based on 100% germination. If the seed germinates only 90%, for example, then divide the pounds of seed by 0.90 to get the planting rate. Do the same with other germination percentages.

Example: 30 seeds/oz to be planted in 22-in. rows at 1-in. spacing between seeds.

$$\frac{595}{0.90} = 661 \text{ lb/acre}$$

Only precision planting equipment begins to approach as exact a job of spacing as this table indicates. Moreover, field conditions such as soil structure, temperature, and moisture affect germination and final stand.

TABLE 3.8. PLANTING RATES FOR LARGE SEEDS

No. of Seeds/oz	Spacing Between Rows (in.) 18						20						22					
	Spacing Between Seeds in Row (in.)																	
	1	2	3	4	5	6	1	2	3	4	5	6	1	2	3	4	5	6
	Seed Needed (lb/acre)																	
30	726	364	242	182	146	121	655	328	218	164	131	109	595	298	198	149	119	98
40	545	273	182	136	110	90	491	246	163	123	99	82	446	223	148	112	90	74
50	440	220	146	110	88	74	396	198	132	99	79	66	361	180	120	90	72	60
60	354	178	118	90	76	59	318	159	106	80	64	53	289	145	97	73	58	48
70	312	156	104	78	62	56	281	140	94	70	56	47	256	128	85	64	51	43
80	272	136	90	68	54	46	245	123	82	62	49	41	223	112	74	56	45	37
90	242	120	82	60	48	40	218	109	73	55	44	37	198	99	66	50	40	33
100	216	108	72	54	42	38	198	99	66	50	39	33	181	90	60	45	35	30
110	198	99	66	50	40	34	173	89	59	44	35	30	161	80	54	40	32	27
120	180	90	60	45	36	30	162	81	54	40	33	27	148	74	49	37	30	25
130	168	84	56	42	34	28	152	76	51	38	31	25	138	69	46	34	28	23
140	156	78	52	38	30	26	141	70	47	35	28	24	128	64	43	32	25	22
150	146	73	49	36	28	24	131	66	44	33	26	22	119	60	40	30	24	20

No. of Seeds/oz	Spacing Between Rows (in.) 24						30						36					
	Spacing Between Seeds in Row (in.) 1	2	3	4	5	6	1	2	3	4	5	6	1	2	3	4	5	6
	Seed Needed (lb/acre)																	
30	545	273	182	136	109	91	437	219	146	109	88	73	363	182	121	91	73	61
40	408	204	136	102	82	68	328	164	106	82	66	54	272	136	91	68	55	45
50	330	165	110	82	66	55	265	132	88	66	53	44	220	110	73	55	44	37
60	265	133	88	67	57	44	212	106	71	59	43	35	177	89	59	45	38	29
70	234	117	78	59	47	39	188	94	63	47	38	31	156	78	52	39	31	26
80	204	102	68	51	41	34	164	82	53	41	33	27	136	68	45	34	27	23
90	181	90	61	45	36	30	146	73	49	37	29	25	121	60	41	30	24	20
100	162	81	55	40	32	28	131	67	44	33	27	22	108	54	37	27	21	19
110	148	74	49	37	30	25	119	60	40	30	24	20	99	49	33	25	20	17
120	135	68	45	34	27	23	108	54	36	27	22	18	90	45	30	23	18	15
130	126	63	42	32	25	21	101	51	34	25	20	17	84	42	28	21	17	14
140	117	58	39	29	23	20	94	47	32	23	19	16	78	39	26	19	15	13
150	109	55	38	27	22	18	88	44	29	22	18	15	73	37	24	18	14	12

06
SPACING OF VEGETABLES

SPACING OF VEGETABLES AND PLANT POPULATIONS

Spacing for vegetables is determined by the equipment used to plant, maintain, and harvest the crop as well as by the area required for growth of the plant without undue competition from neighboring plants. Previously, row spacings were dictated almost entirely by the space requirement of cultivating equipment. Many of the traditional row spacings can be traced to the horse cultivator.

Modern herbicides have largely eliminated the need for extensive cultivation in many crops; thus, row spacings need not be related to cultivation equipment. Instead, the plant's space requirement can be used as the determining factor. In addition, profitability is related to maximum use of field growing space.

Invariably, plant populations increase when this approach is used. A more uniform product with a higher proportion of marketable vegetables as well as higher total yields result from the closer plant spacings. The term *high-density production* has been developed to describe vegetable spacings designed to satisfy the plant's space requirement.

TABLE 3.9. HIGH-DENSITY SPACING OF VEGETABLES

Vegetable	Spacing (in.)	Plant Population (plants/acre)
Bean, snap	3 × 12	174,000
Beet	2 × 12	261,000
Carrot	1½ × 12	349,000
Cauliflower	12 × 18	29,000
Cabbage	12 × 18	29,000
Cucumber (processing)	3 × 20	104,000
Lettuce	12 × 18	29,000
Onion	1 × 12	523,000

TABLE 3.10. TRADITIONAL PLANT AND ROW SPACINGS FOR VEGETABLES

Vegetable	Between Plants in Row (in.)	Between Rows (in.)
Artichoke	48–72	84–96
Asparagus	9–15	48–72
Bean, broad	8–10	20–48
Bean, snap	2–4	18–36
Bean, lima, bush	3–6	18–36
Bean, lima, pole	8–12	36–48
Bean, pole	6–9	36–48
Beet	2–4	12–30
Broccoli[1]	12–24	18–36
Broccoli raab	3–4	24–36
Brussels sprouts	18–24	24–40
Cabbage[1]	12–24	24–36
Cantaloupe and other melons	12	60–84
Cardoon	12–18	30–42
Carrot	1–3	16–30
Cauliflower[1]	14–24	24–36
Celeriac	4–6	24–36
Celery	6–12	18–40
Chard, Swiss	12–15	24–36
Chervil	6–10	12–18
Chicory	4–10	18–24
Chinese cabbage	10–18	18–36
Chive	12–18	24–36
Collards	12–24	24–36
Corn	8–12	30–42
Cress	2–4	12–18
Cucumber[1]	8–12	36–72
Dandelion	3–6	14–24
Dasheen (taro)	24–30	42–48
Eggplant	18–30	24–48
Endive	8–12	18–24
Florence fennel	4–12	24–42
Garlic	1–3	12–24
Horseradish	12–18	30–36

TABLE 3.10. TRADITIONAL PLANT AND ROW SPACINGS FOR VEGETABLES (*Continued*)

Vegetable	Between Plants in Row (in.)	Between Rows (in.)
Jerusalem artichoke	15–18	42–48
Kale	18–24	24–36
Kohlrabi	3–6	12–36
Leek	2–6	12–36
Lettuce, cos	10–14	16–24
Lettuce, head[1]	10–15	16–24
Lettuce, leaf	8–12	12–24
Mustard	5–10	12–36
New Zealand spinach	10–20	36–60
Okra	8–24	42–60
Onion	1–4	16–24
Parsley	4–12	12–36
Parsley, Hamburg	1–3	18–36
Parsnip	2–4	18–36
Pea	1–3	24–48
Pepper[1]	12–24	18–36
Potato	6–12	30–42
Pumpkin	36–60	72–96
Radish	½–1	8–18
Radish, storage type	4–6	18–36
Rhubarb	24–48	36–60
Roselle	24–46	60–72
Rutabaga	5–8	18–36
Salsify	2–4	18–36
Scolymus	2–4	18–36
Scorzonera	2–4	18–36
Shallot	4–8	36–48
Sorrel	½–1	12–18
Southern pea	3–6	18–42
Spinach	2–6	12–36
Squash, bush[1]	24–48	36–60
Squash, vining	36–96	72–96
Strawberry[1]	10–24	24–64
Sweet potato	10–18	36–48

TABLE 3.10. TRADITIONAL PLANT AND ROW SPACINGS FOR VEGETABLES (*Continued*)

Vegetable	Between Plants in Row (in.)	Between Rows (in.)
Tomato, flat	18–48	36–60
Tomato, staked	12–24	36–48
Tomato, processing	2–10	42–60
Turnip	2–6	12–36
Turnip greens	1–4	6–12
Watercress	1–3	6–12
Watermelon	24–36	72–96

[1] Some crops can be grown double-row fashion on polyethylene mulched beds with 10–20 in. between rows.

TABLE 3.11. LENGTH OF ROW PER ACRE AT VARIOUS ROW SPACINGS

Distance Between Rows (in.)	Row Length (ft/acre)	Distance Between Rows (in.)	Row Length (ft/acre)
6	87,120	40	13,068
12	43,560	42	12,445
15	34,848	48	10,890
18	29,040	60	8,712
20	26,136	72	7,260
21	24,891	84	6,223
24	21,780	96	5,445
30	17,424	108	4,840
36	14,520	120	4,356

TABLE 3.12. NUMBER OF PLANTS PER ACRE AT VARIOUS SPACINGS

In order to obtain other spacings, divide 43,560, the number of square feet per acre, by the product of the between-rows and in-the-row spacings, each expressed as feet—that is, 43,560 divided by 0.75 (36 × 3 in. or 3 × 0.25 ft) = 58,080.

Spacing (in.)	Plants	Spacing (in.)	Plants	Spacing (ft)	Plants
12 × 1	522,720	30 × 3	69,696	6 × 1	7,260
12 × 3	174,240	30 × 6	34,848	6 × 2	3,630
12 × 6	87,120	30 × 12	17,424	6 × 3	2,420
12 × 12	43,560	30 × 15	13,939	6 × 4	1,815
		30 × 18	11,616	6 × 5	1,452
15[1] × 1	418,176	30 × 24	8,712	6 × 6	1,210
15 × 3	139,392				
15 × 6	69,696	36 × 3	58,080	7 × 1	6,223
15 × 12	34,848	36 × 6	29,040	7 × 2	3,111
		36 × 12	14,520	7 × 3	2,074
18[1] × 3	116,160	36 × 18	9,680	7 × 4	1,556
18 × 6	58,080	36 × 24	7,260	7 × 5	1,244
18 × 12	29,040	36 × 36	4,840	7 × 6	1,037
18 × 14	24,891			7 × 7	889
18 × 18	19,360	40 × 6	26,136		
		40 × 12	13,068	8 × 1	5,445
20[1] × 3	104,544	40 × 18	8,712	8 × 2	2,722
20 × 6	52,272	40 × 24	6,534	8 × 3	1,815
20 × 12	26,136			8 × 4	1,361
20 × 14	22,402	42 × 6	24,891	8 × 5	1,089
20 × 18	17,424	42 × 12	12,445	8 × 6	907
		42 × 18	8,297	8 × 8	680
21[1] × 3	99,564	42 × 24	6,223		
21 × 6	49,782	42 × 36	4,148	10 × 2	2,178
21 × 12	24,891			10 × 4	1,089
21 × 14	21,336	48 × 6	21,780	10 × 6	726
21 × 18	16,594	48 × 12	10,890	10 × 8	544
		48 × 18	7,260	10 × 10	435
24 × 3	87,120	48 × 24	5,445		

TABLE 3.12. NUMBER OF PLANTS PER ACRE AT VARIOUS SPACINGS (*Continued*)

Spacing (in.)	Plants	Spacing (in.)	Plants	Spacing (ft)	Plants
24 × 6	43,560	48 × 36	3,630		
24 × 12	21,780	48 × 48	2,722		
24 × 18	14,520				
24 × 24	10,890	60 × 12	8,712		
		60 × 18	5,808		
		60 × 24	4,356		
		60 × 36	2,904		
		60 × 48	2,178		
		60 × 60	1,742		

[1] Equivalent to double rows on beds at 30-, 36-, 40-, and 42-in. centers respectively.

07
PRECISION SEEDING

High-density plantings, high costs of hand thinning, and erratic performance of mechanical thinners have resulted in the development of precision seeding techniques. The success of precision seeding depends on having seeds with nearly 100% germination and on exact placement of each seed.

Some of the advantages of precision seeding are:

- Reduced seed costs. Only the seed that is needed is sown.
- Greater crop uniformity. Each seed is spaced equally, fewer harvests are necessary, and/or greater yield is obtained at harvest.
- Improved yields. Each plant has an equal chance to mature; yields can increase 20% to 50%.
- Improved plant stands. Seeds are dropped shorter distances, resulting in less scatter and a uniform depth of planting.
- Thinning can be reduced or eliminated.

Some precautions must be taken to ensure the proper performance of precision seeding equipment:

1. A fine, smooth seedbed is required for uniform seeding depth.
2. Seed must have high germination.
3. Seed must be uniform in size; this can be achieved by seed sizing or seed coating.
4. Seed must be of regular shape; irregular seeds such as carrot, lettuce, and onion must be coated for satisfactory precision seeding. Seed size is increased 2 to 5 times with clay or proprietary coatings.

Several types of equipment are available for precision seeding of vegetables.

Belt type—represented by the StanHay seeder. Circular holes punched in a belt accommodate the seed size. Holes are spaced along the belt at specified intervals. Coated seed usually improves the uniformity obtained with this type of seeder.

Plate type—represented by the John Deere 33 or Earth Way. Seeds drop into a notch in a horizontal plate and are transported to the drop point. The plate is vertical in the Earth Way and catches seed in a pocket in a plastic plate. Most spacing is achieved by gearing the rate of turn of the plate.

Vacuum type—represented by the Gaspardo, Heath, Monosem, StanHay, and several other seeders. Seed is drawn against holes in a vertical plate and agitated to remove excess seed. Various spacings are achieved through a combination of gears and number of holes per plate. Coated seed should not be used in these planters.

Spoon type—represented by the Nibex. Seed is scooped up out of a reservoir by small spoons (sized for the seed) and then carried to a drop shoot, where the spoon turns and drops the seed. Spacing is achieved by spoon number and gearing.

Pneumatic type—represented by the International Harvester cyclo planter. Seed is held in place against a drum until the air pressure is broken. Then it drops in tubes and is blown to the soil. This planter is recommended only for larger vegetable seed.

Grooved cylinder type—represented by the Gramor seeder. This seeder requires round seed or seed that is made round by coating. Seven seeds fall from a supply tube into a slot at the top of a metal case into a metal cylinder. The cylinder turns slowly. As it reaches the bottom of the case, the seed drops out of a diagonal slot. The seed is placed in desired increments by a combination of forward speed and turning rate. This planter can be used with seed as small as pepper seed, but it works best with coated seed.

Guidelines for Operation and Maintenance of Equipment

1. Check the planter for proper operation and replace worn parts during the off season.
2. Thoroughly understand the contents of the manufacturer's manual.
3. Make certain that the operator is trained to use the equipment and check its performance.
4. Double-check settings to obtain desired spacing and depth.
5. Make a trial run before moving to the field.
6. Operate the equipment at the recommended tractor speed.
7. Check the seed drop of each unit periodically during the planting operation.

Adapted in part from D. C. Sanders, "Precision Seeding for Vegetable Crops," North Carolina Cooperative Extension Service Publication HIL-36 (1997).

TABLE 3.13. NUMBER OF SEEDS PLANTED PER MINUTE AT VARIOUS SPEEDS AND SPACINGS[1]

Planter Speed (mph)	In-row Spacing (in.)			
	2	3	4	6
2.5	1,320	880	660	440
3.0	1,584	1,056	792	528
4.0	2,112	1,408	1,056	704
5.0	2,640	1,760	1,320	880

Adapted from Precision Planting Program, Asgrow Seed Co., Kalamazoo, Mich.

[1] For most conditions, a planter speed of 2–3 mph results in the greatest precision.

08
SEED PRIMING

Seed priming is a physiology-based seed enhancement technique designed to improve the germination characteristics of seeds. Germination speed, uniformity, and seedling vigor are all improved by priming. These benefits are especially pronounced under adverse temperature and/or moisture conditions.

The commercial applications of seed priming have been expanding rapidly in recent years. Important vegetable crops now enhanced through priming include brassicas, carrot, celery, cucurbits, lettuce, onion, pepper, and tomato. More crop species are being added on an ongoing basis.

Priming is accomplished by partially hydrating seed and maintaining it under defined moisture, temperature, and aeration conditions for a prescribed period. In this state, the seed is metabolically active. In an optimally hydrated, metabolically active state, important germination steps can be accomplished within the seed. These include repair of membranes and/or genetic material, development of immature embryos, alteration of tissues covering the embryo, and destruction or removal of dormancy blocks.

At the conclusion of the process, the seed is redried to its storage moisture level. The gains made in priming are not lost during storage. Primed seed is physiologically closer to germination than nonprimed seed. When planted at a later date, primed seed starts at this advanced state and moves directly into the final stages of germination and growth.

There are several commercial methods of seed priming. All are based on the basic principles of hydrated seed physiology. They differ in the methods used to control hydration, aeration, temperature, and dehydration. The most important commercial priming methods include:

Liquid osmotic. In this approach, seed is bubbled in a solution of known osmotic concentration (accomplished with various salts or organic osmotic agents). The osmotic properties of the solution control water uptake by the seed. The bubbling is necessary to provide sufficient oxygen to keep the seed alive during the process. The temperature of the solution is controlled throughout the process. After priming is completed, the seeds are removed, washed, and dried.

Membrane and/or flat media osmotic. This method is a variation of liquid osmotic priming. With this method, the seed is placed on a porous membrane suspended on the surface of the osmotic solution. This method addresses some of the aeration concerns associated with liquid osmotic priming but is limited by practical considerations to smaller seed lots.

Drum hydration. With this method, seeds are placed in a rotating drum and controlled quantities of water are sprayed onto the seed, bringing it to the desired moisture level. Drum rotation provides the necessary aeration to the seeds, and temperature and air flow are controlled throughout the process. After the priming period, the seed is dried by flushing air through the drum. Drum priming is a patented technology.

Solid matrix priming (SMP). With the SMP method, water uptake is controlled by suspending seed in a defined medium (or matrix) of solids (organic and/or inorganic) of known water-holding properties. The seed and matrix compete for available water, coming to equilibrium at precisely the right point for priming to occur. Aeration and temperature are precisely controlled throughout the process. After the process is complete, the seed and matrix are separated. The seed is dried to its original moisture. The SMP method is a patented technology.

In maintaining processing conditions during priming, it is important to prevent the seed from progressing too far through the germination process. If germination is allowed to progress beyond the early stages, it is too late to return to a resting state. The seed is committed to growth and cannot be redried without damage and/or reduced shelf life.

Priming alters many basic characteristics of germination and seedling emergence, as indicated below:

Germination speed. Primed seed has already accomplished the early stages of germination and begins growing much more rapidly. The total time required is cut approximately in half. This is especially important with slow-germinating species such as celery and carrot.

Increased temperature range. Primed seed emerges under both cooler and warmer temperatures than unprimed seed. Generally, the temperature range is extended 5–8°F in both directions.

More uniform emergence. The distribution of germination times within most seed lots is greatly reduced, resulting in improved uniformity.

Germination at reduced seed water content. Primed seed germinates at a lower seed water content than unprimed seed.

Control of dormancy mechanisms. In many cases, priming overcomes dormancy mechanisms that slow germination.

Germination percentages. An increase in the germination percentage occurs in many instances with individual seed lots as a result of the priming process. The increase is generally due to repair of weak or abnormal seeds within the lot.

Considerations with Primed Seed

Shelf Life of Primed Seed

Shelf life is a complicated subject and is influenced by many factors. The most important factors are crop species, seed lot quality, seed moisture content in storage, transportation and storage conditions (especially temperature), the degree to which a lot is primed, and subsequent seed treatments (fungicides, film coating, pelleting).

Assuming proper transportation and storage conditions and no other complicating factors (such as coating), deterioration in seed lot performance is rarely experienced during the growing season for which a lot was primed (generally 4 months). In most cases (assuming the same qualifiers listed above), lot performance is maintained for much longer.

As storage time increases, the risk of loss also increases. Most lots are stable, but a percentage deteriorate rapidly. Not only is the priming effect lost, but generally a significant percentage of the lot dies. Screening methods to predict high-risk lots are needed. The results of research in this area are promising, but a usable method of predicting deterioration is not yet available.

Seed should be primed for planting during the immediate growing season only. Priming seed for planting in subsequent years is discouraged. In cases where primed seed must be held for extended periods, the seed should be retested before planting to assess whether or not deterioration occurred.

Treating, Coating, and Pelleting Primed Seed

The compatibility of primed seed with any subsequent seed treatment, coating, or pelleting must be determined on a case-by-case basis. The germination characteristics may be influenced. In some cases, priming is performed to improve the vigor of lots that would otherwise not tolerate the stress of coating or pelleting. In other cases, primed seeds may be more sensitive than unprimed seeds and experience deterioration. Combinations must be tested after priming, on a case-by-case basis, before other commercial treatments are performed.

Transport and Storage Conditions

Exposure to high temperatures, even for brief periods, can induce rapid deterioration of all seeds. The risk is greater with primed seeds. In storage and transport, it is important to maintain seeds that have been enhanced under dry, cool conditions (temperatures of 70°F or lower are recommended). Unfavorable conditions may negatively influence shelf life.

Adapted from John A. Eastin and John S. Vendeland. Kamterter Products, Inc., Lincoln, Neb. "Seed Priming" Presented at Florida Seed Association Seminar (1996), and C. Parera and D. Cantliffe, "Seed Priming," *Horticultural Reviews* 16 (1994):109–141.

TABLE 3.14. STORAGE OF PLANT PARTS USED FOR VEGETATIVE PROPAGATION

Plant Part	Temperature (°F)	Relative Humidity (%)	Comments
Asparagus crowns	30–32	85–90	Roots may be trimmed to 8 in. Prevent heating and excessive drying.
Garlic bulbs	50	50–65	Fumigate for mites, if present. Hot-water-treat (120°F for 20 min) for control of stem and bulb nematode immediately before planting.
Horseradish roots	32	85–90	Pit storage is used in cold climates.
Onion sets	32	70–75	Sets may be cured naturally in the field, in trays, or artificially with warm, dry air.
Potato tubers	36–40 (extended storage), 45–50 (short storage)	90	Cure at 60–65°F and 90–95% relative humidity for 10–14 days. Move to 60–65°F 10–14 days before planting.
Rhubarb crowns	32–35	80–85	Field storage is satisfactory in cold climates.
Strawberry plants	30–32	85–90	Store in crates lined with 1.5-mil polyethylene.

TABLE 3.14. STORAGE OF PLANT PARTS USED FOR VEGETATIVE PROPAGATION (*Continued*)

Plant Part	Temperature (°F)	Relative Humidity (%)	Comments
Sweet potato roots	55–60	85–90	Cure roots at 85°F and 85–90% relative humidity for 6–8 days before storage.
Witloof chicory roots	32	90–95	Prevent excessive drying.

TABLE 3.15. FIELD REQUIREMENTS FOR VEGETATIVELY PROPAGATED CROPS

Vegetable	Plant Parts	Quantity/acre[1]
Artichoke	Root sections	807–1,261
Asparagus	Crowns	5,808–10,890
Dasheen	Corms (2–5 oz)	9–18 cwt
Garlic	Cloves	8–20 cwt
Jerusalem artichoke	Tubers (2 oz)	10–12 cwt
Horseradish	Root cuttings	9,000–11,000
Onion	Sets	5–10 cwt
Potato	Tubers or tuber sections	13–26 cwt
Rhubarb	Crown divisions	4,000–5,000
Strawberry	Plants	6,000–50,000
Sweet potato	Roots for bedding	5–6 cwt

[1] Varies with field spacing, size of individual units, and vigor of stock.

TABLE 3.16. SEED POTATOES REQUIRED PER ACRE, WITH VARIOUS PLANTING DISTANCES AND SIZES OF SEED PIECE

Spacing of Rows and Seed Pieces	Seed Piece Weights				
	1 oz	$1\frac{1}{4}$ oz	$1\frac{1}{2}$ oz	$1\frac{3}{4}$ oz	2 oz
	(Pounds of Seed / Acre)				
Rows 30 in. Apart					
8-in. spacing	1632	2040	2448	2856	3270
10-in. spacing	1308	1638	1956	2286	2614
12-in. spacing	1089	1361	1632	1908	2178
14-in. spacing	936	1164	1398	1632	1868
16-in. spacing	816	1020	1224	1428	1632
Rows 32 in. Apart					
8-in. spacing	1530	1914	2298	2682	3066
10-in. spacing	1224	1530	1836	2142	2448
12-in. spacing	1020	1278	1536	1788	2040
14-in. spacing	876	1092	1314	1530	1752
16-in. spacing	768	960	1152	1344	1536
Rows 34 in. Apart					
8-in. spacing	1440	1800	2160	2520	2880
10-in. spacing	1152	1440	1728	2016	2304
12-in. spacing	960	1200	1440	1680	1920
14-in. spacing	822	1026	1236	1440	1644
16-in. spacing	720	900	1080	1260	1440
Rows 36 in. Apart					
8-in. spacing	1362	1704	2040	2382	2724
10-in. spacing	1086	1362	1632	1902	2178
12-in. spacing	906	1134	1362	1590	1812
14-in. spacing	780	972	1164	1362	1554
16-in. spacing	678	852	1020	1188	1362
18-in. spacing	606	756	906	1056	1212

TABLE 3.16. SEED POTATOES REQUIRED PER ACRE, WITH VARIOUS PLANTING DISTANCES AND SIZES OF SEED PIECE

Spacing of Rows and Seed Pieces	Seed Piece Weights				
	1 oz	1¼ oz	1½ oz	1¾ oz	2 oz
Rows 42 in. Apart					
18-in. spacing	516	648	780	906	1038
24-in. spacing	390	486	582	678	780
30-in. spacing	312	390	468	546	624
36-in. spacing	258	324	390	456	516
Rows 48 in. Apart					
18-in. spacing	456	570	678	792	906
24-in. spacing	342	426	510	594	678
30-in. spacing	270	342	408	474	546
36-in. spacing	228	282	342	396	456

10
POLYETHYLENE MULCHES

Polyethylene mulch has been used commercially on vegetables since the early 1960s. Currently, it is used on thousands of acres of vegetables in the United States. Florida and California lead in use, with about 100,000 acres of mulched vegetables in each state.

Types of Mulch

Basically, three major colors of mulch are used commercially: black, clear, and white (or white-on-black). Black mulch is used most widely because it suppresses weed growth, resulting in less chemical usage. Further, it is useful for cool seasons because it warms the soil by contact. Clear polyethylene is used widely in the northern United States because it promotes warmer soil temperatures (by the greenhouse effect) than does black mulch. Clear mulch requires use of labeled fumigants or herbicides underneath to prevent weed growth. White or white-on-black mulch is used for fall crops established under hot summer conditions. Soils under white mulch or white-on-black mulch remain cooler because less radiant energy is absorbed by the mulch. Some growers create their own white mulch by painting the surface of black-mulched beds with white latex paint. There are some other specialized mulches, such as red or metallized mulch, used for specialized circumstances, such as weed control, plant growth regulation, and insect repelling. Films can be made of thinner gauge and high density compared with low-density polyethylenes.

Benefits of Mulch

Increases early yields. The largest benefit from polyethylene mulch is the increase in soil temperature in the bed, which promotes faster crop development and earlier yields.

Aids moisture retention. Mulch reduces evaporation from the bed soil surface. As a result, a more uniform soil moisture regime is maintained and the frequency of irrigation is reduced slightly. Irrigation is still mandatory for mulched crops so that the soil under the mulch doesn't dry out excessively. Tensiometers placed in the bed between plants can help indicate when irrigation is needed.

Inhibits weed growth. Black and white-on-black mulches greatly inhibit light penetration to the soil. Therefore, weed seedlings cannot survive under the mulch. Nutgrass can still be a problem, however. The nuts provide enough energy for the young nutgrass to puncture the mulch and emerge. Other pests, such as soilborne pathogens, insects, and nematodes, are not reduced by most mulches. Some benefit has been shown from high temperatures under clear mulch (solarization). Currently, the best measure for nutgrass and pest control under the mulch is labeled fumigation.

Reduces fertilizer leaching. Fertilizer placed in the bed under the mulch is less subject to leaching by rainfall. As a result, the fertilizer program is more efficient, and the potential exists for reducing traditional amounts of fertilizer. Heavy rainfall that floods the bed can still result in fertilizer leaching. This fertilizer can be replaced if the grower is using drip irrigation, or it can be replaced with a liquid fertilizer injection wheel.

Decreases soil compaction. Mulch acts as a barrier to the action of rainfall, which can cause soil crusting, compaction, and erosion. Less compacted soil provides a better environment for seedling emergence and root growth.

Protects fruits. Mulch reduces rain-splashed soil deposits on fruits. In addition, mulch reduces fruit rot caused by soil-inhabiting organisms because it provides a protective barrier between the fruit and the organisms.

Aids fumigation. Mulches increase the effectiveness of soil fumigant chemicals. Acting as a barrier to gas escape, mulches help keep gaseous fumigants in the soil. Recently, virtually impermeable films (VIF) have been developed to help trap fumigants better and reduce amounts of fumigants needed.

Negative Aspects of Mulch

Mulch removal and disposal. The biggest problems associated with mulch use are removal and disposal. Because most mulches are not biodegradable, they must be removed from the field after use. This usually involves some hand labor, although mulch lifting and removal machines are available. Some growers burn the mulch, but the buried edges still must be removed by hand. Disposal also presents a problem because of the quantity of waste generated.

Specialized equipment. The mulch cultural system requires a small investment in specialized equipment, including a bed press, mulch layer, and mulch transplanter or plug-mix seeder. Vacuum seeders are also available for seeding through mulch. This equipment is inexpensive and easily obtained, and some can even be manufactured on the farm.

Mulch Application

Mulch is applied by machine for commercial operations. Machines that prepare beds, fertilize, fumigate, and mulch in separate operations or in combination are available. The best option is to complete all of these operations in one pass across the field. In general, all chemicals and fertilizers are applied to the soil before mulching. Nitrogen and potassium fertilizers can be injected through a drip irrigation system under the mulch.

When laying mulch, be sure the bed is pressed firmly and that the mulch is in tight contact with the bed. This helps transfer heat from mulch to bed and reduces flapping in the wind, which results in tears and blowing of mulch from the bed. The mulch layer should be adjusted so that the edges are buried sufficiently to prevent uplifting by wind.

Degradable Mulches

Degradable plastic mulches have many of the properties and provide the usual benefits of standard polyethylene mulches. One important difference is that degradable mulches begin to break down after the film has received a predetermined amount of ultraviolet (UV) light.

When a film has received sufficient UV light, it becomes brittle and develops cracks, tears, and holes. Small sections of film may tear off and be blown around by the wind. Finally, the film breaks down into small flakes and disintegrates into the soil. The edges covered by the soil retain their strength and break down only after being disked to the surface, where they are exposed to UV light.

The use of long-lasting degradable mulches formulated for long-season crops, such as peppers, results in some plastic residue fragments remaining in the soil for the next crop. This residue is primarily the edges of film that were covered with soil. Seeding early crops in a field that had a long-term, degradable mulch the previous season should be avoided. Most plastic fragments should break down and disappear into the soil by the end of the growing season after the mulch was used.

Factors affecting the time and rate of breakdown:

- The formulation and manufacturing of the film—that is, short-, intermediate-, or long-lasting film.
- Factors that influence the amount of UV light received by the mulch film and, thus, the breakdown include the growth habit of the crop (vine or upright), the time of year the film is applied, the time between application and planting, crop vigor, and double- or single-

row planting. Weed growth, mowing off the crop, and length of time the mulch is left in the field after harvest also influence time and extent of breakdown.

- High temperatures can increase the rate of breakdown, and wind can rapidly enlarge tears and holes in film that is breaking down.
- Other factors, including depressions in the bed, footprints, animal and tire tracks, trickle irrigation tubes under the film, and stress on the plastic resulting from making holes for plants and planting, all weaken the film and increase the rate of breakdown.

Suggestions for using degradable mulches:

- Select the proper mulch formulation for the crop. Consult the company representative.
- Make uniform beds, free from depressions and footprints. Apply long-term mulches 1–2 weeks before planting. This allows the mulch to receive UV light and initiate the breakdown process. Apply short-duration films a few days to immediately before planting.
- Minimize damage to the film and avoid unnecessary footprints, especially during planting and early in the growing season.
- Maintain clean weed control between mulch strips. Shading from weed growth can slow the rate of mulch breakdown.
- Lift the soil-covered edges before final harvest or as soon as possible after harvest. This exposes some of the covered edges to UV light and starts the breakdown process.
- Mow down crop immediately after the last harvest to allow UV light to continue the breakdown process.
- When film is brittle, disk the beds. Then, angle or cross-disk to break the mulch (especially the edges) into small fragments.
- Plant a cover crop to trap larger fragments and prevent them from blowing around. Plant a border strip of a tall-growing grass around the field to prevent fragments from blowing into neighboring areas.

Adapted from G. J. Hochmuth, R.C. Hochmuth, and S. M. Olson, "Polyethylene Mulching for Early Vegetable Production in North Florida," Florida Cooperative Extension Service Circular 805 (2001), http://edis.ifas.ufl.edu/CV213, E. R. Kee, P. Mulrooney, D. Caron, M. VanGessel, and J. Whalen, "Commercial Vegetable Production," Delaware Cooperative Extension Bulletin 137 (2005); W. L. Schrader, "Plasticulture in California," Publ. 8016 (2001), http://anrcatalog.ucdavis.edu, and The Center for Plasticulture at Penn State, http://plasticulture.cas.psu.edu.

11
ROW COVERS

Row covers have been used for many years for early growth enhancement of certain vegetables in a few production areas such as San Diego County, California. New materials and methods have been developed recently that make the use of row covers a viable production practice wherever vegetables are seeded or transplanted when temperatures are below optimum and early production is desired. Row covers, when properly used, result in earlier harvest and, perhaps, greater total production. There are two general types of row covers—supported and floating; many variations of the row cover concept are possible, depending on the needs of the individual grower. Row covers generally work best when used in conjunction with black polyethylene-mulched rows or beds.

Supported Row Covers

Clear polyethylene, 5–6 ft wide and 1–1½ mils thick, is the most convenient material to use and is generally used just once. Slitted row covers have slits 5 in. long and ¾ in. apart in two rows. The slits, arranged at the upper sides of the constructed supported row cover, provide ventilation; otherwise the cover would have to be manually opened and closed each day. Hoops of no. 8 or no. 9 wire are cut 63 in. long for 5-ft-wide polyethylene.

Hoops are installed over the polyethylene-mulched crop so that the center of the hoop is 14–16 in. above the row. The slitted row cover can be mechanically applied over the hoops with a high-clearance tractor and a modified mulch applicator.

Floating Row Covers

Floating row covers are made of spun-bonded polyester and polypropylene. The material is similar to the fabrics used in the clothing industry for interlining, interfacing, and other uses. It is white or off-white, porous to air and water, lightweight (0.6 oz/sq yd) and transmits about 80% of the light. The material comes in rolls 67 in. wide and 250–2,500 ft long. One-piece blankets are also available. With care, the spun-bonded fabrics can be used two to three or more times.

Immediately after planting (seeds or transplants), the spun-bonded fabric is laid directly over the row and the edges secured with soil, boards, bricks, or wire pins. Because the material is of such light weight, the plants push it up as they grow. Accordingly, enough slack should be provided to allow for the plants to reach maximum size during the time the material is left over the plants. For bean or tomato, about 12 in. slack should be left. For a crop such as cucumber, 8 in. is sufficient. Supports should be considered in windy growing areas so plants are not damaged by cover abrasion.

Floating covers can be left over vegetables for 3–8 weeks, depending on the crop and the weather. For tomato and pepper, it can be left on for about 1 month but should be removed (at least partially) when the temperature under the covers reaches 86°F and is likely to remain that high for several hours.

Cantaloupe blossoms can withstand high temperatures, but the cover must be removed when the first female flowers appear so bees can begin pollination.

Frost Protection

Frost protection with slitted and floating covers is not as good as with solid plastic covers. A maximum of 3–4°F is all that can be expected, whereas with solid covers, frost protection of 5–7°F has been attained. Polypropylene floating row covers can be used for frost protection of vegetables and strawberries. Heavier covers 1.0–1.5 oz/yd can protect strawberries to 23–25°F. Row covers should not be viewed merely as a frost protection system but as a growth-intensifying system during cool spring weather. Therefore, do not attempt to plant very early and hope to be protected against heavy frosts. An earlier planting date of 10 days to 2 weeks is more reasonable. The purpose of row covers is to increase productivity through an economical increase of early and perhaps total production per unit area.

Adapted from O. S. Wells and J. B. Loy, "Row Covers for Intensive Vegetable Production," New Hampshire Cooperative Extension Service (1985); G. Hochmuth and R. Hochmuth, "Row Covers for Growth Enhancement," Florida Cooperative Extension Service Fact Sheet HS716 (2004), http://edis.ifas.ufl.edu/cv106.

W. L. Schrader, "Plasticulture in California," Publ. 8016 (2000), http://anrcatalog.ucdavis.edu; and The Center for Plasticulture at Penn State, http://plasticulture.cas.psu.edu.

12
WINDBREAKS

Windbreaks are important considerations in an intensive vegetable production system. Use of windbreaks can result in increased yield and earlier crop production.

Young plants are most susceptible to wind damage and sand blasting. Rye or other tall-growing grass strips between rows can provide protection from wind and windborne sand. Windbreaks can improve early plant growth and earlier crop production, particularly with melons, cucumbers, squash, peppers, eggplant, tomatoes, and okra.

A major benefit of a windbreak is improved use of moisture. Reducing the wind speed reaching the crop reduces both direct evaporation from the soil and the moisture transpired by the crop. This moisture advantage also improves conditions for seed germination. Seeds germinate more rapidly, and young plants establish root systems more quickly. Improved moisture conditions continue to enhance crop growth and development throughout the growing season.

The type and height of the windbreak determine its effect. Windbreaks can be living or nonliving. Rye strips are suggested for intensive vegetable production based on economics. In general, windbreaks should be as close as

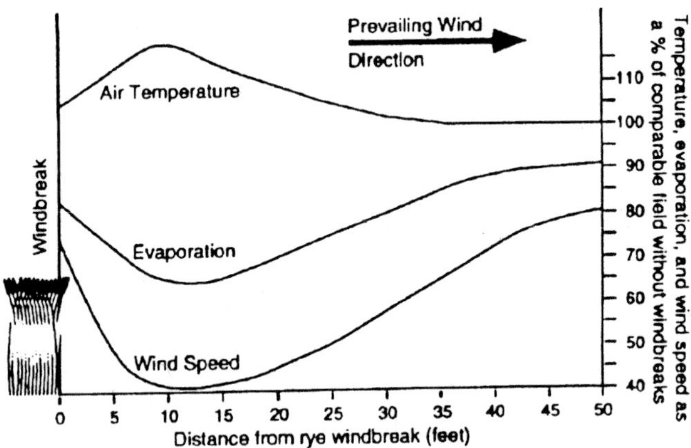

Figure 3.1. Air temperature, evaporation rate, and wind speed changes with distance from windbreak. Variables are expressed as percent of their level if the windbreak was not present.

economically viable—for example, every three or four beds of melons. The windbreak should be planted perpendicular to the prevailing wind direction. Rye strips should be planted prior to the crop to be protected so as to obtain good plant establishment and to provide adequate time for plant growth prior to beginning the next production season. Fertilization and pest management of rye windbreaks may be necessary to encourage growth to the desired height.

Adapted from J. R. Schultheis, D. C. Sanders, and K. B. Perry, "Windbreaks and Drive Rows," in D. C. Sanders (ed.), *A Guide to Intensive Vegetable Systems* (North Carolina Cooperative Extension AG-502, 1993, 9, and from R. Rouse and L. Hodges, "Windbreaks," in W. Lamont (ed.), *Production of Vegetables, Strawberries, and Cut Flowers Using Plasticulture,* (Ithaca, N.Y.: NRAES, Cooperative Extension Service, 2004), 57–66.

Vegetable Production in High Tunnels

Growing vegetables in large, walk-in, plastic-covered structures is popular in many parts of the world and is becoming more popular in the United States. Benefits of vegetable production in high-tunnels include earlier and extended-season production, tunnels can be sited on field soil, protection from rain, reduced disease, and insect pressure, less costly system than greenhouses, and less technology required. Most vegetables can be grown in a high-tunnel but the most popular crops are melons, cucumber, tomato, pepper, strawberry, salad crops, and herbs. More information on high-tunnel production can be found at The Center for Plasticulture at Penn State cited below.

http://plasticulture.cas.psu.edu/H-tunnels.html

13
ADDITIONAL SOURCES OF INFORMATION ON PLASTICULTURE

W. L. Schrader, "Plasticulture in California: Vegetable Production," University of California. Publ. 8016 (2000), http://anardatalog.ucdavis.edu.

H. Taber and V. Lawson, "Melon Row Covers," Iowa State University, http://www.public.iastate.edu/~taber/Extension/Melon/melonrc.html.

The Center for Plasticulture at Penn State, http://plasticulture.cas.psu.edu.

J. Brandle and L. Hodges, "Field Windbreaks," University of Nebraska Cooperative Extension Service EC-00-1778x.

PART 4

SOILS AND FERTILIZERS

01
NUTRIENT BEST MANAGEMENT PRACTICES (BMPs)

With the passage of the Federal Clean Water Act in 1972, states were required to assess the impact of nonpoint sources of pollution on surface and ground waters and to establish programs to minimize these sources of pollution. This Act also requires states to identify impaired water bodies and establish total maximum daily loads (TMDLs) for pollutants entering those water bodies. TMDLs are the maximum amounts of pollutants that can enter a water body and still allow it to meet its designated uses, such as swimming, potable water, fishing, etc. States have implemented various programs to address TMDLs. For example, Florida has adopted a best management practice (BMP) approach to addressing TMDLs whereby nutrient BMPs are adopted by state rule. The following definition of a BMP is taken from the Florida Department of Agriculture and Consumer Services handbook, *Water Quality/Quantity Best Management Practices for Florida Vegetable and Agronomic Crops*. BMPs are a practice or combination of practices determined by state agencies, based on research, field testing, and expert review, to be the most effective and practical on-location means, including economic and technical considerations, for improving water quality in agricultural and urban discharges. BMPs must be technically feasible, economically viable, socially acceptable, and based on sound science. Some states' programs involve incentive measures for adopting BMPs, such as cost-share for certain management practices on the farm, and other technical assistance. Agricultural producers who adopt approved BMPs, depending on the state and the program, may be "presumed to be in compliance" with state water quality standards and are eligible for cost-share funds to implement certain BMPS on the farm. States designate agencies for implementing the BMP programs and for verifying that the BMPs are effective at reducing pollutant loads.

Some information on BMP programs can be found at:

- USDA Natural Resources Conservation Service field office technical guide, http://www.nrcs.usda.gov.
- Florida Department of Agriculture and Consumer Services Water Quality/Quantity Best Management Practices for Florida Vegetable and Agronomic Crops, http://www.floridaagwaterpolicy.com/PDFs/BMPs/vegetable&agronomicCrops.pdf.
- *Farming for Clean Water in South Carolina: A Handbook of Conservation Practices* (S.C.: NRCS), http://www.sc.nrcs.usda.gov/pubs.html.

- T. K. Hartz, *Efficient Nitrogen Management for Cool-season Vegetables* (University of California Vegetable Research and Information Center), http://vric.ucdavis.edu/veginfo/topics/.
- G. Hochmuth, *Nitrogen Management Practices for Vegetable Production in Florida,* http://edis.ifas.ufl.edu/CV237.
- *Maryland Nutrient Management Manual,* http://www.mda.state.md.us/resource_conservation/nutrient_management/manual/index.php.
- *Irrigation Management Practices: Checklist for Oregon,* http://biosys.bre.orst.edu/bre/docs/irrigation.htm.
- *Nutrient and Pesticide Management* (Pacific Northwest Regional Water Program), http://www.pnwwaterweb.com/National/nut_pest.htm.
- *Nutrient Management: NRCS Conservation Practice Standard* (Wis.), http://www.dnr.state.wi.us/org/water/wm/nps/rules/.

02
ORGANIC MATTER

FUNCTION OF ORGANIC MATTER IN SOIL

Rapid decomposition of fresh organic matter contributes most effectively to the physical condition of a soil. Plenty of moisture, nitrogen, and a warm temperature speed the rate of decomposition.

Organic matter serves as a source of energy for soil microorganisms and as a source of nutrients for plants. Organic matter holds the minerals absorbed from the soil against loss by leaching until they are released for plant uptake by the action of microorganisms. Bacteria thriving on the organic matter produce complex carbohydrates that cement soil particles into aggregates. Acids produced in the decomposition of organic matter may make available mineral nutrients of the soil to crop plants. The entrance and percolation of water into and through the soil are facilitated. This reduces losses of soil by erosion. Penetration of roots through the soil is improved by good structure brought about by the decomposition of organic matter. The water-holding capacity of sands and sandy soils may be increased by the incorporation of organic matter. Aggregation in heavy soils may improve drainage. It is seldom possible to make a large permanent increase in the organic matter content of a soil.

ORGANIC SOIL AMENDMENTS

Animal manures, sludges, and plant materials have been used commercially for decades for vegetable production. Today, society demands efficient use of natural materials, so recycling of wastes into agriculture is viewed as important. Many municipalities are producing solid waste materials that can be used on the farm as soil amendments and sources of nutrients for plants. The technology of compost production and utilization is still developing. One challenge for the grower is to locate compost sources that yield consistent chemical and physical qualities. Incompletely composted waste, sometimes called *green compost,* can reduce crop growth because nitrogen is *robbed,* or used by the microorganisms to decompose the organic matter in the compost. Growers contemplating use of soil amendments should thoroughly investigate the quality of the product, including testing for nutrient content.

ENVIRONMENTAL ASPECTS OF ORGANIC SOIL AMENDMENTS

Although the addition of organic matter, such as manures, to the soil can have beneficial effects on crop performance, there are some potential negative effects. As the nitrogen is released from the organic matter, it can be subject to leaching. Heavy applications of manure can contribute to groundwater pollution unless a crop is planted soon to utilize the nitrogen. This potential can be especially great in southern climates, where nitrogen release can be rapid and most nitrogen is released in the first season after application. Plastic mulch placed over the manured soil reduces the potential for nitrate leaching. In today's environmentally aware world, manures must be used carefully to manage the released nutrients. Growers contemplating using manures as soil amendment or crop nutrient source should have the manure tested for nutrient content. The results from such tests can help determine the best rate of application so that excess nutrients such as N or P are not available for leaching or losses to erosion.

03
SOIL-IMPROVING CROPS

TABLE 4.1. SEED REQUIREMENTS OF SOIL-IMPROVING CROPS AND AREAS OF ADAPTATION

Soil-Improving Crops	Seed (lb/acre)	U.S. Area Where Crop Is Adapted
Winter Cover Crops		
Legumes		
Berseem (*Trifolium alexandrinum*)	15	West and southeast
Black medic (*Medicago lupulina*)	15	All
Black lupine (*Lupinus hirsutus*)	70	All
Clover		
Crimson (*Trifolium incarnatum*)	15	South and southeast
Bur, California (*Medicago hispida*)	25	South
Southern (*M. arabica*) unhulled	100	Southeast
Tifton (*M. rigidula*) unhulled	100	Southeast
Sour (*Melilotus indica*)	20	South
Sweet, hubam (*Melilotus alba*)	20	All
Fenugreek (*Trigonella foenumgraecum*)	30	Southwest
Field pea (*Pisum sativum*)		
Canada	80	All
Austrian winter	70	All
Horse bean (*Vicia faba*)	100	Southwest and southeast
Rough pea (*Lathyrus hirsutus*)	60	Southwest and southeast
Vetch		
Bitter (*Vicia ervilia*)	30	West and southeast
Common (*V. sativa*)	50	West and southeast
Hairy (*V. villosa*)	30	All
Hungarian (*V. pannonica*)	50	West and southeast
Monantha (*V. articulata*)	40	West and southeast
Purple (*V. bengalensis*)	40	West and southeast
Smooth (*V. villosa* var. *glabrescens*)	30	All
Woollypod (*V. dasycarpa*)	30	Southeast
Nonlegumes		
Barley (*Hordeum vulgare*)	75	All
Mustard (*Brassica nigra*)	20	All
Oat (*Avena sativa*)	75	All

TABLE 4.1. SEED REQUIREMENTS OF SOIL-IMPROVING CROPS AND AREAS OF ADAPTATION (*Continued*)

Soil-Improving Crops	Seed (lb/acre)	U.S. Area Where Crop is Adapted
Rape (*Brassica napus*)	20	All
Rye (*Secale cereale*)	75	All
Wheat (*Triticum sativum*)	75	All
Summer Cover Crops		
Legumes		
Alfalfa (*Medicago sativa*)	20	All
Beggarweed (*Desmodium purpureum*)	10	Southeast
Clover		
Alyce (*Alysicarpus vaginalis*)	20	Southeast
Crimson (*Trifolium incartum*)	15	Southeast
Red (*T. pratense*)	10	All
Cowpea (*Vigna sinensis*)	90	South and southwest
Hairy indigo (*Indigofera hirsuta*)	10	Southern tier
Lezpedeza		
Common (*Lezpedeza striata*)	25	Southeast
Korean (*L. stipulacea*)	20	Southeast
Sesbania (*Sesbania exaltata*)	30	Southwest
Soybean (*Glycine max*)	75	All
Sweet clover, white (*Melilotus alba*)	20	All
Sweet clover (*M. officinalis*)	20	All
Velvet bean (*Stizolobium deeringianum*)	100	Southeast
Nonlegumes		
Buckwheat (*Fagopyrum esculentum*)	75	All
Pearl millet (*Pennisetum glaucum*)	25	Southern and southeast
Sorghum, Hegari (*Sorghum vulgare*)	40	Western half
Sudan grass (*Sorghum vulgare* var. *sudanese*)	25	All

Adapted from *Growing Summer Cover Crops,* USDA Farmer's Bulletin 2182 (1967); P. R. Henson and E. A. Hollowell, *Winter Annual Legumes for the South,* USDA Farmers Bulletin 2146 (1960); P. R. Miller, W. A. Williams, and B. A. Madson, *Covercrops for California Agriculture,* University of California Division of Agriculture and Natural Resources Publication 21471 (1989), and *Cover Cropping in Vineyards: A Growers Handbook,* University of California Publication 3338.

DECOMPOSITION OF SOIL-IMPROVING CROPS

The normal carbon-nitrogen (C:N) ratio in soils is about 10:1. Turning under organic matter alters this ratio because most organic matter is richer in carbon than in nitrogen. Unless the residue contains at least 1.5% nitrogen, the decomposing organisms will utilize soil nitrogen as the energy source for the decomposition process. Soil organisms can tie up as much as 25 lb nitrogen per acre from the soil in the process of decomposition of carbon-rich fresh organic matter.

A soil-improving crop should be fertilized adequately with nitrogen to increase the nitrogen content somewhat and improve later decomposition. Nitrogen may have to be added as the soil-improving crop is incorporated into the soil. This speeds the decomposition and prevents a temporary shortage of nitrogen for the succeeding vegetable crop.

As a general rule, about 20 lb nitrogen should be added for each ton of dry matter for a nonlegume green-manure crop.

TABLE 4.2. APPROXIMATE CARBON-TO-NITROGEN RATIOS OF COMMON ORGANIC MATERIALS

Material	C:N Ratio
Alfalfa	12:1
Sweet clover, young	12:1
Sweet clover, mature	24:1
Rotted manure	20:1
Oat straw	75:1
Corn stalks	80:1
Timothy straw	80:1
Sawdust	300:1

04
MANURES

TYPICAL COMPOSITION OF MANURES

Manures vary greatly in their nutrient content. The kind of feed used, the percentage and type of litter or bedding, the moisture content, and the age and degree of decomposition or drying all affect the composition. Some nitrogen is lost in the process of producing commercially dried, pulverized manures. The following data are representative analyses from widely scattered reports.

TABLE 4.3. COMPOSITION OF MANURES

		Approximate Composition (% dry weight)		
Source	Dry Matter (%)	N	P_2O_5	K_2O
Dairy	15–25	0.6–2.1	0.7–1.1	2.4–3.6
Feedlot	20–40	1.0–2.5	0.9–1.6	2.4–3.6
Horse	15–25	1.7–3.0	0.7–1.2	1.2–2.2
Poultry	20–30	2.0–4.5	4.5–6.0	1.2–2.4
Sheep	25–35	3.0–4.0	1.2–1.6	3.0–4.0
Swine	20–30	3.0–4.0	0.4–0.6	0.5–1.0

TABLE 4.4. NITROGEN LOSSES FROM ANIMAL MANURE TO THE AIR BY METHOD OF APPLICATION

Application Method	Type of Manure	Nitrogen Loss (%)[1]
Broadcast without incorporation	Solid	15–30
	Liquid	10–25
Broadcast with incorporation	Solid	1–5
	Liquid	1–5
Injection (knifing)	Liquid	0–2
Irrigation	Liquid	30–40

Adapted from D. E. Chaney, L. E. Drinkwater, and G. S. Pettygrove. *Organic Soil Amendments and Fertilizers,* University of California Division of Agriculture and Natural Resources Publication 21505 (1992).

[1] Loss within 3 days of application

TYPICAL COMPOSITION OF SOME ORGANIC FERTILIZER MATERIALS

Under most environments, the nutrients in organic materials become available to plants slowly. However, mineralization of nutrients in organic matter can be hastened under warm, humid conditions. For example, in Florida, most usable nitrogen can be made available from poultry manure during one season. There is considerable variation in nutrient content among samples of organic soil amendments. Commercial manure products should have a summary of the chemical analyses on the container. Growers should have any organic soil amendment tested for nutrient content so fertilization programs can be planned. The data below are representative of many analyses noted in the literature and in reports of state analytical laboratories.

TABLE 4.5. COMPOSITION OF ORGANIC MATERIALS

Organic Materials	Percentage on a Dry Weight Basis		
	N	P_2O_5	K_2O
Bat guano	10.0	4.0	2.0
Blood	13.0	2.0	1.0
Bone meal, raw	3.0	22.0	—
Bone meal, steamed	1.0	15.0	—
Castor bean meal	5.5	2.0	1.0
Cottonseed meal	6.6	3.0	1.5
Fish meal	10.0	6.0	—
Garbage tankage	2.5	2.0	1.0
Peanut meal	7.0	1.5	1.2
Sewage sludge	1.5	1.3	0.4
Sewage sludge, activated	6.0	3.0	0.2
Soybean meal	7.0	1.2	1.5
Tankage	7.0	10.0	1.5

TABLE 4.6. COMPOSITION OF ORGANIC MATERIALS

Materials	Approximate Pounds per Ton of Dry Material			
	Moisture (%)	N	P_2O_5	K_2O
Alfalfa hay	10	50	11	50
Alfalfa straw	7	28	7	36
Barley hay	9	23	11	33
Barley straw	10	12	5	32
Bean straw	11	20	6	25
Beggarweed hay	9	50	12	56
Buckwheat straw	11	14	2	48
Clover hay				
Alyce	11	35	—	—
Bur	8	60	21	70
Crimson	11	45	11	67
Ladino	12	60	13	67
Sweet	8	60	12	38
Cowpea hay	10	60	13	36
Cowpea straw	9	20	5	38
Field pea hay	11	28	11	30
Field pea straw	10	20	5	26
Horse bean hay	9	43	—	—
Lezpedeza hay	11	41	8	22
Lezpedeza straw	10	21	—	—
Oat hay	12	26	9	20
Oat straw	10	13	5	33
Ryegrass hay	11	26	11	25
Rye hay	9	21	8	25
Rye straw	7	11	4	22
Sorghum stover, Hegari	13	18	4	—
Soybean hay	12	46	11	20
Soybean straw	11	13	6	15
Sudan grass hay	11	28	12	31
Sweet corn fodder	12	30	8	24
Velvet bean hay	7	50	11	53
Vetch hay				
Common	11	43	15	53
Hairy	12	62	15	47

TABLE 4.6. COMPOSITION OF ORGANIC MATERIALS (*Continued*)

Materials	Approximate Pounds per Ton of Dry Material			
	Moisture (%)	N	P_2O_5	K_2O
Wheat hay	10	20	8	35
Wheat straw	8	12	3	19

Adapted from *Morrison Feeds and Feeding* (Ithaca, N.Y.: Morrison, 1948).

05
SOIL TEXTURE

The particles of a soil are classified by size into sand, silt, and clay.

TABLE 4.7. CLASSIFICATION OF SOIL-PARTICLE SIZES

Soil Particle Size Classes (diameter, mm)			
2.0	0.02	0.002	0
Gravel	Sand	Silt	Clay
Particles visible with the naked eye	Particles visible under microscope	Particles visible under electron microscope	

SOIL TEXTURAL TRIANGLE

The percentage of sand, silt, and clay may be plotted on the diagram to determine the textural class of that soil.

Example: A soil containing 13% clay, 41% silt, and 46% sand would have a loam texture.

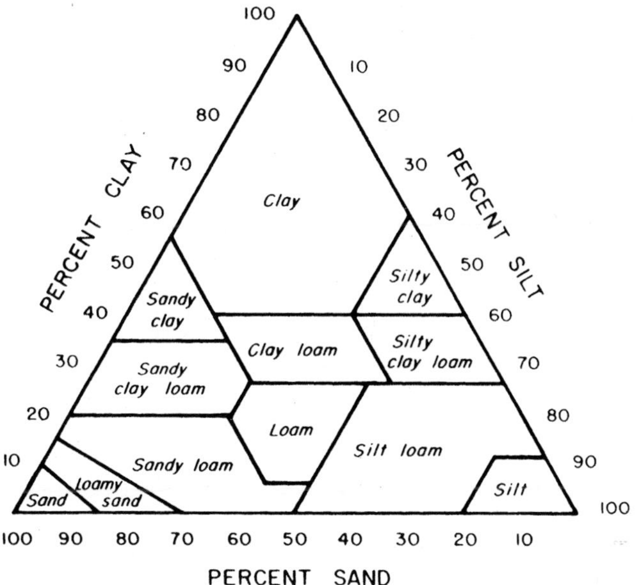

Figure 4.1. Soil textural triangle. From Soil Conservation Service, *Soil Survey Manual,* USDA Agricultural Handbook 18 (1951).

06
SOIL REACTION

RELATIVE TOLERANCE OF VEGETABLE CROPS TO SOIL ACIDITY

Vegetables in the slightly tolerant group can be grown successfully on soils that are on the alkaline side of neutrality. They do well up to pH 7.6 if there is no deficiency of essential nutrients. Vegetables in the very tolerant group grow satisfactorily at a soil pH as low as 5.0. For the most part, even the most tolerant crops grow better at pH 6.0–6.8 than in more acidic soils. Calcium, phosphorus, magnesium, and molybdenum are the nutrients most likely to be deficient in acidic soils.

TABLE 4.8. TOLERANCE OF VEGETABLES TO SOIL ACIDITY

Slightly Tolerant (pH 6.8–6.0)	Moderately Tolerant (pH 6.8–5.5)	Very Tolerant (pH 6.8–5.0)
Asparagus	Bean	Chicory
Beet	Bean, lima	Dandelion
Broccoli	Brussels sprouts	Endive
Cabbage	Carrot	Fennel
Cantaloupe	Collards	Potato
Cauliflower	Cucumber	Rhubarb
Celery	Eggplant	Shallot
Chard, Swiss	Garlic	Sorrel
Chinese cabbage	Gherkin	Sweet potato
Cress	Horseradish	Watermelon
Leek	Kale	
Lettuce	Kohlrabi	
New Zealand spinach	Mustard	
Okra	Parsley	
Onion	Pea	
Orach	Pepper	
Parsnip	Pumpkin	
Salsify	Radish	
Soybean	Rutabaga	
Spinach	Squash	
Watercress	Strawberry	
	Sweet corn	
	Tomato	
	Turnip	

EFFECT OF SOIL REACTION ON AVAILABILITY OF NUTRIENTS

Soil reaction affects plants by influencing the availability of nutrients. Changes in soil reaction caused by liming or by the use of sulfur and acid-forming fertilizers may increase or decrease the supply of the nutrients available to the plants.

The general relationship between soil reaction and availability of plant nutrients in organic soils differs from that in mineral soils. The diagrams

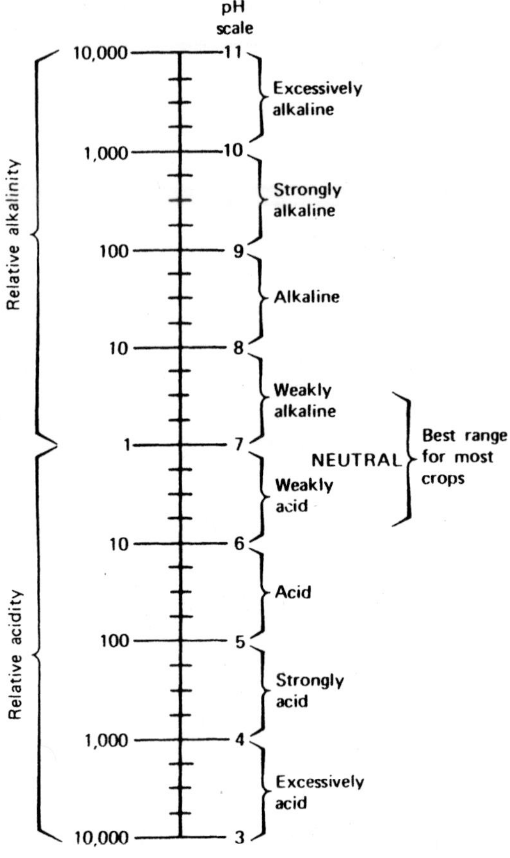

Figure 4.2. Relation Between pH, alkalinity, acidity, and plant growth.

depict nutrient availability for both mineral and organic soils. The width of the band indicates the availability of the nutrient. It does not indicate the actual amount present.

CORRECTION OF SOIL ACIDITY

Liming materials are used to change an unfavorable acidic soil reaction to a pH more favorable for crop production. However, soil types differ in their

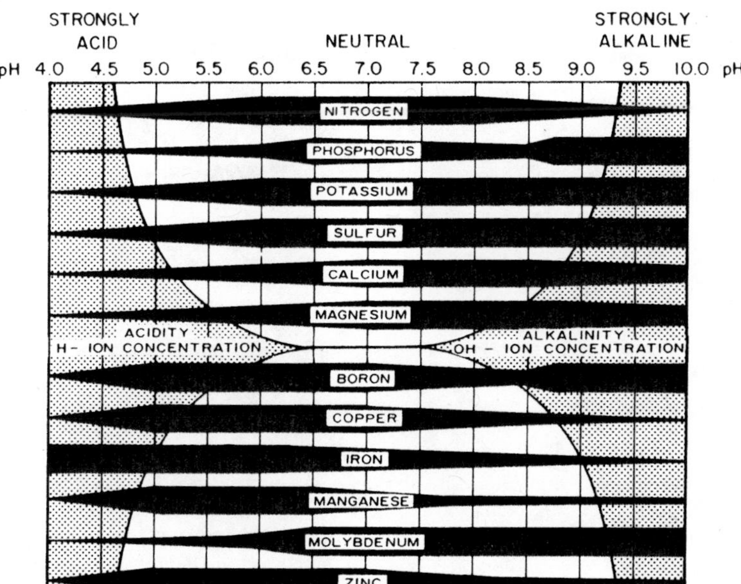

Figure 4.3. Influence of pH on the availability of plant nutrients in organic soils; widest parts of the shaded areas indicate maximum availability of each element.

Adapted from R. E. Lucas and J. F. Davis, "Relationships Between pH Values of Organic Soils and Availability of 12 Plant Nutrients," *Soil Science* 92(1961): 177–182.

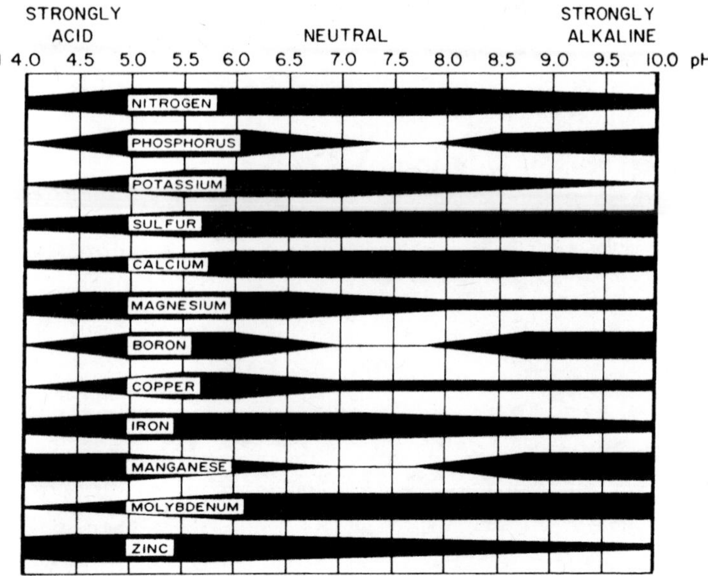

Figure 4.4. Influence of pH on the availability of plant nutrients in mineral soils; widest parts of the shaded areas indicate maximum availability of each element.

Adapted from L. B. Nelson (ed.), *Changing Patterns in Fertilizer Use* (Madison, Wis.: Soil Science Society of America, 1968).

response to liming, a property referred to as the soil's *pH buffering capacity.* Acidic soil reaction is caused by hydrogen ions present in the soil solution (*active acidity*) and attached to soil particles or organic matter (*potential acidity*). Active acidity can be neutralized rapidly, whereas potential acidity is neutralized over time as it is released. Soils vary in their relative content of these sources of acidity. Due to this complexity in soil pH, it is difficult to provide a rule of thumb for rates of liming materials. Most soil testing laboratories now use a lime requirement test to estimate the potential acidity and therefore provide a more accurate liming recommendation than could be done before. The lime requirement test treats the soil sample with a buffer solution to estimate the potential acidity, and thus provides a more accurate lime recommendation than can usually be obtained by treating the soil sample with water only. Soils with similar amounts of active acidity might have different amounts of potential acidity and thus require different lime recommendations even though the rule-of-thumb approach might have given similar lime recommendations. Soils with large potential acidity (clays and mucks) require more lime than sandy soils with a similar water pH.

TABLE 4.9. COMMON LIMING MATERIALS

Material	Chemical Formula	Pure $CaCO_3$ Equivalent (%)	Liming Material (lb) Necessary to Equal 100 lb Limestone
Burned lime	CaO	150	64
Hydrated lime	$Ca(OH)_2$	120	82
Dolomitic limestone	$CaCO_3$, $MgCO_3$	104	86
Limestone	$CaCO_3$	95	100
Marl	$CaCO_3$	95	100
Shell, oyster, etc.	$CaCO_3$	95	100

TABLE 4.10. COMMON ACIDIFYING MATERIALS[1]

Material	Chemical Formula	Sulfur (%)	Acidifying Material (lb) Necessary to Equal 100 lb Soil Sulfur
Soil sulfur	S	99.0	100
Sulfuric acid (98%)	H_2SO_4	32.0	306
Sulfur dioxide	SO_2	50.0	198
Lime-sulfur solution (32° Baumé)	CaS_x + water	24.0	417
Iron sulfate	$FeSO_4 \cdot 7H_2O$	11.5	896
Aluminum sulfate	$Al_2(SO_4)_3$	14.4	694

[1] Certain fertilizer materials also markedly increase soil acidity when used in large quantities (see page 165).

TABLE 4.11. APPROXIMATE QUANTITY OF SOIL SULFUR NEEDED TO INCREASE SOIL ACIDITY TO ABOUT pH 6.5

	Sulfur (lb/acre)		
Change in pH Desired	Sands	Loams	Clays
8.5–6.5	2,000	2,500	3,000
8.0–6.5	1,200	1,500	2,000
7.5–6.5	500	800	1,000
7.0–6.5	100	150	300

TABLE 4.12. EFFECT OF SOME FERTILIZER MATERIALS ON THE SOIL REACTION

Materials	N (%)	Pounds Limestone ($CaCO_3$)	
		Per lb N	Per 100 lb Fertilizer Material
		Needed to Counteract the Acidity Produced	
Acidity-Forming			
Ammonium nitrate	33.5	1.80	60
Monoammonium phosphate	11	5.35	59
Ammonium phosphate sulfate	16	5.35	88
Ammonium sulfate	21	5.35	110
Anhydrous ammonia	82	1.80	148
Aqua ammonia	24	1.80	44
Aqua ammonia	30	1.80	54
Diammonium phosphate	16–18	1.80	70
Liquid phosphoric acid	52 (P_2O_5)	—	110
Urea	46	1.80	84
		Equivalents Produced	
Alkalinity-Forming			
Calcium cyanamide	22	2.85	63
Calcium nitrate	15.5	1.35	20
Potassium nitrate	13	1.80	23
Sodium nitrate	16	1.80	29
Neutral			
Ammonium nitrate-lime	Potassium sulfate		
Calcium sulfate (gypsum)	Superphosphate		
Potassium chloride			

Based on the method of W. H. Pierre, "Determination of Equivalent Acidity and Basicity of Fertilizers," *Industrial Engineering Chemical Analytical Edition*, 5 (1933): 229–234.

RELATIVE SALT EFFECTS OF FERTILIZER MATERIALS ON THE SOIL SOLUTION

When fertilizer materials are placed close to seeds or plants, they may increase the osmotic pressure of the soil solution and cause injury to the crop. The term *salt index* refers to the effect of a material in relation to that produced by sodium nitrate, which is given a rating of 100. The *partial index* shows the relationships per unit (20 lb) of the actual nutrient supplied. Any material with a high salt index must be used with great care.

TABLE 4.13. SALT INDEX OF SEVERAL FERTILIZER MATERIALS

Material	Salt Index	Partial Salt Index per Unit of Plant Food
Anhydrous ammonia	47.1	0.572
Ammonium nitrate	104.7	2.990
Ammonium nitrate-lime (Cal-Nitro)	61.1	2.982
Ammonium sulfate	69.0	3.253
Calcium carbonate (limestone)	4.7	0.083
Calcium nitrate	52.5	4.409
Calcium sulfate (gypsum)	8.1	0.247
Diammonium phosphate	29.9	1.614[1]
		0.637[2]
Dolomite (calcium and magnesium carbonates)	0.8	0.042
Monoammonium phosphate	34.2	2.453[1]
		0.485[2]
Monocalcium phosphate	15.4	0.274
Nitrogen solution, 37%	77.8	2.104
Potassium chloride, 50%	109.4	2.189
Potassium chloride, 60%	116.3	1.936
Potassium nitrate	73.6	5.336[1]
		1.580[3]
Potassium sulfate	46.1	0.853
Sodium chloride	153.8	2.899
Sodium nitrate	100.0	6.060
Sulfate of potash-magnesia	43.2	1.971
Superphosphate, 20%	7.8	0.390
Superphosphate, 45%	10.1	0.224
Urea	75.4	1.618

Adapted from L. F. Rader, L. M. White, and C. W. Whittaker, "The Salt Index: A Measure of the Effect of Fertilizers on the Concentration of the Soil Solution," *Soil Science* 55 (1943):201–218.

[1] N [2] P_2O_5 [3] K_2O

TABLE 4.14. RELATIVE SALT TOLERANCE OF VEGETABLES
The indicated salt tolerances are based on growth rather than yield. With most crops, there is little difference in salt tolerance among varieties. Boron tolerances may vary depending on climate, soil condition, and crop variety.

Vegetable	Maximum Soil Salinity Without Yield Loss (Threshold) (dS/m)	Decrease in Yield at Soil Salinities Above the Threshold (% per dS/m)
Sensitive crops		
Bean	1.0	19
Carrot	1.0	14
Strawberry	1.0	33
Onion	1.2	16
Moderately sensitive		
Turnip	0.9	9
Radish	1.2	13
Lettuce	1.3	13
Pepper	1.5	14
Sweet potato	1.5	11
Broad bean	1.6	10
Corn	1.7	12
Potato	1.7	12
Cabbage	1.8	10
Celery	1.8	6
Spinach	2.0	8
Cucumber	2.5	13
Tomato	2.5	10
Broccoli	2.8	9
Squash, scallop	3.2	16
Moderately tolerant		
Beet	4.0	9
Squash, zucchini	4.7	9

Adapted from E. V. Maas, "Crop Tolerance," *California Agriculture* (October 1984).

Note: 1 decisiemens per meter (dS/m) = 1 mmho/cm = approximately 640 mg/L salt

07
SALINITY

SOIL SALINITY

With an increase in soil salinity, plant roots extract water less easily from the soil solution. This situation is more critical under hot and dry than under humid conditions. High soil salinity may result also in toxic concentrations of ions in plants. Soil salinity is determined by finding the electrical conductivity of the soil saturation extract (ECe). The electrical conductivity is measured in millimhos per centimeter (mmho/cm). One mmho/cm is equivalent to 1 decisiemens per meter (dS/m) and, on the average, to 640 ppm salt.

TABLE 4.15. CROP RESPONSE TO SALINITY

Salinity (expressed as ECe, mmho/cm, or dS/m)	Crop Responses
0–2	Salinity effects mostly negligible.
2–4	Yields of very sensitive crops may be restricted.
4–8	Yields of many crops restricted.
8–16	Only tolerant crops yield satisfactorily.
Above 16	Only a few very tolerant crops yield satisfactorily.

Adapted from Leon Bernstein, *Salt Tolerance of Plants*, USDA Agricultural Information Bulletin 283 (1970).

08
FERTILIZERS

FERTILIZER DEFINITIONS

Grade or *analysis* means the minimum guarantee of the percentage of total nitrogen (N), available phosphoric acid (P_2O_5), and water-soluble potash (K_2O) in the fertilizer.

Example: 20-0-20 or 5-15-5

Ratio is the grade reduced to its simplest terms.

Example: A 20-0-20 has a ratio of 1-0-1, as does a 10-0-10.

Formula shows the actual pound and percentage composition of the ingredients or compounds that are mixed to make up a ton of fertilizer.

An *open-formula mix* carries the formula as well as the grade on the tag attached to each bag.

Carrier, simple, or *source* is the material or compound in which a given plant nutrient is found or supplied.

Example: Ammonium nitrate and urea are sources or carriers that supply nitrogen.

Unit means 1% of 1 ton or 20 lb. On the basis of a ton, the units per ton are equal to the percentage composition or the pounds per 100 lb.

Example: Ammonium sulfate contains 21% nitrogen, or 21 lb nitrogen/100 lb, or 21 units nitrogen in a ton.

Primary nutrient refers to nitrogen, phosphorus, and potassium, which are used in considerable quantities by crops.

Secondary nutrient refers to calcium, magnesium, and sulfur, which are used in moderate quantities by crops.

Micronutrient, trace, or *minor element* refers to iron, boron, manganese, zinc, copper, and molybdenum, the essential plant nutrients used in relatively small quantities.

TABLE 4.16. APPROXIMATE COMPOSITION OF SOME CHEMICAL FERTILIZER MATERIALS[1]

Fertilizer Material	Total Nitrogen (% N)	Available Phosphorus (% P_2O_5)	Water-soluble Potassium (% K_2O)
Nitrogen			
Ammonium nitrate	33.5	—	—
Ammonium nitrate-lime (A-N-L, Cal-Nitro)	20.5	—	—
Monoammonium phosphate	11.0	48.0	—
Ammonium phosphate-sulfate	16.0	20.0	—
Ammonium sulfate	21.0	—	—
Anhydrous ammonia	82.0	—	—
Aqua ammonia	20.0	—	—
Calcium cyanamide	21.0	—	—
Calcium nitrate	15.5	—	—
Calcium ammonium nitrate	17.0	—	—
Diammonium phosphate	16–18	46.0–48.0	—
Potassium nitrate	13.0	—	44.0
Sodium nitrate	16.0	—	—
Urea	46.0	—	—
Urea formaldehyde	38.0	—	—
Phosphorus			
Phosphoric acid solution	—	52.0–54.0	—
Normal (single) superphosphate	—	18.0–20.0	—
Concentrated (triple or treble) superphosphate	—	45.0–46.0	—
Monopotassium phosphate	—	53.0	—
Potassium			
Potassium chloride	—	—	60.0–62.0
Potassium nitrate	13.0	—	44.0
Potassium sulfate	—	—	50.0–53.0
Sulfate of potash-magnesia	—	—	26.0
Monopotassium phosphate	—	—	34.0

[1] See page 165 for effect of these materials on soil reaction.

TABLE 4.17. SOLUBILITY OF FERTILIZER MATERIALS

Solubility of fertilizer materials is an important factor in preparing starter solutions, foliar sprays, and solutions to be knifed into the soil or injected into an irrigation system. Hot water may be needed to dissolve the chemicals.

Material	Solubility in Cold Water (lb/100 gal)
Primary Nutrients	
Ammonium nitrate	984
Ammonium sulfate	592
Calcium cyanamide	Decomposes
Calcium nitrate	851
Diammonium phosphate	358
Monoammonium phosphate	192
Potassium nitrate	108
Sodium nitrate	608
Superphosphate, single	17
Superphosphate, treble	33
Urea	651
Secondary Nutrients and Micronutrients	
Ammonium molybdate	Decomposes
Borax	8
Calcium chloride	500
Copper oxide	Insoluble
Copper sulfate	183
Ferrous sulfate	242
Magnesium sulfate	592
Manganese sulfate	876
Sodium chloride	300
Sodium molybdate	467
Zinc sulfate	625

TABLE 4.18. AMOUNT OF CARRIERS NEEDED TO SUPPLY A CERTAIN AMOUNT OF NUTRIENT PER ACRE[1]

	Nutrients (lb/acre)							
	20	40	60	80	100	120	160	200
Nutrient in Carrier (%)	Carriers Needed (lb)							
3	667	1,333	2,000					
4	500	1,000	1,500	2,000				
5	400	800	1,200	1,600	2,000			
6	333	667	1,000	1,333	1,667	2,000		
7	286	571	857	1,142	1,429	1,714		
8	250	500	750	1,000	1,250	1,500	2,000	
9	222	444	667	889	1,111	1,333	1,778	
10	200	400	600	800	1,000	1,200	1,600	2,000
11	182	364	545	727	909	1,091	1,455	1,818
12	166	333	500	666	833	1,000	1,333	1,666
13	154	308	462	615	769	923	1,231	1,538
15	133	267	400	533	667	800	1,067	1,333
16	125	250	375	500	625	750	1,000	1,250
18	111	222	333	444	555	666	888	1,111
20	100	200	300	400	500	600	800	1,000
21	95	190	286	381	476	571	762	952
25	80	160	240	320	400	480	640	800
30	67	133	200	267	333	400	533	667
34	59	118	177	235	294	353	471	588
42	48	95	143	190	238	286	381	476
45	44	89	133	178	222	267	356	444
48	42	83	125	167	208	250	333	417
50	40	80	120	160	200	240	320	400
60	33	67	100	133	167	200	267	333

[1] This table can be used in determining the acre rate for applying a material in order to supply a certain number of pounds of a nutrient.

Example: A carrier provides 34% of a nutrient. To get 200 lb of the nutrient, 588 lb of the material is needed, and for 60 lb of the nutrient, 177 lb of carrier is required.

TABLE 4.19. APPROXIMATE RATES OF MATERIALS TO PROVIDE CERTAIN QUANTITIES OF NITROGEN PER ACRE

	N (lb/acre):	15	30	45	60	75	100
Fertilizer Material	% N	Material to Apply (lb/acre)					
Solids							
Ammonium nitrate	33	45	90	135	180	225	300
Ammonium phosphate (48% P_2O_5)	11	135	270	410	545	680	910
Ammonium phosphate-sulfate (20% P_2O_5)	16	95	190	280	375	470	625
Ammonium sulfate	21	70	140	215	285	355	475
Calcium nitrate	15.5	95	195	290	390	485	645
Potassium nitrate	13	115	230	345	460	575	770
Sodium nitrate	16	95	190	280	375	470	625
Urea	46	35	65	100	130	165	215
Liquids							
Anhydrous ammonia (approx. 5 lb/gal)[1]	82	20	35	55	75	90	120
Aqua ammonium phosphate (24% P_2O_5; approx. 10 lb/gal)	8	190	375	560	750	940	1250
Aqua ammonia (approx. 7½ lb/gal)[1]	20	75	150	225	300	375	500
Nitrogen solution (approx. 11 lb/gal)	32	50	100	150	200	250	330

[1] To avoid burning, especially on alkaline soils, these materials must be placed deeper and farther away from the plant row than dry fertilizers are placed.

TABLE 4.20. RATES OF APPLICATION FOR SOME NITROGEN SOLUTIONS

Nitrogen (lb/acre)	Nitrogen Solution Needed (gal/acre)		
	21% Solution	32% Solution	41% Solution
20	8.9	5.6	5.1
25	11.1	7.1	6.4
30	13.3	8.5	7.7
35	15.6	9.9	9.0
40	17.8	11.3	10.3
45	20.0	12.7	11.5
50	22.2	14.1	12.8
55	24.4	15.5	14.1
60	26.7	16.5	15.4
65	28.9	18.4	16.7
70	31.1	19.8	17.9
75	33.3	21.2	19.2
80	35.6	22.6	20.5
85	37.8	24.0	21.8
90	40.0	25.4	23.1
95	42.2	26.8	24.4
100	44.4	28.2	25.6
110	48.9	31.1	28.2
120	53.3	33.9	30.8
130	57.8	36.7	33.3
140	62.2	39.6	35.9
150	66.7	42.4	38.5
200	88.9	56.5	51.3

Adapted from C. W. Gandt, W. C. Hulburt, and H. D. Brown, *Hose Pump for Applying Nitrogen Solutions,* USDA Farmer's Bulletin 2096 (1956).

09
FERTILIZER CONVERSION FACTORS

TABLE 4.21. CONVERSION FACTORS FOR FERTILIZER MATERIALS

Multiply	By	To Obtain Equivalent Nutrient
Ammonia—NH_3	4.700	Ammonium nitrate—NH_4NO_3
Ammonia—NH_3	3.879	Ammonium sulfate—$(NH_4)_2SO_4$
Ammonia—NH_3	0.823	Nitrogen—N
Ammonium nitrate—NH_4NO_3	0.350	Nitrogen—N
Ammonium sulfate—$(NH_4)_2SO_4$	0.212	Nitrogen—N
Borax—$Na_2B_4O_7 \cdot 10H_2O$	0.114	Boron—B
Boric acid—H_3BO_3	0.177	Boron—B
Boron—B	8.813	Borax—$Na_2B_4O_7 \cdot 10H_2O$
Boron—B	5.716	Boric acid—H_3BO_3
Calcium—Ca	1.399	Calcium oxide—CaO
Calcium—Ca	2.498	Calcium carbonate—$CaCO_3$
Calcium—Ca	1.849	Calcium hydroxide—$Ca(OH)_2$
Calcium—Ca	4.296	Calcium sulfate—$CaSO_4 \cdot 2H_2O$ (gypsum)
Calcium carbonate—$CaCO_3$	0.400	Calcium—Ca
Calcium carbonate—$CaCO_3$	0.741	Calcium hydroxide—$Ca(OH)_2$
Calcium carbonate—$CaCO_3$	0.560	Calcium oxide—CaO
Calcium carbonate—$CaCO_3$	0.403	Magnesia—MgO
Calcium carbonate—$CaCO_3$	0.842	Magnesium carbonate—$MgCO_3$
Calcium hydroxide—$Ca(OH)_2$	0.541	Calcium—Ca
Calcium hydroxide—$Ca(OH)_2$	1.351	Calcium carbonate—$CaCO_3$
Calcium hydroxide—$Ca(OH)_2$	0.756	Calcium oxide—CaO
Calcium oxide—CaO	0.715	Calcium—Ca
Calcium oxide—CaO	1.785	Calcium carbonate—$CaCO_3$
Calcium oxide—CaO	1.323	Calcium hydroxide—$Ca(OH)_2$
Calcium oxide—CaO	3.071	Calcium sulfate—$CaSO_4 \cdot 2H_2O$ (gypsum)
Gypsum—$CaSO_4 \cdot 2H_2O$	0.326	Calcium oxide—CaO

TABLE 4.21. CONVERSION FACTORS FOR FERTILIZER MATERIALS (*Continued*)

Multiply	By	To Obtain Equivalent Nutrient
Gypsum—$CaSO_4 \cdot 2H_2O$	0.186	Sulfur—S
Magnesia—MgO	2.480	Calcium carbonate—$CaCO_3$
Magnesia—MgO	0.603	Magnesium—Mg
Magnesia—MgO	2.092	Magnesium carbonate—$MgCO_3$
Magnesia—MgO	2.986	Magnesium sulfate—$MgSO_4$
Magnesia—MgO	6.114	Magnesium sulfate— $MgSO_4 \cdot 7H_2O$ (Epsom salts)
Magnesium—Mg	4.116	Calcium carbonate—$CaCO_3$
Magnesium—Mg	1.658	Magnesia—MgO
Magnesium—Mg	3.466	Magnesium carbonate—$MgCO_3$
Magnesium—Mg	4.951	Magnesium sulfate—$MgSO_4$
Magnesium—Mg	10.136	Magnesium sulfate— $MgSO_4 \cdot 7H_2O$ (Epsom salts)
Magnesium carbonate—$MgCO_3$	1.187	Calcium carbonate—$CaCO_3$
Magnesium carbonate—$MgCO_3$	0.478	Magnesia—MgO
Magnesium carbonate—$MgCO_3$	0.289	Magnesium—Mg
Magnesium sulfate—$MgSO_4$	0.335	Magnesia—MgO
Magnesium sulfate—$MgSO_4$	0.202	Magnesium—Mg
Magnesium sulfate—$MgSO_4 \cdot 7H_2O$ (Epsom salts)	0.164	Magnesia—MgO
Magnesium sulfate—$MgSO_4 \cdot 7H_2O$ (Epsom salts)	0.099	Magnesium—Mg
Manganese—Mn	2.749	Manganese(ous) sulfate—$MnSO_4$
Manganese—Mn	4.060	Manganese(ous) sulfate—$MnSO_4 \cdot 4H_2O$
Manganese(ous) sulfate—$MnSO_4$	0.364	Manganese—Mn
Manganese(ous) sulfate—$MnSO_4 \cdot 4H_2O$	0.246	Manganese—Mn
Nitrate—NO_3	0.226	Nitrogen—N
Nitrogen—N	1.216	Ammonia—NH_3
Nitrogen—N	2.856	Ammonium nitrate—NH_4NO_3

TABLE 4.21. CONVERSION FACTORS FOR FERTILIZER MATERIALS (*Continued*)

Multiply	By	To Obtain Equivalent Nutrient
Nitrogen—N	4.716	Ammonium sulfate—$(NH_4)_2SO_4$
Nitrogen—N	4.426	Nitrate—NO_3
Nitrogen—N	6.068	Sodium nitrate—$NaNO_3$
Nitrogen—N	6.250	Protein
Phosphoric acid—P_2O_5	0.437	Phosphorus—P
Phosphorus—P	2.291	Phosphoric acid—P_2O_5
Potash—K_2O	1.583	Potassium chloride—KCl
Potash—K_2O	2.146	Potassium nitrate—KNO_3
Potash—K_2O	0.830	Potassium—K
Potash—K_2O	1.850	Potassium sulfate—K_2SO_4
Potassium—K	1.907	Potassium chloride—KCl
Potassium—K	1.205	Potash—K_2O
Potassium—K	2.229	Potassium sulfate—K_2SO_4
Potassium chloride—KCl	0.632	Potash—K_2O
Potassium chloride—KCl	0.524	Potassium—K
Potassium nitrate—KNO_3	0.466	Potash—K_2O
Potassium nitrate—KNO_3	0.387	Potassium—K
Potassium sulfate—K_2SO_4	0.540	Potash—K_2O
Potassium sulfate—K_2SO_4	0.449	Potassium—K
Sodium nitrate—$NaNO_3$	0.165	Nitrogen—N
Sulfur—S	5.368	Calcium sulfate—$CaSO_4 \cdot 2H_2O$ (gypsum)
Sulfur—S	2.497	Sulfur trioxide—SO_3
Sulfur—S	3.059	Sulfuric acid—H_2SO_4
Sulfur trioxide—SO_3	0.401	Sulfur—S
Sulfuric acid—H_2SO_4	0.327	Sulfur—S

Examples: 80 lb ammonia (NH_3) contains the same amount of N as 310 lb ammonium sulfate [$(NH_4)_2SO_4$], 80 × 3.88 = 310. Likewise, 1000 lb calcium carbonate multiplied by 0.400 equals 400 lb calcium. A material contains 20% phosphoric acid. This percentage (20) multiplied by 0.437 equals 8.74% phosphorus.

10
NUTRIENT ABSORPTION

APPROXIMATE CROP CONTENT OF NUTRIENT ELEMENTS

Sometimes crop removal values are used to estimate fertilizer needs by crops. Removal values are obtained by analyzing plants and fruits for nutrient content and then expressing the results on an acre basis. It is risky to relate fertilizer requirements on specific soils to generalized listings of crop removal values. A major problem is that crop removal values are usually derived from analyzing plants grown on fertile soils where much of the nutrient content of the crop is supplied from soil reserves rather than from fertilizer application. Because plants can absorb larger amounts of specific nutrients than they require, crop removal values can overestimate the true crop nutrient requirement of a crop. Crop removal values can estimate the nutrient supply capacity on an unfertilized soil. The crop content (removal) values presented in the table are presented for information purposes and are not suggested for use in formulating fertilizer recommendations. The values were derived from various sources and publications. For example, a similar table was published in M. McVicker and W. Walker, *Using Commercial Fertilizer* (Danville, Ill.: Interstate Printers and Publishers, 1978).

TABLE 4.22. APPROXIMATE ACCUMULATION OF NUTRIENTS BY SOME VEGETABLE CROPS

Vegetable	Yield (cwt/acre)	Nutrient Absorption (lb/acre)		
		N	P	K
Broccoli	100 heads	20	2	45
	Other	145	8	165
		165	10	210
Brussels sprouts	160 sprouts	150	20	125
	Other	85	9	110
		235	29	235
Cantaloupe	225 fruits	95	17	120
	Vines	60	8	35
		155	25	155
Carrot	500 roots	80	20	200
	Tops	65	5	145
		145	25	345
Celery	1000 tops	170	35	380
	Roots	25	15	55
		195	50	435
Honeydew melon	290 fruits	70	8	65
	Vines	135	15	95
		205	23	160
Lettuce	350 plants	95	12	170
Onion	400 bulbs	110	20	110
	Tops	35	5	45
		145	25	155

TABLE 4.22. APPROXIMATE ACCUMULATION OF NUTRIENTS BY SOME VEGETABLE CROPS (*Continued*)

Vegetable	Yield (cwt/acre)	Nutrient Absorption (lb/acre)		
		N	P	K
Pea, shelled	40 peas	100	10	30
	Vines	70	12	50
		170	22	80
Pepper	225 fruits	45	6	50
	Plants	95	6	90
		140	12	140
Potato	400 tubers	150	19	200
	Vines	60	11	75
		210	30	275
Snap bean	100 beans	120	10	55
	Plants	50	6	45
		170	16	100
Spinach	200 plants	100	12	100
Sweet corn	130 ears	55	8	30
	Plants	100	12	75
		155	20	105
Sweet potato	300 roots	80	16	160
	Vines	60	4	40
		140	20	200
Tomato	600 fruits	100	10	180
	Vines	80	11	100
		180	21	280

11
PLANT ANALYSIS

TABLE 4.23. PLANT ANALYSIS GUIDE FOR SAMPLING TIME, PLANT PART, AND NUTRIENT CONCENTRATION OF VEGETABLE CROPS (DRY WEIGHT BASIS)[1]

Crop	Time of Sampling	Plant Part	Source	Nutrient Concentration[2]	Nutrient Level	
					Deficient	Sufficient
Asparagus	Midgrowth of fern	4-in. tip section of new fern branch	NO_3	N, ppm	100	500
			PO_4	P, ppm	800	1,600
				K, %	1	3
Bean, bush snap	Midgrowth	Petiole of 4th leaf from tip	NO_3	N, ppm	2,000	3,000
			PO_4	P, ppm	1,000	2,000
				K, %	3	5
	Early bloom	Petiole of 4th leaf from tip	NO_3	N, ppm	1,000	1,500
			PO_4	P, ppm	800	1,500
				K, %	2	4
Broccoli	Midgrowth	Midrib of young, mature leaf	NO_3	N, ppm	7,000	9,000
			PO_4	P, ppm	2,500	4,000
				K, %	3	5
	First buds	Midrib of young, mature leaf	NO_3	N, ppm	5,000	7,000
			PO_4	P, ppm	2,500	4,000
				K, %	2	4

Brussels sprouts	Midgrowth	Midrib of young, mature leaf	NO_3	N, ppm	5,000	7,000
			PO_4	P, ppm	2,000	3,500
				K, %	3	5
	Late growth	Midrib of young, mature leaf	NO_3	N, ppm	2,000	3,000
			PO_4	P, ppm	1,000	3,000
				K, %	2	4
Cabbage	At heading	Midrib of wrapper leaf	NO_3	N, ppm	5,000	7,000
			PO_4	P, ppm	2,500	3,500
				K, %	2	4
Cantaloupe	Early growth (short runners)	Petiole of 6th leaf from growing tip	NO_3	N, ppm	8,000	12,000
			PO_4	P, ppm	2,000	3,000
				K, %	4	6
	Early fruit	Petiole of 6th leaf from growing tip	NO_3	N, ppm	5,000	8,000
			PO_4	P, ppm	1,500	2,500
				K, %	3	5
	First mature fruit	Petiole of 6th leaf from growing tip	NO_3	N, ppm	2,000	3,000
			PO_4	P, ppm	1,000	2,000
				K, %	2	4
	Early growth	Blade of 6th leaf from growing tip	NO_3	N, ppm	2,000	3,000
			PO_4	P, ppm	1,500	2,300
				K, %	1	2.5
	Early fruit	Blade of 6th leaf from growing tip	NO_3	N, ppm	1,000	1,500
			PO_4	P, ppm	1,300	1,700
				K, %	1	2.0
	First mature fruit	Blade of 6th leaf from growing tip	NO_3	N, ppm	500	800
			PO_4	P, ppm	1,000	1,500
				K, %	1	1.8

TABLE 4.23. PLANT ANALYSIS GUIDE FOR SAMPLING TIME, PLANT PART, AND NUTRIENT CONCENTRATION OF VEGETABLE CROPS (DRY-WEIGHT BASIS) (*Continued*)

Crop	Time of Sampling	Plant Part	Source	Nutrient[1] Concentration	Nutrient Level: Deficient	Nutrient Level: Sufficient
Chinese cabbage	At heading	Midrib of wrapper leaf	NO_3	N, ppm	8,000	10,000
			PO_4	P, ppm	2,000	3,000
				K, %	4	7
Carrot	Midgrowth	Petiole of young, mature leaf	NO_3	N, ppm	5,000	7,500
			PO_4	P, ppm	2,000	3,000
				K, %	4	6
Cauliflower	Buttoning	Midrib of young, mature leaf	NO_3	N, ppm	5,000	7,000
			PO_4	P, ppm	2,500	3,500
				K, %	2	4
Celery	Midgrowth	Petiole of newest fully elongated leaf	NO_3	N, ppm	5,000	7,000
			PO_4	P, ppm	2,500	3,000
				K, %	4	7
	Near maturity	Petiole of newest fully elongated leaf	NO_3	N, ppm	4,000	6,000
			PO_4	P, ppm	2,000	3,000
				K, %	3	5
Cucumber, pickling	Early fruit set	Petiole of 6th leaf from tip	NO_3	N, ppm	5,000	7,500
			PO_4	P, ppm	1,500	2,500
				K, %	3	5

Cucumber, slicing	Early harvest period	Petiole of 6th leaf from growing tip	NO_3	N, ppm	5,000	7,500
			PO_4	P, ppm	1,500	2,500
				K, %	4	7
Eggplant	At first harvest	Petiole of young, mature leaf	NO_3	N, ppm	5,000	7,500
			PO_4	P, ppm	2,000	3,000
				K, %	4	7
Garlic	Early growth (prebulbing)	Newest fully elongated leaf	PO_4	P, ppm	2,000	3,000
				K, %	3	4
	Midseason (bulbing)	Newest fully elongated leaf	PO_4	P, ppm	2,000	3,000
				K, %	2	3
	Late season (postbulbing)	Newest fully elongated leaf	PO_4	P, ppm	2,000	3,000
				K, %	1	2
Lettuce	At heading	Midrib of wrapper leaf	NO_3	N, ppm	4,000	6,000
			PO_4	P, ppm	2,000	3,000
				K, %	2	4
	At harvest	Midrib of wrapper leaf	NO_3	N, ppm	3,000	5,000
			PO_4	P, ppm	1,500	2,500
				K, %	1.5	2.5
Onion	Early season	Tallest leaf	PO_4	P, ppm	1,000	2,000
				K, %	3	4.5
	Midseason	Tallest leaf	PO_4	P, ppm	1,000	2,000
				K, %	2	4
	Late season	Tallest leaf	PO_4	P, ppm	1,000	2,000
				K, %	2	3
Pepper, chile	Early growth first bloom	Petiole of young, mature leaf	NO_3	N, ppm	5,000	7,000
			PO_4	P, ppm	2,000	2,500
				K, %	3	5

TABLE 4.23. PLANT ANALYSIS GUIDE FOR SAMPLING TIME, PLANT PART, AND NUTRIENT CONCENTRATION OF VEGETABLE CROPS (DRY-WEIGHT BASIS) (*Continued*)

Crop	Time of Sampling	Plant Part	Source	Nutrient[1] Concentration	Nutrient Level	
					Deficient	Sufficient
	Early fruit set	Petiole of young, mature leaf	NO_3	N, ppm	1,000	1,500
			PO_4	P, ppm	1,500	2,000
				K, %	2	4
	Fruits, full size	Petiole of young, mature leaf	NO_3	N, ppm	750	1,000
			PO_4	P, ppm	1,500	2,000
				K, %	1.5	3
	Early growth first bloom	Blade of young, mature leaf	NO_3	N, ppm	1,500	2,000
			PO_4	P, ppm	1,500	2,000
				K, %	3	5
	Early fruit set	Blade of young, mature leaf	NO_3	N, ppm	500	800
			PO_4	P, ppm	1,500	2,000
				K, %	2	4
Pepper, sweet	Early growth, first flower	Petiole of young, mature leaf	NO_3	N, ppm	8,000	10,000
			PO_4	P, ppm	2,000	3,000
				K, %	4	6
	Early fruit set, 1 in. diameter	Petiole of young, mature leaf	NO_3	N, ppm	5,000	7,000
			PO_4	P, ppm	1,500	2,500
				K, %	3	5

	Fruit ¾ size	Petiole of young, mature leaf	NO_3	N, ppm	3,000	5,000
			PO_4	P, ppm	1,200	2,000
				K, %	2	4
	Early growth, first flower	Blade of young, mature leaf	NO_3	N, ppm	2,000	3,000
			PO_4	P, ppm	1,800	2,500
				K, %	3	5
	Early fruit set, 1 in. diameter	Blade of young, mature leaf	NO_3	N, ppm	1,500	2,000
			PO_4	P, ppm	1,500	2,000
				K, %	2	4
Potato	Early season	Petiole of 4th leaf from growing tip	NO_3	N, ppm	8,000	12,000
			PO_4	P, ppm	1,200	2,000
				K, %	9	11
	Midseason	Petiole of 4th leaf from growing tip	NO_3	N, ppm	6,000	9,000
			PO_4	P, ppm	800	1,600
				K, %	7	9
	Late season	Petiole of 4th leaf from growing tip	NO_3	N, ppm	3,000	5,000
			PO_4	P, ppm	500	1,000
				K, %	4	6
Spinach	Midgrowth	Petiole of young, mature leaf	NO_3	N, ppm	4,000	6,000
			PO_4	P, ppm	2,000	3,000
				K, %	2	4
Summer squash (zucchini)	Early bloom	Petiole of young, mature leaf	NO_3	N, ppm	12,000	15,000
			PO_4	P, ppm	4,000	6,000
				K, %	6	10
Sweet corn	Tasseling	Midrib of 1st leaf above primary ear	NO_3	N, ppm	500	1,000
			PO_4	P, ppm	500	1,000
				K, %	2	4

TABLE 4.23. PLANT ANALYSIS GUIDE FOR SAMPLING TIME, PLANT PART, AND NUTRIENT CONCENTRATION OF VEGETABLE CROPS (DRY-WEIGHT BASIS) (*Continued*)

Crop	Time of Sampling	Plant Part	Source	Nutrient[1] Concentration	Nutrient Level	
					Deficient	Sufficient
Sweet potato	Midgrowth	Petiole of 6th leaf	NO_3	N, ppm	1,500	2,500
		from the	PO_4	P, ppm	1,000	2,000
		growing tip		K, %	3	5
Tomato, cherry	Early fruit set	Petiole of 4th leaf	NO_3	N, ppm	8,000	10,000
		from the	PO_4	P, ppm	2,000	3,000
		growing tip		K, %	4	7
	Fruit ½ in. diameter	Petiole of 4th leaf	NO_3	N, ppm	5,000	7,000
		from growing	PO_4	P, ppm	2,000	3,000
		tip		K, %	3	5
	At first harvest	Petiole of 4th leaf	NO_3	N, ppm	1,000	2,000
		from growing	PO_4	P, ppm	2,000	3,000
		tip		K, %	2	4
Tomato, processing and determinate, fresh market	Early bloom	Petiole of 4th leaf	NO_3	N, ppm	8,000	12,000
		from growing	PO_4	P, ppm	2,000	3,000
		tip		K, %	3	6

	Fruit 1 in. diameter	Petiole of 4th leaf from growing tip	NO_3	N, ppm	4,000	6,000
			PO_4	P, ppm	1,500	2,500
				K, %	2	4
	First color	Petiole of 4th leaf from growing tip	NO_3	N, ppm	2,000	3,000
			PO_4	P, ppm	1,000	2,000
				K, %	1	3
Tomato, fresh market indeterminate	Early bloom	Petiole of 4th leaf from growing tip	NO_3	N, ppm	10,000	14,000
			PO_4	P, ppm	2,500	3,000
				K, %	4	7
	Fruit 1 in. diameter	Petiole of 4th leaf from growing tip	NO_3	N, ppm	8,000	12,000
			PO_4	P, ppm	2,500	3,000
				K, %	3	5
	Full ripe fruit	Petiole of 4th leaf from growing tip	NO_3	N, ppm	4,000	6,000
			PO_4	P, ppm	2,000	2,500
				K, %	2	4
Watermelon	Early fruit set	Petiole of 6th leaf from growing tip	NO_3	N, ppm	5,000	7,500
			PO_4	P, ppm	1,500	2,500
				K, %	3	5

[1]Adapted from H. M. Reisenauer (ed.), *Soil and Plant Tissue Testing in California,* University of California Division of Agricultural Science Bulletin 1879 (1983).

[2]Two percent acetic acid-soluble NO_3–N and PO_4–P and total K (dry weight basis). Updated 1995, personal communication, T. K. Hartz, University of California, Davis. O.A. Lorenz and K.B. Tyler, *Plant Tissue Analysis of Vegetable Crops* (University of California—Davis Vegetable Research and Information Center), http://vric.ucdavis.edu/veginfo/topics/fertilizer/tissueanalysis.pdf. Values represent conventionally fertilized crops. Organically managed crops may show lower petiole-nitrate (NO_3–N) concentrations. Total macronutrient concentrations of whole leaves is the preferred method of evaluating nutrient sufficiency under organic fertility management.

TABLE 4.24. TOTAL NUTRIENT CONCENTRATION FOR DIAGNOSIS OF THE NUTRIENT LEVEL OF VEGETABLE CROPS

				Nutrient Level (% dry weight)	
Crop	Time of Sampling	Plant Part	Nutrient	Deficient	Sufficient
Asparagus	Early fern growth	4-in. tip section of new fern branch	N	4.00	5.00
			P	0.20	0.40
			K	2.00	4.00
	Mature fern	4-in. tip section of new fern branch	N	3.00	4.00
			P	0.20	0.40
			K	1.00	3.00
Bean, bush snap	Full bloom	Petiole: recent fully exposed trifoliate leaf	N	1.50	2.25
			P	0.15	0.30
			K	1.00	2.50
	Full bloom	Blade: recent fully exposed trifoliate leaf	N	1.25	2.25
			P	0.25	0.40
			K	0.75	1.50
Bean, lima	Full bloom	Oldest trifoliate leaf	N	2.50	3.50
			P	0.20	0.30
			K	1.50	2.25
Celery	Midgrowth	Petiole	N	1.00	1.50
			P	0.25	0.55
			K	4.00	5.00

Cantaloupe	Early growth	Petiole of 6th leaf from growing tip	N	2.50	3.50
			P	0.30	0.60
			K	4.00	6.00
	Early fruit	Petiole of 6th leaf from growing tip	N	2.00	3.00
			P	0.20	0.35
			K	3.00	5.00
	First mature fruit	Petiole of 6th leaf from growing tip	N	1.50	2.00
			P	0.15	0.30
			K	2.00	4.00
Garlic	Early season (prebulbing)	Newest fully elongated leaf	N	4.00	5.00
			P	0.20	0.30
			K	3.00	4.00
	Midseason (bulbing)	Newest fully elongated leaf	N	3.00	4.00
			P	0.20	0.30
			K	2.00	3.00
	Late season (postbulbing)	Newest fully elongated leaf	N	2.00	3.00
			P	0.20	0.30
			K	1.00	2.00
Lettuce	At heading	Leaves	N	1.50	3.00
			P	0.20	0.35
			K	2.50	5.00
	Nearly mature	Leaves	N	1.25	2.50
			P	0.15	0.30
			K	2.50	5.00
Onion	Early season	Tallest leaf	N	3.00	4.00
			P	0.10	0.20
			K	3.00	4.00

TABLE 4.24. TOTAL NUTRIENT CONCENTRATION FOR DIAGNOSIS OF THE NUTRIENT LEVEL OF VEGETABLE CROPS (*Continued*)

				Nutrient Level (% dry weight)	
Crop	Time of Sampling	Plant Part	Nutrient	Deficient	Sufficient
	Midseason	Tallest leaf	N	2.50	3.00
			P	0.10	0.20
			K	2.50	4.00
	Late season	Tallest leaf	N	2.00	2.50
			P	0.10	0.20
			K	2.00	3.00
Pepper, sweet	Full bloom	Blade and petiole	N	3.00	4.00
			P	0.15	0.25
			K	1.50	2.50
	Full bloom, fruit 3/4 size	Blade and petiole	N	2.50	3.50
			P	0.12	0.20
			K	1.00	2.00
Potato	Early, plants 12 in. tall	Petiole of 4th leaf from tip	N	2.50	3.50
			P	0.20	0.30
			K	9.00	11.00
	Midseason	Petiole of 4th leaf from tip	N	2.25	2.75
			P	0.10	0.20
			K	7.00	9.00

	Late, nearly mature	Petiole of 4th leaf from tip	N	1.50	2.25
			P	0.08	0.15
			K	4.00	6.00
	Early, plants 12 in. tall	Blade of 4th leaf from tip	N	4.00	6.00
			P	0.30	0.60
			K	3.50	5.00
	Midseason	Blade of 4th leaf from tip	N	3.00	5.00
			P	0.20	0.40
			K	2.50	3.50
	Late, nearly mature	Blade of 4th leaf from tip	N	2.00	4.00
			P	0.10	0.20
			K	1.50	2.50
Southern pea (cowpea)	Full bloom	Blade and petiole	N	2.00	3.50
			P	0.20	0.30
			K	1.00	2.00
Spinach	Midgrowth	Mature leaf blade and petiole	N	2.00	4.00
			P	0.20	0.40
			K	3.00	6.00
	At harvest	Mature leaf blade and petiole	N	1.50	3.00
			P	0.20	0.35
			K	2.00	5.00
Sweet corn	Tasseling	Sixth leaf from base of plant	N	2.75	3.50
			P	0.18	0.28
			K	1.75	2.25
	Silking	Leaf opposite first ear	N	1.50	2.00
			P	0.20	0.30
			K	1.00	2.00

TABLE 4.24. TOTAL NUTRIENT CONCENTRATION FOR DIAGNOSIS OF THE NUTRIENT LEVEL OF VEGETABLE CROPS (*Continued*)

Crop	Time of Sampling	Plant Part	Nutrient	Nutrient Level (% dry weight)	
				Deficient	Sufficient
Tomato (determinate)	Flowering	Leaf blade and petiole	N	2.50	3.50
			P	0.20	0.30
			K	1.50	2.50
	First ripe fruit	Leaf blade and petiole	N	1.50	2.50
			P	0.15	0.25
			K	1.00	2.00

Adapted from H. M. Reisenauer (ed.), *Soil and Plant Tissue Testing in California*, University of California Division of Agricultural Science Bulletin 1879 (1983). Updated 1995, personal communication, T. K. Hartz, University of California—Davis.

TABLE 4.25. CRITICAL (DEFICIENCY) VALUES, ADEQUATE RANGES, HIGH VALUES, AND TOXICITY VALUES FOR PLANT NUTRIENT CONCENTRATION OF VEGETABLES

				%						ppm					
Crop	Plant Part[1]	Time of Sampling	Status	N	P	K	Ca	Mg	S	Fe	Mn	Zn	B	Cu	Mo
Bean, snap	MRM	Before bloom	Deficient	<3.0	0.25	2.0	0.8	0.20	0.20	25	20	20	15	5	—
	trifoliate		Adequate	3.0	0.25	2.0	0.8	0.20	0.40	25	20	20	15	5	0.4
	leaf		range	4.0	0.45	3.0	1.5	0.45	0.40	200	100	40	40	10	1.0
			High	>4.1	0.46	3.1	1.6	0.45	0.40	200	100	40	40	10	—
			Toxic (>)	—	—	—	—	—	—	—	1000	—	150	—	—
	MRM	First bloom	Deficient	<3.0	0.25	2.0	0.8	0.25	0.20	25	20	20	15	5	—
	trifoliate		Adequate	3.0	0.25	2.0	0.8	0.26	0.21	25	20	20	15	5	0.4
	leaf		range	4.0	0.45	3.0	1.5	0.45	0.40	200	100	40	40	10	1.0
			High	>4.1	0.46	3.1	1.6	0.45	0.40	200	100	40	40	10	—
			Toxic (>)	—	—	—	—	—	—	—	1000	—	150	—	—
	MRM	Full bloom	Deficient	<2.5	0.20	1.5	0.8	0.25	0.20	25	20	20	15	5	—
	trifoliate		Adequate	2.5	0.20	1.6	0.8	0.26	0.21	25	20	20	15	5	0.4
			range	4.0	0.40	2.5	1.5	0.45	0.40	200	100	40	40	10	1.0
			High	>4.1	0.41	2.5	1.6	0.45	0.40	200	100	40	40	10	—
			Toxic (>)	—	—	—	—	—	—	—	1000	—	150	—	—
Beet, table	Leaf blades	5 weeks	Deficient	<3.0	0.22	2.0	1.5	0.25	—	40	30	15	30	5	0.05
		after seeding	Adequate	3.0	0.25	2.0	1.5	0.25	0.60	40	30	15	30	5	0.20
			range	5.0	0.40	6.0	2.0	1.00	0.80	200	200	30	80	10	0.60
			High	>5.0	0.40	6.0	2.0	1.00	—	—	—	—	80	10	—

TABLE 4.25. CRITICAL (DEFICIENCY) VALUES, ADEQUATE RANGES, HIGH VALUES, AND TOXICITY VALUES FOR PLANT NUTRIENT CONCENTRATION OF VEGETABLES (*Continued*)

Crop	Plant Part[1]	Time of Sampling	Status	%						ppm					
				N	P	K	Ca	Mg	S	Fe	Mn	Zn	B	Cu	Mo
			Toxic (>)	—	—	—	—	—	—	—	—	—	650	—	—
	Leaf blades	9 weeks after seeding	Deficient	<2.5	0.20	1.7	1.5	0.30	—	—	—	15	30	5	0.05
			Adequate	2.6	0.20	1.7	1.5	0.30	0.60	—	70	15	60	5	0.60
			range	4.0	0.30	4.0	3.0	1.00	0.80	—	200	30	80	10	—
			High	>4.0	0.30	4.0	3.0	1.00	—	—	—	—	80	10	—
			Toxic (>)	—	—	—	—	—	—	—	—	—	650	—	—
Broccoli	MRM leaf	Heading	Deficient	<3.0	0.30	1.1	0.8	0.23	0.20	40	20	25	20	3	0.04
			Adequate	3.0	0.30	1.5	1.2	0.23	—	40	25	45	30	5	0.04
			range	4.5	0.50	4.0	2.5	0.40	—	300	150	95	50	10	0.16
			High	>4.5	0.50	4.0	2.5	0.40	—	300	150	100	100	10	—
Brussels sprouts	MRM leaf	At early sprouts	Deficient	<2.2	0.20	2.4	0.4	0.20	0.20	50	20	20	20	4	0.04
			Adequate	2.2	0.20	2.4	0.4	0.20	0.20	50	20	20	30	5	0.16
			range	5.0	0.60	3.5	2.0	0.40	0.80	150	200	80	70	10	0.16
			High	>5.0	0.60	3.5	2.0	0.40	0.80	150	200	80	70	—	—
Cabbage	MRM leaf	5 weeks after transplanting	Deficient	<3.2	0.30	2.8	0.5	0.25	—	30	20	30	20	3	0.3
			Adequate	3.2	0.30	2.8	1.1	0.25	0.30	30	20	30	20	3	0.3
			range	6.0	0.60	5.0	2.0	0.60	—	60	40	50	40	7	0.6
			High	>6.0	0.60	5.0	2.0	0.60	—	100	40	50	40	10	—

	MRM	8 weeks	Deficient	<3.0	0.30	2.0	0.5	0.20	—	30	20	30	20	3	0.3
	leaf	after	Adequate	3.0	0.30	2.0	1.5	0.25	0.30	30	20	30	20	3	0.3
		transplanting	range	6.0	0.60	4.0	2.0	0.60	—	60	40	50	40	7	0.6
			High	>6.0	0.60	4.0	2.0	0.60	—	100	40	50	40	10	—
	Wrapper	Heads	Deficient	<3.0	0.30	1.7	0.5	0.25	—	20	20	20	30	4	0.3
	leaf	½ grown	Adequate	3.0	0.30	2.3	1.5	0.25	0.30	20	20	20	30	4	0.3
			range	4.0	0.50	4.0	2.0	0.45	—	40	40	30	50	8	0.6
			High	>4.0	0.50	4.0	2.0	0.45	—	100	40	40	50	10	—
	Wrapper	At harvest	Deficient	<1.8	0.26	1.2	0.5	0.25	—	20	20	20	30	4	0.3
	leaf		Adequate	1.8	0.26	1.5	1.5	0.25	0.30	20	20	20	30	4	0.3
			range	3.0	0.40	3.0	2.0	0.45	—	40	40	30	50	8	0.6
			High	>3.0	0.40	3.0	2.0	0.45	—	100	40	40	50	10	—
Cantaloupe	MRM	12-in.	Deficient	<4.0	0.40	5.0	3.0	0.35	—	40	20	20	20	5	0.6
	leaf	vines	Adequate	4.0	0.40	5.0	3.0	0.35	0.20	40	20	20	20	5	0.6
			range	5.0	0.70	7.0	5.0	0.45	0.50	100	100	60	80	10	1.0
			High	>5.0	0.70	7.0	5.0	0.45	—	100	100	60	80	10	1.0
			Toxic (>)	—	—	—	—	—	—	—	900	—	150	—	—
	MRM	Early fruit	Deficient	<3.5	0.25	1.8	1.8	0.30	—	40	20	20	20	5	0.6
	leaf	set	Adequate	3.5	0.25	1.8	1.8	0.30	0.20	40	20	20	20	5	0.6
			range	4.5	0.40	4.0	5.0	0.40	0.50	100	100	60	80	10	1.0
			High	>4.5	0.40	4.0	5.0	0.40	—	100	100	60	80	10	1.0
			Toxic (>)	—	—	—	—	—	—	—	900	—	150	—	—
Carrot	MRM	60 days	Deficient	<1.8	0.20	2.0	1.0	0.15	—	30	30	20	20	4	—
	leaf	after seeding	Adequate	1.8	0.20	2.0	2.0	0.20	—	30	30	20	20	4	—
			range	2.5	0.40	4.0	3.5	0.50	—	60	60	60	40	10	—
			High	>2.5	0.40	4.0	3.5	0.50	—	60	100	60	40	10	—
	MRM	Harvest	Deficient	<1.5	0.18	1.0	1.0	0.25	—	20	30	20	20	4	—

TABLE 4.25. CRITICAL (DEFICIENCY) VALUES, ADEQUATE RANGES, HIGH VALUES, AND TOXICITY VALUES FOR PLANT NUTRIENT CONCENTRATION OF VEGETABLES (*Continued*)

				%						ppm					
Crop	Plant Part[1]	Time of Sampling	Status	N	P	K	Ca	Mg	S	Fe	Mn	Zn	B	Cu	Mo
	leaf		Adequate	1.5	0.18	1.4	1.0	0.40	—	20	30	20	20	4	—
			range	2.5	0.40	4.0	1.5	0.50	—	30	60	60	40	10	—
			High	>2.5	0.40	4.0	1.5	0.50	—	60	100	60	40	10	—
Cauliflower	MRM	Buttoning	Deficient	<3.0	0.40	2.0	0.8	0.25	0.60	30	30	30	30	5	—
	leaf		Adequate	3.0	0.40	2.0	0.8	0.25	0.60	30	30	30	30	5	—
			range	5.0	0.70	4.0	2.0	0.60	1.00	60	80	50	50	10	—
			High	>5.0	0.70	4.0	2.0	0.60	—	100	100	50	50	10	—
	MRM	Heading	Deficient	<2.2	0.30	1.5	1.0	0.25	—	30	50	30	30	5	—
	leaf		Adequate	2.2	0.30	1.5	1.0	0.25	—	30	50	30	30	5	—
			range	4.0	0.70	3.0	2.0	0.60	—	60	80	50	50	10	—
			High	>4.0	0.70	3.0	2.0	0.60	—	100	100	50	50	10	—
Celery	Outer	6 weeks	Deficient	<1.5	0.30	6.0	1.3	0.30	—	20	5	20	15	4	—
	petiole	after	Adequate	1.5	0.30	6.0	1.3	0.30	—	20	5	20	15	4	—
		transplanting	range	1.7	0.60	8.0	2.0	0.60	—	30	10	40	25	6	—
			High	>1.7	0.60	8.0	2.0	0.60	—	100	20	60	25	—	—
	Outer	At maturity	Deficient	<1.5	0.30	5.0	1.3	0.30	—	20	5	20	20	1	—
	petiole		Adequate	1.5	0.30	5.0	1.3	0.30	—	20	5	20	20	1	—
			range	1.7	0.60	7.0	2.0	0.60	—	30	10	40	40	3	—
			High	>1.7	0.60	7.0	2.0	0.60	—	100	20	60	40	3	—

Chinese	Oldest	8-leaf	Deficient	<4.5	0.50	7.5	4.5	0.35	—	—	8	30	15	5	—
cabbage	undamaged	stage	Adequate	4.5	0.50	7.5	4.5	0.35	—	—	14	30	15	5	—
(heading)	leaf		range	5.0	0.60	8.5	5.0	0.45	—	—	20	50	25	10	—
			High	>5.0	0.60	8.5	5.0	0.45	—	—	20	50	25	10	—
	Oldest	At maturity	Deficient	<3.5	0.30	3.0	—	0.40	—	—	7	20	30	4	—
	undamaged		Adequate	3.5	0.30	3.0	3.7	0.40	—	—	13	20	30	4	—
	leaf		range	4.0	0.60	6.5	6.0	0.50	—	—	19	40	50	6	—
			High	>4.0	0.60	6.5	6.0	0.50	—	—	20	40	50	6	—
Collards	Tops	Young plants	Deficient	<4.0	0.30	3.0	1.0	0.40	—	40	40	25	25	5	—
			Adequate	4.0	0.30	3.0	1.0	0.40	—	40	40	25	25	5	—
			range	5.0	0.60	5.0	2.0	1.00	—	100	100	50	50	10	—
			High	>5.0	0.60	5.0	2.0	1.00	—	100	100	50	50	10	—
	MRM	Harvest	Deficient	<3.0	0.25	2.5	1.0	0.35	—	40	40	20	25	5	—
	leaf		Adequate	3.0	0.25	2.5	1.0	0.35	—	40	40	20	25	5	—
			range	5.0	0.50	4.0	2.0	0.10	—	100	100	40	50	10	—
			High	>5.0	0.50	4.0	2.0	0.10	—	100	100	40	50	10	—
Cucumber	MRM	Before bloom	Deficient	<3.5	0.30	1.6	2.0	0.58	0.30	40	30	20	20	5	0.2
	leaf		Adequate	3.5	0.30	1.6	2.0	0.58	0.30	40	30	20	20	5	0.3
			range	6.0	0.60	3.0	4.0	0.70	0.80	100	100	50	60	20	1.0
			High	>6.0	0.60	3.0	4.0	0.70	0.80	100	100	50	60	20	2.0
	MRM	Early bloom	Deficient	<2.5	0.25	1.6	1.3	0.30	0.30	40	30	20	20	5	0.2
	leaf		Adequate	2.5	0.25	1.6	1.3	0.30	0.30	40	30	20	20	5	0.3
			range	5.0	0.60	3.0	3.5	0.60	0.80	100	100	50	60	20	1.0
			High	>5.0	0.60	3.0	3.5	0.60	0.80	100	100	50	60	20	2.0
			Toxic (>)	—	—	—	—	—	—	—	900	950	150	—	—
Eggplant	MRM	Early fruit	Deficient	<4.2	0.30	3.5	0.8	0.25	0.40	50	50	20	20	5	0.5
	leaf	set	Adequate	4.2	0.30	3.5	0.8	0.25	0.40	50	50	20	20	5	0.5
			range	5.0	0.60	5.0	1.5	0.60	0.60	100	100	40	40	10	0.8

TABLE 4.25. CRITICAL (DEFICIENCY) VALUES, ADEQUATE RANGES, HIGH VALUES, AND TOXICITY VALUES FOR PLANT NUTRIENT CONCENTRATION OF VEGETABLES (*Continued*)

Crop	Plant Part[1]	Time of Sampling	Status	%						ppm					
				N	P	K	Ca	Mg	S	Fe	Mn	Zn	B	Cu	Mo
			High	>6.0	0.60	5.0	1.5	0.60	0.60	100	100	40	40	10	0.8
Endive	Oldest undamaged leaf	8-leaf stage	Deficient	<4.5	0.45	4.5	2.0	0.25	—	—	15	30	25	5	—
			Adequate range	4.5	0.45	4.5	2.0	0.25	—	—	15	30	25	5	—
				6.0	0.80	6.0	4.0	0.60	—	—	25	50	35	10	—
			High	>6.0	0.80	6.0	4.0	0.60	—	—	25	50	35	10	—
	Oldest undamaged leaf	Maturity	Deficient	<3.5	0.40	4.0	1.8	0.30	—	—	15	20	30	5	—
			Adequate range	3.5	0.40	4.0	1.8	0.30	—	—	15	20	30	5	—
				3.5	0.60	6.0	3.0	0.40	—	—	20	40	40	10	—
			High	>4.2	0.60	6.0	3.0	0.40	—	—	20	40	40	10	—
Escarole	Oldest undamaged leaf	8-leaf stage	Deficient	<4.2	0.45	5.7	1.7	0.25	—	—	15	30	20	4	—
			Adequate range	4.2	0.45	5.7	1.7	0.25	—	—	15	30	20	4	—
				5.0	0.60	6.5	2.2	0.35	—	—	25	50	30	6	—
			High	>5.0	0.60	6.5	2.2	0.35	—	—	25	50	30	6	—
	Oldest undamaged leaf	Maturity	Deficient	<3.0	0.35	5.5	2.0	0.25	—	—	15	20	30	4	—
			Adequate range	3.0	0.35	5.5	2.0	0.25	—	—	15	20	30	4	—
				4.5	0.45	6.5	3.0	0.35	—	—	25	50	45	6	—
			High	>4.5	0.45	6.5	3.0	0.35	—	—	25	50	45	6	—
Lettuce, Boston	Oldest undamaged	8-leaf stage	Deficient	<4.0	0.40	5.0	1.0	0.40	—	50	10	40	15	5	0.1
			Adequate	4.0	0.40	5.0	1.7	0.40	—	50	10	40	15	5	0.1

	leaf		range	6.0	0.60	6.0	2.0	0.60	—	100	20	60	25	10	0.2
			High	>6.0	0.60	6.0	2.0	0.60	—	100	20	60	25	10	0.4
			Toxic (>)	—	—	—	—	—	—	—	250	—	100	—	—
	Oldest	Maturity	Deficient	<3.0	0.35	5.0	1.0	0.30	—	50	10	20	15	5	0.1
	undamaged		Adequate	3.0	0.35	5.0	1.7	0.30	—	50	10	20	15	5	0.1
	leaf		range	4.0	0.45	6.0	2.0	0.60	—	100	20	40	25	10	0.2
			High	>4.0	0.45	6.0	2.0	0.60	—	100	20	40	25	10	0.4
			Toxic (>)	—	—	—	—	—	—	—	250	—	100	—	—
Lettuce,	Oldest	8-leaf	Deficient	<4.0	0.50	4.0	1.7	0.30	—	40	10	40	20	5	—
cos	undamaged	stage	Adequate	4.0	0.50	4.0	1.7	0.30	—	40	10	40	20	5	—
	leaf		range	5.0	0.60	6.0	2.0	1.70	—	100	20	60	40	10	—
			High	>5.0	0.60	6.0	2.0	1.70	—	100	20	60	40	10	—
	Oldest	Maturity	Deficient	<3.0	0.40	4.0	1.7	0.30	—	20	10	20	20	5	—
	undamaged		Adequate	3.0	0.40	4.0	1.7	0.30	—	20	10	20	20	5	—
	leaf		range	4.0	0.60	6.0	2.0	0.70	—	50	20	40	40	10	—
			High	>4.0	0.60	6.0	2.0	0.70	—	50	20	40	40	10	—
Lettuce,	MRM	8-leaf	Deficient	<4.0	0.40	5.0	1.0	0.30	0.30	50	20	25	15	5	—
crisphead		stage	Adequate	4.0	0.40	5.0	1.0	0.30	—	50	20	25	15	5	—
			range	5.0	0.60	7.0	2.0	0.50	0.50	150	40	50	30	10	—
			High	>5.0	0.60	7.0	2.0	0.50	—	150	40	50	30	10	—
	Wrapper	Heads	Deficient	<2.5	0.40	4.5	1.4	0.30	—	50	20	25	15	5	—
	leaf	½ size	Adequate	2.5	0.40	4.5	1.4	0.30	0.30	50	20	25	15	5	—
			range	4.0	0.60	8.0	2.0	0.70	0.50	150	40	50	30	10	—
			High	>4.0	0.60	8.0	2.0	0.70	—	150	40	50	30	10	—
	Wrapper	Maturity	Deficient	<2.0	0.25	2.5	1.4	0.30	—	50	20	25	15	5	—
	leaf		Adequate	2.0	0.25	2.5	1.4	0.30	0.30	50	20	25	15	5	—
			range	3.0	0.50	5.0	2.0	0.70	0.50	150	40	50	30	10	—
			High	>3.0	0.50	5.0	2.0	0.70	—	150	40	50	30	10	—

TABLE 4.25. CRITICAL (DEFICIENCY) VALUES, ADEQUATE RANGES, HIGH VALUES, AND TOXICITY VALUES FOR PLANT NUTRIENT CONCENTRATION OF VEGETABLES (*Continued*)

				%						ppm					
Crop	Plant Part[1]	Time of Sampling	Status	N	P	K	Ca	Mg	S	Fe	Mn	Zn	B	Cu	Mo
Lettuce, romaine	Oldest undamaged leaf	8-leaf stage	Deficient	<5.0	0.35	5.0	2.0	0.25	—	—	15	20	30	5	—
			Adequate range	5.0	0.35	5.0	2.0	0.25	—	—	15	20	30	5	—
				6.0	0.80	6.0	3.0	0.35	—	—	25	50	45	10	—
			High	>6.0	0.80	6.0	3.0	0.35	—	—	25	50	45	10	—
	Oldest undamaged leaf	Maturity	Deficient	<3.5	0.35	5.0	2.0	0.25	—	—	15	20	30	5	0.1
			Adequate range	3.5	0.35	5.0	2.0	0.25	—	—	15	20	30	5	0.1
				4.5	0.60	6.0	3.0	0.40	—	—	25	50	45	10	0.4
			High	>4.5	0.60	6.0	3.0	0.40	—	—	25	50	45	10	—
Okra	MRM leaf	30 days after seeding	Deficient	<3.5	0.30	2.0	0.5	0.25	—	50	30	30	25	5	—
			Adequate range	3.5	0.30	2.0	0.5	0.25	—	50	30	30	25	5	—
				5.0	0.60	3.0	0.8	0.50	—	100	100	50	50	10	—
			High	>5.0	0.60	3.0	0.8	0.50	—	100	100	50	50	10	—
	MRM leaf	Prior to harvest	Deficient	<2.5	0.30	2.0	1.0	0.25	—	50	30	30	25	5	—
			Adequate range	2.5	0.30	2.0	1.0	0.25	—	50	30	30	25	5	—
				3.0	0.60	3.0	1.5	0.50	—	100	100	50	50	10	—
			High	>3.0	0.60	3.0	1.5	0.50	—	100	100	50	50	10	—
Onion, sweet	MRM leaf	Just prior to bulb initiation	Deficient	<2.0	0.20	1.5	0.6	0.15	0.20	—	10	15	10	5	—
			Adequate range	2.0	0.20	1.5	0.6	0.15	0.20	—	10	15	10	5	—
				3.0	0.50	3.0	0.8	0.30	0.60	—	20	20	25	10	—

			High	>3.0	0.50	3.0	0.8	0.30	0.60	—	20	20	25	10	—
			Toxic (>)	—	—	—	—	—	—	—	—	—	100	—	—
Pepper	MRM leaf	Prior to blossoming	Deficient	<4.0	0.30	5.0	0.9	0.35	0.30	30	30	25	20	5	—
			Adequate	4.0	0.30	5.0	0.9	0.35	0.30	30	30	25	20	5	—
			range	5.0	0.50	6.0	1.5	0.60	0.60	150	100	80	50	10	—
			High	>5.0	0.50	6.0	1.5	0.60	0.60	150	100	80	50	10	—
			Toxic (>)	—	—	—	—	—	—	—	—	—	350	—	—
	MRM leaf	First blossoms open	Deficient	<3.0	0.30	2.5	0.9	0.30	0.30	30	30	25	20	5	—
			Adequate	3.0	0.30	2.5	0.9	0.30	0.30	30	30	25	20	5	—
			range	5.0	0.50	5.0	1.5	0.50	0.60	150	100	80	50	10	—
			High	>5.0	0.50	5.0	1.5	0.50	0.60	150	100	80	50	10	—
			Toxic (>)	—	—	—	—	—	—	—	1000	—	350	—	—
	MRM leaf	Early fruit set	Deficient	<2.9	0.25	2.5	1.0	0.30	0.30	30	30	25	20	5	—
			Adequate	2.9	0.25	2.5	1.0	0.30	0.30	30	30	25	20	5	—
			range	4.0	0.40	4.0	1.5	0.40	0.40	150	100	80	50	10	—
			High	>4.0	0.40	4.0	1.5	0.40	0.40	150	100	80	50	10	—
			Toxic (>)	—	—	—	—	—	—	—	—	—	350	—	—
	MRM leaf	Early harvest	Deficient	<2.5	0.20	2.0	1.0	0.30	0.30	30	30	25	20	5	0.1
			Adequate	2.5	0.20	2.0	1.0	0.30	0.30	30	30	25	20	5	0.1
			range	3.0	0.40	3.0	1.5	0.40	0.40	150	100	80	50	10	0.2
			High	>3.0	0.40	3.0	1.5	0.40	0.40	150	100	80	50	10	—
			Toxic (>)	—	—	—	—	—	—	—	—	—	350	—	—
Potato	MRM leaf	Plants 8–10 in. tall	Deficient	<3.0	0.20	3.5	0.6	0.30	0.25	40	30	30	20	5	0.1
			Adequate	3.0	0.20	3.5	0.6	0.30	0.25	40	30	30	20	5	0.1
			range	6.0	0.80	6.0	2.0	0.60	0.50	150	60	60	60	10	0.2
			High	>6.0	0.80	6.0	2.0	0.60	0.50	150	60	60	60	10	—
	MRM	First	Deficient	<3.0	0.20	3.0	0.6	0.25	0.20	40	30	30	20	5	0.1

TABLE 4.25. CRITICAL (DEFICIENCY) VALUES, ADEQUATE RANGES, HIGH VALUES, AND TOXICITY VALUES FOR PLANT NUTRIENT CONCENTRATION OF VEGETABLES (*Continued*)

				%						ppm					
Crop	Plant Part[1]	Time of Sampling	Status	N	P	K	Ca	Mg	S	Fe	Mn	Zn	B	Cu	Mo
	leaf	blossom	Adequate	3.0	0.20	3.0	0.6	0.25	0.20	40	30	30	20	5	0.1
			range	4.0	0.50	5.0	2.0	0.60	0.50	150	100	60	30	10	0.2
			High	>4.0	0.50	5.0	2.0	0.60	0.50	150	100	60	30	10	—
	MRM	Tubers	Deficient	<2.0	0.20	2.5	0.6	0.25	0.20	40	20	30	20	5	0.1
	leaf	½ grown	Adequate	2.0	0.20	2.5	0.6	0.25	0.20	40	20	30	20	5	0.1
			range	4.0	0.40	4.0	2.0	0.60	0.50	150	100	60	30	10	0.2
			High	>4.0	0.40	4.0	2.0	0.60	0.50	150	100	60	30	10	—
	MRM	At tops-down	Deficient	<2.0	0.16	1.5	0.6	0.20	0.20	40	20	30	20	5	0.1
	leaf		Adequate	2.0	0.16	1.5	0.6	0.20	0.20	40	20	30	20	5	0.1
			range	3.0	0.40	3.0	2.0	0.50	0.50	150	100	60	30	10	0.2
			High	>3.0	0.40	3.0	2.0	0.50	0.50	150	100	60	30	10	—
Pumpkin	MRM	5 weeks	Deficient	<3.0	0.30	2.3	0.9	0.35	0.20	40	40	20	25	5	0.3
	leaf	after seeding	Adequate	3.0	0.30	2.3	0.9	0.35	0.20	40	40	20	25	5	0.3
			range	6.0	0.50	4.0	1.5	0.60	0.40	100	100	50	40	10	0.5
			High	>6.0	0.50	4.0	1.5	0.60	0.40	100	100	50	40	10	—
	MRM	8 weeks from	Deficient	<3.0	0.25	2.0	0.9	0.30	0.20	40	40	20	20	5	0.3
	leaf	seeding	Adequate	3.0	0.25	2.0	0.9	0.30	0.20	40	40	20	20	5	0.3
			range	4.0	0.40	3.0	1.5	0.50	0.40	100	100	50	40	10	0.5

			High	>4.0	0.40	3.0	1.5	0.50	0.40	100	100	50	40	10	—
Radish	MRM leaf	At harvest	Deficient	<3.0	0.25	1.5	1.0	0.30	—	30	20	30	15	3	0.1
			Adequate	3.0	0.25	1.5	1.0	0.30	—	30	20	30	15	3	0.1
			range	4.5	0.40	3.0	2.0	0.50	—	50	40	50	30	10	2.0
			High	>4.5	0.40	3.0	2.0	0.50	—	50	40	50	30	10	2.0
			Toxic (>)	—	—	—	—	—	—	—	—	—	85	—	—
Southern pea	MRM leaf	Before bloom	Deficient	<3.5	0.30	2.0	1.0	0.30	—	30	30	20	15	5	—
			Adequate	3.5	0.30	2.0	1.0	0.30	—	30	30	20	15	5	—
			range	5.0	0.80	4.0	1.5	0.50	—	100	100	40	25	10	—
			High	>5.0	0.80	4.0	1.5	0.50	—	100	100	40	25	10	—
	MRM leaf	First bloom	Deficient	<2.5	0.20	2.0	1.0	0.30	—	30	30	20	15	5	4.0
			Adequate	2.5	0.20	2.0	1.0	0.30	—	30	30	20	15	5	4.0
			range	4.0	0.40	4.0	1.5	0.50	—	100	100	40	25	10	6.0
			High	>4.0	0.40	4.0	1.5	0.50	—	100	100	40	25	10	6.0
Spinach	MRM leaf	30 days after seeding	Deficient	<3.0	0.30	3.0	0.6	1.00	—	—	50	50	20	5	0.1
			Adequate	3.0	0.30	3.0	0.6	1.00	—	—	50	50	20	5	0.1
			range	4.5	0.50	4.0	1.0	1.60	—	—	100	70	40	7	1.0
			High	>5.0	0.50	4.0	1.0	1.60	—	—	100	70	40	7	1.0
	MRM leaf	Harvest	Deficient	<3.0	0.25	2.5	0.6	1.00	—	—	30	50	20	5	0.1
			Adequate	3.0	0.25	2.5	0.6	1.00	—	—	30	50	20	5	0.1
			range	4.0	0.50	3.5	1.0	1.60	—	—	50	70	40	7	1.0
			High	>4.0	0.50	4.0	1.0	1.60	—	—	80	70	40	7	1.0
Squash	MRM	Early fruit	Deficient	<3.0	0.25	2.0	1.0	0.30	0.20	40	40	20	25	5	0.3
			Adequate	3.0	0.25	2.0	1.0	0.30	0.20	40	40	20	25	5	0.3
			range	5.0	0.50	3.0	2.0	0.50	0.50	100	100	50	40	20	0.5
			High	>5.0	0.50	3.0	2.0	0.50	0.50	100	100	50	40	20	0.5
Strawberry	MRM leaf	Transplants	Deficient	<2.8	0.25	1.5	0.3	0.30	—	50	30	25	25	5	—
			Adequate	2.8	0.25	1.5	0.3	0.30	—	50	30	25	25	5	—
			range	3.5	0.40	3.0	1.5	0.60	—	100	100	40	40	10	—

TABLE 4.25. CRITICAL (DEFICIENCY) VALUES, ADEQUATE RANGES, HIGH VALUES, AND TOXICITY VALUES FOR PLANT NUTRIENT CONCENTRATION OF VEGETABLES (*Continued*)

				%						ppm					
Crop	Plant Part[1]	Time of Sampling	Status	N	P	K	Ca	Mg	S	Fe	Mn	Zn	B	Cu	Mo
			High	>3.5	0.40	3.0	1.5	0.60	—	100	100	40	40	10	—
	MRM	Initial	Deficient	<3.0	0.20	1.5	0.4	0.25	—	50	30	20	20	5	—
	leaf	flower	Adequate	3.0	0.20	1.5	0.4	0.25	—	50	30	20	20	5	—
			range	4.0	0.40	3.0	1.5	0.50	—	100	100	40	40	10	—
			High	>4.0	0.40	3.0	1.5	0.50	—	100	100	40	20	10	—
	MRM	Initial	Deficient	<3.0	0.20	1.5	0.4	0.25	—	50	30	20	20	5	—
	leaf	harvest	Adequate	3.0	0.20	1.5	0.4	0.25	—	50	30	20	20	5	—
			range	3.5	0.40	2.5	1.5	0.50	—	100	100	40	40	10	—
			High	>3.5	0.40	2.5	1.5	0.50	—	100	100	40	40	10	—
			Toxic (>)	—	—	—	—	—	—	—	800	—	—	—	—
	MRM	Midseason	Deficient	<2.8	0.20	1.1	0.4	0.20	0.8	50	25	20	20	5	0.5
	leaf		Adequate	2.8	0.20	1.1	0.4	0.20	0.8	50	25	20	20	5	0.5
			range	3.0	0.40	2.5	1.5	0.40	1.0	100	100	40	40	10	0.8
			High	>3.0	0.40	2.5	1.5	0.40	1.0	100	100	40	40	10	0.8
			Toxic (>)	—	—	—	—	—	—	—	800	—	—	—	—
	MRM	End of	Deficient	<2.5	0.20	1.1	0.4	0.20	—	50	25	20	20	5	—
	leaf	season	Adequate	2.5	0.20	1.1	0.4	0.20	—	50	25	20	20	5	—
			range	3.0	0.30	2.0	1.5	0.40	—	100	100	40	40	10	—
			High	>3.0	0.30	2.0	1.5	0.40	—	100	100	40	40	10	—

Sweet corn	Whole	3-leaf	Deficient	<3.0	0.35	2.5	0.6	0.25	0.4	50	40	30	10	5	0.1
	seedlings	stage	Adequate	3.0	0.35	2.5	0.6	0.25	0.4	50	40	30	10	5	0.1
			range	4.0	0.50	4.0	0.8	0.50	0.6	100	100	40	30	10	0.2
			High	>4.0	0.50	4.0	0.8	0.50	0.6	100	100	40	30	10	0.2
			Toxic (>)	—	—	—	—	—	—	—	—	—	100	—	—
	Whole	6-leaf stage	Deficient	<3.0	0.25	2.5	0.5	0.25	0.4	50	40	30	10	5	0.1
	seedlings		Adequate	3.0	0.25	2.5	0.5	0.25	0.4	50	40	30	10	5	0.1
			range	4.0	0.50	4.0	0.8	0.50	0.6	100	100	40	30	10	0.2
			High	>4.0	0.50	4.0	0.8	0.50	0.6	100	100	40	30	10	0.2
			Toxic (>)	—	—	—	—	—	—	—	—	—	100	—	—
	MRM	30 in. tall	Deficient	<2.5	0.20	2.5	0.5	0.20	0.2	40	40	25	10	4	0.1
	leaf		Adequate	2.5	0.20	2.5	0.5	0.20	0.2	40	40	25	10	4	0.1
			range	4.0	0.40	4.0	0.8	0.40	0.4	100	100	40	30	10	0.2
			High	>4.0	0.40	4.0	0.8	0.40	0.4	100	100	40	30	10	0.2
			Toxic (>)	—	—	—	—	—	—	—	—	—	100	—	—
	MRM	Just prior	Deficient	<2.5	0.20	2.0	0.3	0.15	0.2	30	30	20	10	4	0.1
	leaf	to tassel	Adequate	2.5	0.20	2.0	0.3	0.15	0.2	30	30	20	10	4	0.1
			range	4.0	0.40	3.5	0.6	0.40	0.4	100	100	40	20	10	0.2
			High	>4.0	0.40	3.5	0.6	0.40	0.4	100	100	40	20	10	0.2
			Toxic (>)	—	—	—	—	—	—	—	—	—	100	—	—
	Ear leaf	Tasseling	Deficient	<1.5	0.20	1.2	0.3	0.15	0.20	30	20	20	10	4	0.1
			Adequate	1.5	0.20	1.2	0.3	0.15	0.20	30	20	20	10	4	0.1
			range	2.5	0.40	2.0	0.6	0.40	0.40	100	100	40	20	10	0.2
			High	>2.5	0.40	2.0	0.6	0.40	0.40	100	100	40	20	10	0.2
Sweet	MRM	Early	Deficient	<4.0	0.30	2.5	0.8	0.40	0.20	40	40	25	20	5	—
potato	leaf	vining	Adequate	4.0	0.30	2.5	0.8	0.40	0.20	40	40	25	20	5	—
			range	5.0	0.50	4.0	1.6	0.80	0.60	100	100	50	50	10	—

TABLE 4.25. CRITICAL (DEFICIENCY) VALUES, ADEQUATE RANGES, HIGH VALUES, AND TOXICITY VALUES FOR PLANT NUTRIENT CONCENTRATION OF VEGETABLES (*Continued*)

				%						ppm					
Crop	Plant Part[1]	Time of Sampling	Status	N	P	K	Ca	Mg	S	Fe	Mn	Zn	B	Cu	Mo
			High	>5.0	0.50	4.0	1.6	0.80	0.60	100	100	50	50	10	—
	MRM	Midseason	Deficient	<3.0	0.20	2.0	0.8	0.25	0.20	40	40	25	25	5	—
	leaf	before root	Adequate	3.0	0.20	2.0	0.8	0.25	0.20	40	40	25	25	5	—
		enlargement	range	4.0	0.30	4.0	1.8	0.50	0.40	100	100	40	40	10	—
			High	>4.0	0.30	4.0	1.8	0.50	0.40	100	100	40	40	10	—
	MRM	Root	Deficient	<3.0	0.20	2.0	0.8	0.25	0.20	40	40	25	20	5	—
	leaf	enlargement	Adequate	3.0	0.20	2.0	0.8	0.25	0.20	40	40	25	20	5	—
			range	4.0	0.30	4.0	1.6	0.50	0.60	100	100	50	50	10	—
			High	>4.0	0.30	4.0	1.6	0.50	0.60	100	100	50	50	10	—
	MRM	Just before	Deficient	<2.8	0.20	2.0	0.8	0.25	0.20	40	40	25	20	5	—
	leaf	harvest	Adequate	2.8	0.20	2.0	0.8	0.25	0.20	40	40	25	20	5	—
			range	3.5	0.30	4.0	1.6	0.50	0.60	100	100	50	50	10	—
			High	>3.5	0.30	4.0	1.6	0.50	0.60	100	100	50	50	10	—
Tomato	MRM	5-leaf	Deficient	<3.0	0.30	3.0	1.0	0.30	0.30	40	30	25	20	5	0.2
	leaf	stage	Adequate	3.0	0.30	3.0	1.0	0.30	0.30	40	30	25	20	5	0.2
			range	5.0	0.60	5.0	2.0	0.50	0.80	100	100	40	40	15	0.6
			High	>5.0	0.60	5.0	2.0	0.50	0.80	100	100	40	40	15	0.6
	MRM	First	Deficient	<2.8	0.20	2.5	1.00	0.30	0.30	40	30	25	20	5	0.2
	leaf	flower	Adequate	2.8	0.20	2.5	1.00	0.30	0.30	40	30	25	20	5	0.2

			range	4.0	0.40	4.0	2.00	0.50	0.80	100	100	40	40	15	0.6
			High	>4.0	0.40	4.0	2.00	0.50	0.80	100	100	40	40	15	0.6
			Toxic (>)	—	—	—	—	—	—	—	1500	300	250	—	—
	MRM	Early	Deficient	<2.5	0.20	2.5	1.0	0.25	0.30	40	30	20	20	5	0.2
	leaf	fruit set	Adequate	2.5	0.20	2.5	1.0	0.25	0.30	40	30	20	20	5	0.2
			range	4.0	0.40	4.0	2.0	0.50	0.60	100	100	40	40	10	0.6
			High	>4.0	0.40	4.0	2.0	0.50	0.60	100	100	40	40	10	0.6
			Toxic (>)	—	—	—	—	—	—	—	—	—	250	—	—
	MRM	First ripe	Deficient	<2.0	0.20	2.0	1.0	0.25	0.30	40	30	20	20	5	0.2
	leaf	fruit	Adequate	2.0	0.20	2.0	1.0	0.25	0.30	40	30	20	20	5	0.2
			range	3.5	0.40	4.0	2.0	0.50	0.60	100	100	40	40	10	0.6
			High	>3.5	0.40	4.0	2.0	0.50	0.60	100	100	40	40	10	0.6
	MRM	During	Deficient	<2.0	0.20	1.5	1.0	0.25	0.30	40	30	20	20	5	0.2
	leaf	harvest	Adequate	2.0	0.20	1.5	1.0	0.25	0.30	40	30	20	20	5	0.2
		period	range	3.0	0.40	2.5	2.0	0.50	0.60	100	100	40	40	10	0.6
			High	>3.0	0.40	2.5	2.0	0.50	0.60	100	100	40	40	10	0.6
Turnip	MRM	Hypocotyl	Deficient	<3.0	0.25	2.5	0.8	0.25	0.20	30	30	20	20	5	—
greens	leaf	1-in.	Adequate	3.0	0.25	2.5	0.8	0.25	0.20	30	30	20	20	5	—
		diameter	range	5.0	0.80	4.0	1.5	0.60	0.60	100	100	40	40	10	—
			High	>5.0	0.80	4.0	1.5	0.60	0.60	100	100	40	40	10	—
Watermelon	MRM	Layby	Deficient	<3.0	0.25	3.0	1.0	0.25	0.20	30	20	20	20	5	—
	leaf	(last	Adequate	3.0	0.25	3.0	1.0	0.25	0.20	30	20	20	20	5	—
		cultivation)	range	4.0	0.50	4.0	2.0	0.50	0.40	100	100	40	40	10	—
			High	>4.0	0.50	4.0	2.0	0.50	0.40	100	100	40	40	10	—
			Toxic (>)	—	—	—	—	—	—	—	800	—	—	—	—
	MRM	First	Deficient	<2.5	0.25	2.7	1.0	0.25	0.20	30	20	20	20	5	—
	leaf	flower	Adequate	2.5	0.25	2.7	1.0	0.25	0.20	30	20	20	20	5	—

TABLE 4.25. CRITICAL (DEFICIENCY) VALUES, ADEQUATE RANGES, HIGH VALUES, AND TOXICITY VALUES FOR PLANT NUTRIENT CONCENTRATION OF VEGETABLES (*Continued*)

				%						ppm					
Crop	Plant Part[1]	Time of Sampling	Status	N	P	K	Ca	Mg	S	Fe	Mn	Zn	B	Cu	Mo
			range	3.5	0.50	3.5	2.0	0.50	0.40	100	100	40	40	10	—
			High	>3.5	0.50	3.5	2.0	0.50	0.40	100	100	40	40	10	—
	MRM leaf	First fruit	Deficient	<2.0	0.25	2.3	1.0	0.25	0.20	30	20	20	20	5	—
			Adequate range	2.0	0.25	2.3	1.0	0.25	0.20	30	20	20	20	5	—
				3.0	0.50	3.5	2.0	0.50	0.40	100	100	40	40	10	—
			High	>3.0	0.50	3.5	2.0	0.50	0.40	100	100	40	40	10	—
	MRM leaf	Harvest period	Deficient	<2.0	0.25	2.0	1.0	0.25	0.20	30	20	20	20	3	—
			Adequate range	2.0	0.25	2.0	1.0	0.25	0.20	30	20	20	20	3	—
				3.0	0.50	3.0	2.0	0.50	0.40	100	100	40	40	10	—
			High	>3.0	0.50	3.0	2.0	0.50	0.40	100	100	40	40	10	—

Adapted from G. Hochmuth, D. Maynard, C. Vavrina, E. Hanlon, and E. Simonne, *Plant Tissue Analysis and Interpretation for Vegetable Crops in Florida* (Florida Cooperative Extension Service), http://edis.ifas.ufl.edu/hs162.

[1] *MRM leaf* is the most recently matured whole leaf blade plus petiole.

TABLE 4.26. UNIVERSITY OF FLORIDA GUIDELINES FOR LEAF PETIOLE FRESH SAP NITRATE-NITROGEN AND POTASSIUM TESTING

		Fresh Petiole Sap Concentration (ppm)	
Crop	Development Stage/Time	NO_3–N	K
Eggplant	First fruit (2 in. long)	1,200–1,600	4,500–5,000
	First harvest	1,000–1,200	4,000–4,500
	Midharvest	800–1,000	3,500–4,000
Pepper	First flower buds	1,400–1,600	3,200–3,500
	First open flowers	1,400–1,600	3,000–3,200
	Fruits half-grown	1,200–1,400	3,000–3,200
	First harvest	800–1,000	2,400–3,000
	Second harvest	500–800	2,000–2,400
Potato	Plants 8 in. tall	1,200–1,400	4,500–5,000
	First open flowers	1,000–1,400	4,500–5,000
	50% flowers open	1,000–1,200	4,000–4,500
	100% flowers open	900–1,200	3,500–4,000
	Tops falling over	600–900	2,500–3,000
Strawberry[1]	November	800–900	3,000–3,500
	December	600–800	3,000–3,500
	January	600–800	2,500–3,000
	February	300–500	2,000–2,500
	March	200–500	1,800–2,500
	April	200–500	1,500–2,000
Tomato	First buds	1,000–1,200	3,500–4,000
	First open flowers	600–800	3,500–4,000
	Fruits 1-in. diameter	400–600	3,000–3,500
	Fruits 2-in. diameter	400–600	3,000–3,500
	First harvest	300–400	2,500–3,000
	Second harvest	200–400	2,000–2,500
Watermelon	Vines 6 in. long	1,200–1,500	4,000–5,000
	Fruit 2 in. long	1,000–1,200	4,000–5,000
	Fruits one-half mature	800–1,000	3,500–4,000
	At first harvest	600–800	3,000–3,500

Adapted from G. Hochmuth, "Plant Petiole Sap-testing Guide for Vegetable Crops," Florida Cooperative Extension Service Circular 1144 (2003), http://edis.ifas.ufl.edu/pdffiles/cv/cv00400.pdf.

[1] Annual hill production system

12
SOIL TESTS

Analyses for total amounts of nutrients in the soil are of limited value in predicting fertilizer needs. Consequently, various methods and extractants have been developed to estimate available soil nutrients and to serve as a basis for predicting fertilizer needs. Proper interpretation of the results of soil analysis is essential in recommending fertilizer needs. Soil testing procedure and extractants must be correlated with crop response to be of value for predicting crop need for fertilizer. One soil test procedure developed for one soil and crop condition in one area of the country may not apply in another area. Growers should consult their crop and soils experts about soil testing procedures appropriate for their growing area.

DETERMINING THE KIND AND QUANTITY OF FERTILIZER TO USE

Many states issue suggested rates of application of fertilizers for specific vegetables. These recommendations are sometimes made according to the type of soil—that is, light or heavy, sands, loams, clays, peats, and mucks. Other factors often used in establishing these rates are whether manure or soil-improving crops are employed and whether an optimum moisture supply can be maintained. The nutrient requirements of each crop must be considered, as must the past fertilizer and cropping history. The season of the year affects nutrient availability. Broad recommendations are at best only a point from which to make adjustments to suit individual conditions. Each field may require a different fertilizer program for the same vegetable.

Calibrated soil testing can provide an estimate of the concentration of essential elements that will be available to the crop during the season and predict the amount of fertilizer needed to produce a crop. Various extraction solutions are used by soil testing labs around the country to estimate the nutrient-supplying capacity of the soil, and not all solutions are calibrated for all soils. Therefore, growers must exercise care in selecting a lab to analyze soil samples, using only those labs that employ analytical procedures calibrated with yield response in specific soil types and growing regions. Even though several labs might differ in lab procedures, if all procedures are calibrated, then fertilizer recommendations should be similar. If unclear about specific soil testing practices, growers should consult their Cooperative Extension Service and the specific analytical lab.

Phosphorus. This element is not very mobile in most agricultural soils.

Phosphorus is fixed in soils with basic reactions and large quantities of calcium, or in acidic soils containing aluminum or iron. Even though phosphorus can be fixed, if a calibrated soil test predicts no response to phosphorus fertilization, then growers need not add large amounts of phosphorus because enough phosphorus will be made available to the crop during the growing season, even though the soil has a high phosphorus-fixing capacity. Sometimes crops might respond to small amounts of starter phosphorus supplied to high phosphorus testing soils in cold planting seasons.

Potassium. Although not generally considered a mobile element in soils, potassium can leach in coarse, sandy soils. Clay soils and loamy soils often contain adequate amounts of available potassium and may not need fertilization with potassium. Coarse, sandy soils usually test medium or low in extractable potassium, and crops growing on these soils respond to potassium fertilization.

Nitrogen. Most soil testing labs have no calibrated soil test for nitrogen because nitrogen is highly mobile in most soils and predicting a crop's response to nitrogen fertilization from a soil test is risky. However, some labs do predict the nitrogen-supplying capacity of a soil from a determination of soil organic matter. Estimates vary from 20 to 40 lb nitrogen made available during the season for each percent soil organic matter. Another soil nitrogen estimation procedure used by some labs is the pre-sidedress soil nitrate test. This test predicts the likelihood of need for sidedressed nitrogen during the season but is relatively insensitive for predicting exact amounts of sidedress nitrogen.

TABLE 4.27. PREDICTED RESPONSES OF CROPS TO RELATIVE AMOUNTS OF EXTRACTED PLANT NUTRIENTS BY SOIL TEST

Soil Test Interpretation	Predicted Crop Response
Very high	No crop response predicted to fertilization with a particular element.
High	No crop response predicted to fertilization with a particular element.
Medium	75–100% maximum expected yield predicted without fertilization.
Low	50–75% maximum expected yield predicted without fertilization.
Very low	25–50% maximum expected yield predicted without fertilization.

TABLE 4.28. INTERPRETATION OF SOIL TEST RESULTS FOR PHOSPHORUS BY THE OLSEN BICARBONATE EXTRACTION METHOD, FOR POTASSIUM AND MAGNESIUM BY THE AMMONIUM ACETATE EXTRACTION METHOD, AND FOR ZINC BY THE DPTA EXTRACTION METHOD

	Amount in Soil (ppm)			
Nutrient Need	Phosphorus[1] (PO_4–P)	Potassium[2] (K)	Magnesium[2] (Mg)	Zinc[3] (Zn)
Deficient levels for most vegetables	0–10	0–60	0–25	0–0.3
Deficient for susceptible vegetables	10–20	60–120	25–50	0.3–0.6
A few susceptible crops may respond	20–40	120–200	50–100	0.6–1.0
No crop response	Above 40	Above 200	Above 100	Above 1.0
Levels are excessive and could cause problems	Above 150	Above 2000	Above 1000	Above 3.0

Adapted from H. M. Reisenauer (ed.), *Soil and Plant Tissue Testing in California,* University of California Division of Agricultural Science Bulletin 1879 (1978).

[1] Olsen (0.5*M*, pH 8.5) sodium bicarbonate extractant

[2] Exchangeable with 1*N* ammonium acetate extractant

[3] DTPA extractable Zn

TABLE 4.29. INTERPRETATION OF SOIL TEST RESULTS OBTAINED BY THE MEHLICH-1 DOUBLE ACID (0.05*N* HCl, 0.025*N* H_2SO_4) AND MEHLICH-3 SOIL EXTRACTANTS

Relative Level in Soil	Soil Amount (lb/acre)							
	Phosphorus (P)		Potassium (K)		Magnesium (Mg)		Calcium (Ca)	
	Mehlich-1	Mehlich-3	Mehlich-1	Mehlich-3	Mehlich-1	Mehlich-3	Mehlich-1	Mehlich-3
Very low	0–13	0–24	0–29	0–40	0–35	0–45	0–400	0–615
Low	14–27	25–45	30–70	41–81	36–70	46–83	401–800	616–1,007
Medium	28–45	46–71	71–134	82–145	71–125	84–143	801–1,200	1,008–1,400
High	46–89	72–137	135–267	146–277	126–265	144–295	1,201–1,600	1,401–1,790
Very high	90+	138+	268+	278+	266+	1,601+	296+	1,791+

Adapted from *Commercial Vegetable Production Recommendations,* Delaware Cooperative Extension Service Bulletin 137 (2005).

TABLE 4.30. INTERPRETATION OF THE MEHLICH-1 (DOUBLE-ACID) EXTRACTANT USED BY THE UNIVERSITY OF FLORIDA

	ppm (soil)				
Element	Very Low	Low	Medium	High	Very High
P	<10	10–15	16–30	31–60	>60
K	<20	20–35	36–60	61–125	>125
Mg		<15	15–30	>30	

Micronutrients

	Soil pH (mineral soils only)		
	5.5–5.9	6.0–6.4 ppm (soil)	6.5–7.0
Test level below which there may be a crop response to applied *copper*	0.1–0.3	0.3–0.5	0.5
Test level above which *copper* toxicity may occur	2.0–3.0	3.0–5.0	5.0
Test level below which there may be a crop response to applied *manganese*	3.0–5.0	5.0–7.0	7.0–9.0
Test level below which there may be a crop response to applied *zinc*	0.5	0.5–1.0	1.0–3.0

Adapted from G. Hochmuth and E. Hanlon, "IFAS Standardized Fertilization Recommendations for Vegetable Crops" (Florida Cooperative Extension Service Circular 1152, 2000), http://edis.ifas.ufl.edu/CV002.

TABLE 4.31. GUIDE FOR DIAGNOSING NUTRIENT STATUS OF CALIFORNIA SOILS FOR VEGETABLE CROPS[1]

		Vegetable Yield Response to Fertilizer Application	
Vegetable	Nutrient[1]	Likely (soil ppm less than)	Not Likely (soil ppm more than)
Lettuce	P	15	25
	K	50	80
	Zn	0.5	1.0
Cantaloupe	P	8	12
	K	80	100
	Zn	0.4	0.6
Onion	P	8	12
	K	80	100
	Zn	0.5	1.0
Potato (mineral soils)	P	12	25
	K	100	150
	Zn	0.3	0.7
Tomato	P	8	12
	K	100	150
	Zn	0.3	0.7
Warm-season vegetables	P	8	12
	K	50	70
	Zn	0.2	0.5
Cool-season vegetables	P	20	30
	K	50	80
	Zn	0.5	1.0

Adapted from *Soil and Plant Tissue Testing in California*, University of California Division Agricultural Science Bulletin 1879 (1983). Updated 1996, personal communication, T. K. Hartz, University of California—Davis.

[1] Soil extracts:

PO_4–P: 0.5*M* pH 8.5 sodium bicarbonate ($NaHCO_3$)
K: 1.0*M* ammonium acetate (NH_4OAc)
Zn: 0.005*M* diethyienetriaminepentaacetic acid (DTPA)

PRE-SIDEDRESS NITROGEN TEST FOR SWEET CORN

The Pre-sidedress Nitrate Test was developed to aid farmers in the prediction of nitrogen needs by corn at the time when sidedress applications are normally made. This test takes into consideration nitrogen released from organic nutrient sources (such as manure, compost, cover crops, and soil organic matter) in addition to nitrogen fertilizer.

Sampling Procedure for Nitrogen Soil Test

1. Sample soil when corn is 8–12 in. tall.
2. Collect 15–20 soil cores per field to a depth of 12 in., if possible. If not, sample as deeply as possible. Avoid areas where starter fertilizer bands were applied, areas where manure was stacked, and areas where starter fertilizer applications were unusually heavy or light.
3. Combine the cores for each field and mix completely. Take a subsample of approximately 1 cup and dry it immediately. Soil can be dried in an oven at about 200°F. Samples can also be air dried if spread out thinly on a nonabsorbent material in a dry, well-ventilated area. A fan reduces drying time. Do not put wet samples on absorbent material because it will absorb some nitrate. The longer the delay in drying the sample, the less accurate the results will be.

TABLE 4.32. SWEET CORN NITROGEN TEST

Soil Test Results (ppm NO_3–N)	Recommended Nitrogen Sidedressing (lbs actual N/acre)
0–10	130
11–20	100
21–25	50
25+	0

Adapted from Vern Grubinger, University of Vermont Cooperative Extension Service, http://www.uvm.edu/vtvegandberry/factsheets/PSNT.html. Similar research results were obtained with pumpkin by T. F. Morris et al. (University of Connecticut, 1999), http://www.hort.uconn.edu/ipm/veg/htms/presidrs.htm.

Additional references for soil testing for vegetable fertilizer recommendations:

- E. A. Hanlon, *Procedures Used by State Soil Testing Laboratories in the Southern Region of the United States,* (Southern Cooperative Series Bulletin 190-B, 1998), http://www.imok.ufl.edu/hanlon/bull190.pdf.
- H. M. Reisenauer, J. Quick, R. E. Voss, and A. L. Brown, *Chemical Soil Tests for Soil Fertility Evaluation* (University of California Vegetable Research and Information Center), http://vric.ucdavis.edu.

TABLE 4.33. CONVERSION OF FERTILIZER RATES FROM APPLICATION ON A PER-ACRE BASIS TO RATES BASED ON LINEAR BED FEET FOR FULL-BED MULCHED CROPS

Typical bed spacing for mulched vegetables grown in Florida:

Vegetable	Typical Spacing (ft)[1]	Rows of Plants per Bed	Vegetable	Typical Spacing (ft)	Rows of Plants per Bed
Broccoli	6	2	Lettuce	4	2
Cabbage	6	2	Pepper	6	2
Cantaloupe	5	1	Summer squash	6	2
Cauliflower	6	2	Strawberry	4	2
Cucumber	6	2	Tomato	6	1
Eggplant	6	1	Watermelon	8	1

[1] Spacing between the centers of two adjacent beds.

TABLE 4.33. CONVERSION OF FERTILIZER RATES FROM APPLICATION ON A PER-ACRE BASIS TO RATES BASED ON LINEAR BED FEET FOR FULL-BED MULCHED CROPS (*Continued*)

Typical Bed Spacing (ft)	Recommended Fertilizer (N, P_2O_5, or K_2O) (lb/A)								
	20	40	60	80	100	120	140	160	180
	Resulting Fertilizer Rate (N, P_2O_5, or K_2O) (lb/100 LBF)								
3	0.14	0.28	0.41	0.55	0.69	0.83	0.96	1.10	1.24
4	0.18	0.37	0.55	0.73	0.92	1.10	1.29	1.47	1.65
5	0.23	0.46	0.69	0.92	1.15	1.38	1.61	1.84	2.07
6	0.28	0.55	0.83	1.10	1.38	1.65	1.93	2.20	2.48
8	0.37	0.73	1.10	1.47	1.84	2.20	2.57	2.94	3.31

To determine the correct fertilization rate in lb nutrient per 100 linear bed feet (LBF), choose the crop and its typical bed spacing, Locate that typical bed spacing value in the bottom part of the table. Then locate the desired value for recommended fertilizer rate. Read down the column under recommended fertilizer rate until you reach the value in the row for your typical bed spacing.

Adapted from G. J. Hochmuth, "Soil and Fertilizer Management for Vegetable Production in Florida," in D. N. Maynard and G. J. Hochmuth (eds.), *Vegetable Production Guide for Florida* (Florida Cooperative Extension Service Circular SP-170, 1995), 2–17.

TABLE 4.34. RATES OF FERTILIZERS RECOMMENDED FOR VEGETABLE CROPS IN MID-ATLANTIC (DELAWARE, MARYLAND, NEW JERSEY, PENNSYLVANIA, AND VIRGINIA) STATES BASED ON SOIL ANALYSES[1]

		Amount P_2O_5 (lb/acre)			Amount K_2O (lb/acre)		
Vegetable	Amount N (lb/acre)	Low Soil P	Medium Soil P	Optimum Soil P	Low Soil K	Medium Soil K	Optimum Soil K
Asparagus (cutting beds)	50	200	100	50	200	100	50
Bean, snap	40–80	80	60	40	80	60	40
Beet	75–100	150	100	50	150	100	50
Broccoli	150–200	200	100	50	200	100	50
Cabbage	100–150	200	100	50	200	100	50
Cantaloupe	75–100	150	100	50	200	150	100
Carrot	50–80	150	100	50	150	100	50
Cauliflower	100–150	200	100	50	200	100	50
Celery	150–175	250	150	100	250	150	100
Cucumber	100–125	150	100	50	200	150	100
Eggplant	125–150	250	150	100	250	150	100
Lettuce, iceberg	60–80	200	150	100	200	150	100
Leek	100–125	200	150	100	200	150	100
Onion, bulbs	75–100	200	100	50	200	100	50

TABLE 4.34. RATES OF FERTILIZERS RECOMMENDED FOR VEGETABLE CROPS IN MID-ATLANTIC (DELAWARE, MARYLAND, NEW JERSEY, PENNSYLVANIA, AND VIRGINIA) STATES BASED ON SOIL ANALYSES[1] (*Continued*)

		Amount P_2O_5 (lb/acre)			Amount K_2O (lb/acre)		
Vegetable	Amount N (lb/acre)	Low Soil P	Medium Soil P	Optimum Soil P	Low Soil K	Medium Soil K	Optimum Soil K
Pea	40–80	120	80	40	120	80	40
Pepper	100–130	200	150	100	200	150	100
Potato	125–150	200	150	100	300	200	100
Pumpkin	50–100	150	100	50	200	150	100
Spinach	100–195	200	150	100	200	150	100
Squash, summer	75–100	150	100	50	200	150	100
Strawberry (established)	60–110	165	115	65	165	115	65
Sweet corn (fresh)	125–150	160	120	80	160	120	80
Sweet potato	50–75	200	100	50	300	200	100
Tomato (fresh)	80–90	200	150	100	300	200	100
Watermelon	125–150	150	100	50	200	150	100

Adapted from *Commercial Vegetable Production Recommendations,* Delaware Cooperative Extension Service Bulletin 137 (2005).

[1] A common recommendation is to broadcast and work deeply into the soil one-third to one-half of the fertilizer at planting and to apply the balance as a side dressing in one or two applications after the crop is fully established.

TABLE 4.35. RATES OF FERTILIZERS RECOMMENDED FOR VEGETABLE CROPS IN FLORIDA ON SANDY SOILS BASED ON MEHLICH-I SOIL TEST RESULTS

		P_2O_5 (lb/acre)[1] Soil-test P			K_2O (lb/acre)[1] Soil-test K		
Vegetable	N (lb/acre)	Very Low	Low	Medium	Very Low	Low	Medium
Bean	100	120	100	80	120	100	80
Beet	120	120	100	80	120	100	80
Broccoli	175	150	120	100	150	120	100
Cabbage	175	150	120	80	150	120	80
Cantaloupe	150	150	120	80	150	120	80
Carrot	175	150	120	100	150	120	100
Cauliflower	175	150	120	100	150	120	100
Celery	200	200	150	100	250	150	100
Chinese cabbage	150	150	120	80	150	120	80
Collards	150	150	120	80	150	120	80
Cucumber	150	120	100	80	120	100	80
Eggplant	200	150	120	100	160	130	100
Lettuce	200	150	120	80	150	120	80
Mustard	120	150	120	100	150	120	100
Okra	120	150	120	100	150	120	100
Onion	150	150	120	80	150	120	80

TABLE 4.35. RATES OF FERTILIZERS RECOMMENDED FOR VEGETABLE CROPS IN FLORIDA ON SANDY SOILS BASED ON MEHLICH-I SOIL TEST RESULTS (*Continued*)

Vegetable	N (lb/acre)	P_2O_5 (lb/acre)[1] Soil-test P			K_2O (lb/acre)[1] Soil-test K		
		Very Low	Low	Medium	Very Low	Low	Medium
Parsley	120	150	120	100	150	120	100
Pea, southern	60	80	80	60	80	80	60
Pepper	200	150	120	100	160	130	100
Potato	200	120	120	60	140	140	140
Radish	90	120	100	80	120	100	80
Spinach	90	120	100	80	120	100	80
Squash	150	120	100	80	120	100	80
Strawberry	150	150	120	100	150	120	100
Sweet corn	200	150	120	100	120	100	80
Sweet potato	60	120	100	80	120	100	80
Tomato	200	150	120	100	225	150	100
Watermelon	150	150	120	80	150	120	80

Adapted from G. Hochmuth and E. Hanlon, "IFAS Standardized Fertilization Recommendations for Vegetable Crops," Florida Cooperative Extension Service Circular 1152 (2000), http://edis.ifas.ufl.edu/CV002.

[1] No P or K recommended for soils testing high except for potato, which receives no P but receives 140 lb K_2O/acre.

TABLE 4.36. FERTILIZATION RECOMMENDATIONS FOR NEW ENGLAND VEGETABLES

Crop	Nitrogen	Soil Phosphorus					Soil Potassium				
		Very Low	Low	Medium	High	Very High	Very Low	Low	Medium	High	Very High
	(lb/acre)	P_2O_5 (lb/acre)					K_2O (lb/acre)				
Asparagus, established	75	200	175	150	100	50	300	250	200	150	75
Bean	50	100	75	50	25	25	100	75	50	50	25
Beet, Swiss chard	100–130	150	125	100	50	0	300	200	100	50	25
Carrot, Parsnip	110–150	150	100	75	50	0	400	350	250	150	0–75
Cantaloupe	130	150	120	100	80	0	200	150	100	80	0
Celery	180	200	150	100	50	0	300	240	180	120	0–60
Cole crops	160	200	170	130	100	20	175	150	125	50	0
Sweet Corn, early	100–130	110	80	40	20	0	200	160	130	30	0
Sweet Corn, main	100–160	110	80	40	20	0	200	160	130	30	0
Cucumber	130	150	120	100	80	0	200	150	100	80	0
Eggplant	80–110	200	150	100	50	0	200	150	100	50	0
Gourd	90	125	100	75	50	0	150	125	100	75	0

TABLE 4.36. FERTILIZATION RECOMMENDATIONS FOR NEW ENGLAND VEGETABLES (*Continued*)

Crop	Nitrogen	Soil Phosphorus					Soil Potassium				
		Very Low	Low	Medium	High	Very High	Very Low	Low	Medium	High	Very High
	(lb/acre)	P_2O_5 (lb/acre)					K_2O (lb/acre)				
Lettuce, Endive, Escarole	80–125	190	165	140	90	0	190	165	140	90	0
Onion	130	175	150	100	50	0	175	150	100	50	0
Pea	75	150	100	75	50	25	150	100	75	50	0
Pepper	140	200	150	100	50	0	200	150	100	50	0
Potato	120–180	300	250	200	180	150	250	225	200	180	150
Pumpkin, Squash	70–90	150	125	100	0–70	0	200	150	100	70	0
Radish	50	125	100	75	50	0–25	125	100	75	50	25
Rutabaga, Turnip	50	100	75	50	25	0	100	75	50	25	0
Spinach	90–110	150	120	100	60	30	200	150	100	50	0
Tomato	140–160	200	150	100	50	0	250	200	150	100	0
Watermelon	130	150	120	100	80	0	200	150	100	80	0

Adapted from J.C. Howell (ed.), *New England Vegetable Management Guide* (Cooperative Extension Services of New England States, 2004–2005).

TABLE 4.37. FERTILIZER RATES RECOMMENDED FOR VEGETABLE CROPS IN NEW YORK

	Amount (lb/acre)[1]		
Vegetable	N	P_2O_5	K_2O
Asparagus	50	25–75	40–80
Bean	40	40–80	20–60
Beet	150–175	50–150	100–300
Broccoli	100–120	40–120	60–160
Brussels sprouts	100–120	40–120	60–160
Cabbage	100–120	40–120	60–160
Carrot	80–90	40–120	60–160
Cantaloupe	100–120	40–120	40–120
Cauliflower	100–120	40–120	60–160
Celery	130–150	50–150	120–240
Cucumber	100–120	40–120	40–120
Eggplant	120	60–150	60–150
Endive	100	30–120	50–150
Lettuce	100	30–120	50–150
Onion (mineral soil)	90–120	50–150	50–150
Pea	40–50	50–100	40–120
Pepper	125	75–150	75–150
Potato	120–175	120–240	75–240
Pumpkin	100–120	40–120	40–120
Radish	50	50–110	50–150
Spinach	100–125	80–140	50–150
Squash, summer	100–120	40–120	40–120
Squash, winter	100–120	40–120	40–120
Sweet corn	120–140	40–120	40–120
Tomato	100	60–150	60–150
Turnip	50	50–110	50–150
Watermelon	100–120	40–120	40–120

Adapted from S. Reiners and C. Petzuldt (eds.), *Integrated Crop and Pest Management Guidelines for Commercial Vegetable Production* (Cornell Cooperative Extension Service, 2005), http://www.nysaes.cornell.edu/recommends/.

[1] Total amounts are listed; application may be broadcast and plow down, broadcast and disk in, band, or sidedress. Actual rate of fertilization depends on soil type, previous cropping history, and soil test results. Ranges reflect recommendations made for high to low soil test situations. See above document for details.

ADDITIONAL VEGETABLE PRODUCTION GUIDES CONTAINING FERTILIZER RECOMMENDATIONS

New England Vegetable Management Guide, http://www.nevegetable.org

Oregon Vegetable Production Guides, http://oregonstate.edu/Dept/NWREC/vegindex.html

Texas Vegetable Grower's Handbook, http://aggie-horticulture.tamu.edu/extension/veghandbook/index.html

Ohio Vegetable Production Guide, http://ohioline.osu.edu/b672/index.html

Cornell (New York) Integrated Crop and Pest Management Guidelines for Commercial Vegetable Production, http://www.nysaes.cornell.edu/recommends/

13
NUTRIENT DEFICIENCIES

TABLE 4.38. A KEY TO NUTRIENT DEFICIENCY SYMPTOMS

Nutrient	Plant Symptoms	Occurrence
Primary		
Nitrogen	Stems are thin, erect, and hard. Leaves are smaller than normal, pale green or yellow; lower leaves are affected first, but all leaves may be deficient in severe cases. Plants grow slowly.	Excessive leaching on light soils
Phosphorus	Stems are thin and shortened. Leaves develop purple coloration, first on undersides and later throughout. Plants grow slowly, and maturity is delayed.	On acid soils; temporary deficiencies on cold, wet soils
Potassium	Older leaves develop gray or tan areas near the margins. Eventually a scorch around the entire leaf margin may occur. Chlorotic areas may develop throughout leaf.	Excessive leaching on light soils
Secondary Nutrients and Micronutrients		
Boron	Growing points die; stems are shortened and hard; leaves are distorted. Specific deficiencies include browning of cauliflower, cracked stem of celery, blackheart of beet, and internal browning of turnip.	On soils with a pH above 6.8 or on crops with a high boron requirement

TABLE 4.38. A KEY TO NUTRIENT–DEFICIENCY SYMPTOMS (*Continued*)

Nutrient	Plant Symptoms	Occurrence
Calcium	Stem elongation restricted by death of the growing point. Root tips die, and root growth is restricted. Specific deficiencies include blossom-end rot of tomato, brownheart of escarole, celery blackheart, and carrot cavity spot.	On acid soils, following leaching rains, on soils with very high potassium levels, or on very dry soils
Copper	Yellowing of leaves. Leaves may become elongated. Onion bulbs are soft, with thin, pale-yellow scales.	Most cases of copper deficiency occur on muck or peat soils.
Iron	Distinct yellow or white areas appear between the veins on the youngest leaves.	On soils with a pH above 6.8
Magnesium	Initially, older leaves show yellowing between the veins; continued deficiency causes younger leaves to become affected. Older leaves may fall with prolonged deficiency.	On acid soils, on soils with very high potassium levels, or on very light soils subject to leaching
Manganese	Yellow mottled areas, not as intense as with iron deficiency, appear on the youngest leaves. This finally results in an overall pale appearance. In beet, foliage becomes densely red. Onion and corn show narrow striping of yellow.	On soils with a pH above 6.7

TABLE 4.38. A KEY TO NUTRIENT–DEFICIENCY SYMPTOMS (*Continued*)

Nutrient	Plant Symptoms	Occurrence
Molybdenum	Pale, distorted, very narrow leaves with some interveinal yellowing on older leaves. Whiptail of cauliflower; small, open, loose curds.	On very acid soils
Zinc	Small reddish-brown spots on cotyledon leaves of bean. Green and yellow broad striping at base of leaves of corn. Interveinal yellowing with marginal burning on beet.	On wet soils in early spring; often related to heavy phosphorus fertilization

14
MICRONUTRIENTS

TABLE 4.39. INTERPRETATION OF MICRONUTRIENT SOIL TESTS

Element	Method	Range in Critical Level (ppm)[1]
Boron (B)	Hot H_2O	0.1–0.7
Copper (Cu)	$NH_4C_2H_3O_2$ (pH 4.8)	0.2
	0.5*M* EDTA	0.1–0.7
	0.43*N* HNO_3	3–4
	Biological assay	2–3
Iron (Fe)	$NH_4C_2H_3O_2$ (pH 4.8)	2
	DTPA + $CaCl_2$ (pH 7.3)	2.5–4.5
Manganese (Mn)	0.05*N* HCl + 0.025*N* H_2SO_4	5–9
	0.1*N* H_3PO_4 and 3*N* $NH_4H_2PO_4$	15–20
	Hydroquinone + $NH_4C_2H_3O_2$	25–65
	H_2O	2
Molybdenum (Mo)	$(NH_4)_2C_2O_4$ (pH 3.3)	0.04–0.2
Zinc (Zn)	0.1*N* HCl	1.0–7.5
	Dithizone + $NH_4C_2H_3O_2$	0.3–2.3
	EDTA + $(NH_4)_2CO_3$	1.4–3.0
	DTPA + $CaCl_2$ (pH 7.3)	0.5–1.0

Reprinted with permission from S. S. Mortvedt, P. M. Giordano, and W. L. Lindsay (eds.), *Micronutrients in Agriculture* (Madison, Wis.: Soil Science Society of America, 1972).

[1] Deficiencies are likely to occur when concentrations are below the critical level. Consult the state's Extension Service for the latest interpretations.

TABLE 4.40. MANGANESE RECOMMENDATIONS FOR RESPONSIVE CROPS GROWN ON MINERAL SOILS IN MICHIGAN

Soil Test ppm[1]	Soil pH 5.8	6.0	6.2	6.4	6.6	6.8	7.0
	Band-applied Mn (lb/acre)						
2	2	4	5	7	9	10	12
4	1	2	3	5	6	7	8
8	0	1	2	3	5	6	7
12	0	0	1	2	3	4	6
16	0	0	0	1	2	3	4
20	0	0	0	0	0	2	3
24	0	0	0	0	0	0	1

Adapted from D. D. Warncke, D. R. Christenson, L. W. Jacobs, M. L. Vitosh, and B. H. Zandstra, *Fertilizer Recommendations for Vegetable Crops in Michigan* (Michigan Cooperative Extension Service Bulletin E550B, 1994), http://web1.msue.msu.edu/msue/imp/modaf/55092001.html.

[1] 0.1*N* HCl extractant

TABLE 4.41. MANGANESE RECOMMENDATIONS FOR RESPONSIVE CROPS GROWN ON ORGANIC SOILS IN MICHIGAN

Soil Test ppm[1]	Soil pH 5.8	6.0	6.2	6.4	6.6	6.8	7.0
	Band-applied Mn (lb/acre)						
2	2	4	5	7	9	10	12
4	1	3	5	6	8	10	11
8	0	1	3	5	7	8	10
12	0	0	2	4	6	7	9
16	0	0	1	3	4	6	8
20	0	0	0	1	3	5	6
24	0	0	0	0	2	4	5
28	0	0	0	0	1	2	4
32	0	0	0	0	0	1	3
36	0	0	0	0	0	0	1

Adapted from D. D. Warncke, D. R. Christenson, L. W. Jacobs, M. L. Vitosh, and B. H. Zandstra, *Fertilizer Recommendations for Vegetable Crops in Michigan* (Michigan Cooperative Extension Service Bulletin E550B, 1994).

[1] 0.1*N* HCl extractant

TABLE 4.42. ZINC RECOMMENDATIONS FOR RESPONSIVE CROPS GROWN ON MINERAL AND ORGANIC SOILS IN MICHIGAN

Soil Test ppm[1]	Soil pH 6.7	7.0	7.3	7.6
	Band-applied Zn (lb/acre)[2]			
2	1	2	4	5
4	0	1	3	4
6	0	0	2	4
8	0	0	1	3
10	0	0	0	2
12	0	0	0	1

Adapted from D. D. Warncke, D. R. Christenson, L. W. Jacobs, M. L. Vitosh, and B. H. Zandstra, *Fertilizer Recommendations for Vegetable Crops in Michigan* (Michigan Cooperative Extension Service Bulletin E550B, 1994).

[1] 0.1*N* HCl extractant

[2] Rates may be divided by 5 when chelates are used.

TABLE 4.43. COPPER RECOMMENDATIONS FOR CROPS GROWN ON ORGANIC SOILS IN MICHIGAN

Soil Test ppm[1]	Crop Response		
	Low	Medium	High
	Cu (lb/acre)		
1	3	4	6
4	3	4	5
8	2	3	4
12	1	2	3
16	1	2	2
20	1	1	2
24	0	1	1

Adapted from D. D. Warncke, D. R. Christenson, L. W. Jacobs, M. L. Vitosh, and B. H. Zandstra, *Fertilizer Recommendations for Vegetable Crops in Michigan* (Michigan Cooperative Extension Service Bulletin E550B, 1994), http://web1.msue.msu.edu/msue/imp/modaf/55092001.html.

[1] 0.1*N* HCl extractant

TABLE 4.44. RELATIVE RESPONSE OF VEGETABLES TO MICRONUTRIENTS[1]

	Response to Micronutrient					
Vegetable	Manganese	Boron	Copper	Zinc	Molybdenum	Iron
Asparagus	Low	Low	Low	Low	Low	Medium
Bean	High	Low	Low	High	Medium	High
Beet	High	High	High	Medium	High	High
Broccoli	Medium	High	Medium	—	High	High
Cabbage	Medium	Medium	Medium	Low	Medium	Medium
Carrot	Medium	Medium	Medium	Low	Low	—
Cauliflower	Medium	High	Medium	—	High	High
Celery	Medium	High	Medium	—	Low	—
Cucumber	High	Low	Medium	—	—	—
Lettuce	High	Medium	High	Medium	High	—
Onion	High	Low	High	High	High	—
Pea	High	Low	Low	Low	Medium	—
Potato	High	Low	Low	Medium	Low	—
Radish	High	Medium	Medium	Medium	Medium	—
Spinach	High	Medium	High	High	High	High
Sweet corn	High	Medium	Medium	High	Low	Medium
Tomato	Medium	Medium	High	Medium	Medium	High
Turnip	Medium	High	Medium	—	Medium	—

Adapted from M. L. Vitosh, D. D. Warncke, and R. E. Lucas, *Secondary and Micronutrients for Vegetables and Field Crops* (Michigan Extension Bulletin E-486, 1994), http://web1.msue.msu.edu/msue/imp/modf1/05209701.html.

[1] The crops listed respond as indicated to applications of the respective micronutrient when that micronutrient concentration in the soil is low.

TABLE 4.45. BORON REQUIREMENTS OF VEGETABLES ARRANGED IN APPROXIMATE ORDER OF DECREASING REQUIREMENTS

High Requirement (more than 0.5 ppm in soil)	Medium Requirement (0.1–0.5 ppm in soil)	Low Requirement (less than 0.1 ppm in soil)
Beet	Tomato	Corn
Turnip	Lettuce	Pea
Cabbage	Sweet potato	Bean
Broccoli	Carrot	Lima bean
Cauliflower	Onion	Potato
Asparagus		
Radish		
Brussels sprouts		
Celery		
Rutabaga		

Adapted from K. C. Berger, "Boron in Soils and Crops," *Advances in Agronomy*, vol. 1, (New York: Academic Press, 1949), 321–351.

TABLE 4.46. RELATIVE TOLERANCE OF VEGETABLES TO BORON, ARRANGED IN ORDER OF INCREASING SENSITIVITY

Tolerant	Semitolerant	Sensitive
Asparagus	Celery	Jerusalem artichoke
Artichoke	Potato	Bean
Beet	Tomato	
Cantaloupe	Radish	
Broad bean	Corn	
Onion	Pumpkin	
Turnip	Bell pepper	
Cabbage	Sweet potato	
Lettuce	Lima bean	
Carrot		

Adapted from L. V. Wilcox, *Determining the Quality of Irrigation Water,* USDA Agricultural Information Bulletin 197 (1958).

TABLE 4.47. SOIL AND FOLIAR APPLICATION OF SECONDARY AND TRACE ELEMENTS

Vegetables differ in their requirements for these secondary nutrients. Availability in the soil is influenced by soil reaction and soil type. Use higher rates on muck and peat soils than on mineral soils and lower rates for band application than for broadcast. Foliar application is one means of correcting an evident deficiency that appears while the crop is growing.

Element	Application Rate (per acre basis)	Source	Composition
Boron	0.5–3.5 lb (soil)	Borax ($Na_2B_4O_7 \cdot 10H_2O$)	11% B
		Boric acid (H_3BO_3)	17% B
		Sodium pentaborate ($Na_2B_{10}O_{16} \cdot 10H_2O$)	18% B
		Sodium tetraborate ($Na_2B_4O_7$)	21% B
Calcium	2–5 lb (foliar)	Calcium chloride ($CaCl_2$)	36% Ca
		Calcium nitrate ($CaNO_3 \cdot 2H_2O$)	20% Ca
		Liming materials and gypsum supply calcium when used as soil amendments	
Copper	2–6 lb (soil)	Cupric chloride ($CuCl_2$)	47% Cu
		Copper sulfate ($CuSO_4 \cdot H_2O$)	35% Cu
		Copper sulfate ($CuSO_4 \cdot 5H_2O$)	25% Cu
		Cupric oxide (CuO)	80% Cu
		Cuprous oxide (Cu_2O)	89% Cu
		Copper chelates	8–13% Cu
Iron	2–4 lb (soil)	Ferrous sulfate ($FeSO_4 \cdot 7H_2O$)	20% Fe
	0.5–1 lb (foliar)	Ferric sulfate [$Fe_2(SO_4)_3 \cdot 9H_2O$]	20% Fe
		Ferrous carbonate ($FeCO_3 \cdot H_2O$)	42% Fe
		Iron chelates	5–12% Fe
Magnesium	25–30 lb (soil)	Magnesium sulfate ($MgSO_4 \cdot 7H_2O$)	10% Mg

TABLE 4.47. SOIL AND FOLIAR APPLICATION OF SECONDARY AND TRACE ELEMENTS (*Continued*)

Element	Application Rate (per acre basis)	Source	Composition
	2–4 lb (foliar)	Magnesium oxide (MgO)	55% Mg
		Dolomitic limestone	11% Mg
		Magnesium chelates	2–4% Mg
Manganese	20–100 lb (soil)	Manganese sulfate ($MnSO_4 \cdot 3H_2O$)	27% Mn
	2–5 lb (foliar)	Manganous oxide (MnO)	41–68% Mn
		Manganese chelates (Mn EDTA)	12% Mn
Molybdenum	25–400 g (soil)	Ammonium molybdate [$(NH_4)_6MO_7O_{24} \cdot 4H_2O$]	54% Mo
	25 g (foliar)	Sodium molybdate ($Na_2MoO_4 \cdot 2H_2O$)	39% Mo
Sulfur	20–50 lb (soil)	Sulfur (S)	100% S
		Ammonium sulfate [$(NH_4)_2SO_4$]	24% S
		Potassium sulfate (K_2SO_4)	18% S
		Calcium sulfate ($CaSO_4$)	16–18% S
		Ferric sulfate [$Fe_2(SO_4)_3$]	18–19% S
Zinc	2–10 lb (soil)	Zinc oxide (ZnO)	80% Zn
	0.25 lb (foliar)	Zinc sulfate ($ZnSO_4 \cdot 7H_2O$)	23% Zn
		Zinc chelates (Na_2Zn EDTA)	14% Zn

TABLE 4.48. BORON RECOMMENDATIONS BASED ON SOIL TESTS FOR VEGETABLE CROPS

Interpretation of Boron Soil Tests

ppm	lb/acre	Relative Level	Crops That Often Need Additional Boron	Boron Recommendation (lb/acre)
0.0–0.35	0.0–0.70	Low	Beet, broccoli, Brussels sprouts, cabbage, cauliflower, celery, rutabaga, turnip	3
			Asparagus, carrot, eggplant, horseradish, leek, onion, parsnip, radish, squash,	2
			Strawberry, sweet corn, tomato	
			Pepper, sweet potato	1
0.36–0.70	0.71–1.40	Medium	Beet, broccoli, Brussels sprouts, cabbage, cauliflower, celery, rutabaga, turnip	1½
			Asparagus, carrot, eggplant, horseradish, leek, onion, parsnip, radish, squash,	1
			Strawberry, sweet corn, tomato	
>0.70	>1.40	High	All crops	0

Adapted from *Commercial Vegetable Production Recommendations,* Delaware Cooperative Extension Service Bulletin 137 (2005).

TABLE 4.49. TOLERANCE OF VEGETABLES TO A DEFICIENCY OF SOIL MAGNESIUM

Tolerant	Not Tolerant
Bean	Cabbage
Beet	Cantaloupe
Chard	Corn
Lettuce	Cucumber
Pea	Eggplant
Radish	Pepper
Sweet potato	Potato
	Pumpkin
	Rutabaga
	Tomato
	Watermelon

Adapted from W. S. Ritchie and E. B. Holland, *Minerals in Nutrition*, Massachusetts Agricultural Experiment Station Bulletin 374 (1940).

15
FERTILIZER DISTRIBUTORS

ADJUSTMENT OF FERTILIZER DISTRIBUTORS

Each time a distributor is used, it is important to ensure that the proper quantity of fertilizer is being supplied. Fertilizers vary greatly in the way they flow through the equipment. Movement is influenced by the humidity of the atmosphere as well as the degree of granulation of the material.

TABLE 4.50. ADJUSTMENT OF ROW CROP DISTRIBUTOR

1. Disconnect from one hopper the downspout or tube to the furrow opener for a row.
2. Attach a can just below the fertilizer hopper.
3. Fill the hopper under which the can is placed.
4. Engage the fertilizer attachment and drive the tractor the suggested distance according to the number of inches between rows.

Distance Between Rows (in.)	Distance to Pull the Distributor (ft)
20	261
24	218
30	174
36	145
38	138
40	131
42	124

5. Weigh the fertilizer in the can. Each pound in it equals 100 lb/acre. Each tenth of a pound equals 10 lb/acre.
6. Adjust the distributor for the rate of application desired, and then adjust the other distributor or distributors to the same setting.

TABLE 4.51. ADJUSTMENT OF GRAIN-DRILL-TYPE DISTRIBUTOR

1. Remove four downspouts or tubes.
2. Attach a paper bag to each of the four outlets.
3. Fill the part of the drill over the bagged outlets.
4. Engage the distributor and drive the tractor the suggested distance according to the inches between the drill rows.

Distance Between Drill Rows (in.)	Distance to Pull the Drill (ft)
7	187
8	164
10	131
12	109
14	94

5. Weigh total fertilizer in the four bags. Each pound equals 100 lb/acre. Each tenth of a pound equals 10 lb/acre.

TABLE 4.52. CALIBRATION OF FERTILIZER DRILLS

Set drill at opening estimated to give the desired rate of application. Mark level of fertilizer in the hopper. Operate the drill for 100 ft. Weigh a pail full of fertilizer. Refill hopper to marked level and again weigh pail. The difference is the pounds of fertilizer used in 100 ft. Consult the column under the row spacing being used. The left-hand column opposite the amount used shows the rate in pounds per acre at which the fertilizer has been applied. Adjust setting of the drill, if necessary, and recheck.

Rate (lb/acre)	Distance Between Rows (in.)				
	18	20	24	36	48
	Approximate Amount of Fertilizer (lb/100 ft of row)				
250	0.9	1.1	1.4	1.7	2.3
500	1.7	2.3	2.9	3.5	4.6
750	2.6	3.4	4.3	5.2	6.9
1000	3.5	4.6	5.8	6.9	9.2
1500	5.2	6.8	8.6	10.4	13.8
2000	6.8	9.2	11.6	13.0	18.4
3000	10.5	14.0	17.5	21.0	28.0

PART 5

WATER AND IRRIGATION

01
SUGGESTIONS ON SUPPLYING WATER TO VEGETABLES

Plants in hot, dry areas lose more moisture into the air than those in cooler, more humid areas. Vegetables utilize and evaporate more water in the later stages of growth when size and leaf area are greater. The root system becomes deeper and more widespread as the plant ages.

Some vegetables, especially lettuce and sweet corn, have sparse root systems that do not come into contact with all the soil moisture in their root-depth zone. Cool-season vegetables normally root to a shallower depth than do warm-season vegetables and perennials.

When applying water, use enough to bring the soil moisture content of the effective rooting zone of the crop up to field capacity. This is the quantity of water that the soil holds against the pull of gravity.

The frequency of irrigation depends on the total supply of available moisture reached by the roots and the rate of water use. The first is affected by soil type, depth of wetted soil, and the depth and dispersion of roots. The latter is influenced by weather conditions and the age of the crops. Add water when the moisture in the root zone has been used to about the halfway point in the range of available moisture. Do not wait until vegetables show signs of wilting or develop color or texture changes that indicate they are not growing rapidly. A general rule is that vegetables need an average of 1 in. water per week from rain or supplemental irrigation in order to grow vigorously. In arid regions, about 2 in./week is required. These amounts of water may vary from 0.5 in./week early in the season to more than 1 in. later in the season.

IRRIGATION MANAGEMENT AND NUTRIENT LEACHING

Irrigation management is critical to success in nutrient management for mobile nutrients such as nitrogen. Irrigation management is particularly important in vegetable production in sandy soils where nitrogen is highly prone to leaching from the root zone with heavy rainfall or excessive irrigation. Leaching can occur with all irrigation systems if more water is applied than the soil can hold at one time. If the water-holding capacity of the soil is exceeded with any irrigation event, nutrient leaching can occur. Information is provided here to assist growers in understanding the rooting zone for crops and water-holding capacity of soils as well as application rates for various irrigation systems. Optimum irrigation management involves attention to these factors, knowing crop water needs, and keeping an eye on soil moisture levels during the season. These factors vary for the crop being grown, the soil used, the season, and the climate, among other factors. Please consult your local Extension Service for specific information for your production area.

02
ROOTING OF VEGETABLES

ROOTING DEPTH OF VEGETABLES

The depth of rooting of vegetables is influenced by the soil profile. If it is a clay pan, hard pan, compacted layer, or other dense formation, the normal depth of rooting is not possible. Also, some transplanted vegetables may not develop root systems as deep as those of seeded crops. Although vegetables may root as deep as 18–24 in., most of the active root system for water uptake may be between 8 and 12 in.

TABLE 5.1. CHARACTERISTIC MAXIMUM ROOTING DEPTHS OF VARIOUS VEGETABLES

Shallow (18–24 in.)	Moderately Deep (36–48 in.)	Deep (More than 48 in.)
Broccoli	Bean, bush	Artichoke
Brussels sprouts	Bean, pole	Asparagus
Cabbage	Beet	Bean, lima
Cauliflower	Cantaloupe	Parsnip
Celery	Carrot	Pumpkin
Chinese cabbage	Chard	Squash, winter
Corn	Cucumber	Sweet potato
Endive	Eggplant	Tomato
Garlic	Mustard	Watermelon
Leek	Pea	
Lettuce	Pepper	
Onion	Rutabaga	
Parsley	Squash, summer	
Potato	Turnip	
Radish		
Spinach		
Strawberry		

03
SOIL MOISTURE

DETERMINING MOISTURE IN SOIL BY APPEARANCE OR FEEL

A shovel serves to obtain a soil sample from a shallow soil or when a shallow-rooted crop is being grown. A soil auger or soil tube is necessary to draw samples from greater depths in the root zone.

Squeeze the soil sample in the hand and compare its behavior with those of the soils listed in the Practical Soil-Moisture Interpretation Chart to get a rough idea of its moisture content.

TABLE 5.2. PRACTICAL SOIL-MOISTURE INTERPRETATION CHART

	Soil Type			
Amount of Readily Available Moisture Remaining for the Plant	Sand (gritty when moist, almost like beach sand)	Sandy Loam (gritty when moist; dirties fingers; contains some silt and clay)	Clay Loam (sticky and plastic when moist)	Clay (very sticky when moist; behaves like modeling clay)
Close to 0%. Little or no moisture available	Dry, loose, single-grained; flows through fingers.	Dry, loose, flows through fingers.	Dry clods that break down into powdery condition.	Hard, baked, cracked surface. Hard clods difficult to break, sometimes have loose crumbs on surface.
50% or less. Approaching time to irrigate	Still appears to be dry; will not form a ball with pressure.	Still appears to be dry; will not form a ball.	Somewhat crumbly, but will hold together with pressure.	Somewhat pliable; will ball under pressure.
50–75%. Enough available moisture	Same as sand under 50%.	Tends to ball under pressure but seldom holds together.	Forms a ball, somewhat plastic; sometimes sticks slightly with pressure.	Forms a ball; ribbons out between thumb and forefinger.

75% to field capacity. Plenty of available moisture	Tends to stick together slightly, sometimes forms a very weak ball under pressure.	Forms weak ball, breaks easily, does not become slick.	Forms a ball and is very pliable; becomes slick readily if high in clay.	Easily ribbons out between fingers; feels slick.
At field capacity. Soil will not hold any more water (after draining)	Upon squeezing, no free water appears; moisture is left on hand.	Same as sand.	Same as sand.	Same as sand.
Above field capacity. Unless water drains out, soil will be water-logged	Free water appears when soil is bounced in hand.	Free water is released with kneading.	Can squeeze out free water.	Puddles and free water form on surface.

Adapted from R. W. Harris and R. H. Coppock (eds.), "Saving Water in Landscape Irrigation," University of California Division of Agricultural Science Leaflet 2976 (1978). Also from N. Klocke and P. Fischbach, *Estimation, Soil Moisture by Appearance and Feel* (University of Nebraska Cooperative Extension Service, 1998), http://ianrpubs.unl.edu/irrigation/g690.htm.

TABLE 5.3. FIELD DEVICES FOR MONITORING SOIL MOISTURE

Method	Advantages	Disadvantages
Neutron moderation	inexpensive per location large sensing volume not affected by salinity stabile	safety hazard cumbersome expensive slow
Time Domain Reflectometry (TDR)	accurate easily expanded insensitive to normal salinity soil-specific calibration not needed	expensive problems under high salinity small sensing volume
Frequency Domain (FD) Capacitance and FDR	accurate after specific-soil calibration better than TDR in saline soils more flexible in probes than TDR less expensive than TDR (some devices)	small sensing sphere needs careful installation needs specific soil calibration
Tensiometer	direct reading minimal skill inexpensive not affected by salinity	limited suction range slow response time frequent maintenance requires intimate contact with soil
Resistance blocks	no maintenance simple, inexpensive	low resolution slow reaction time not suited for clays block properties change with time
Granular matrix sensors	no maintenance simple, inexpensive reduced problems compared to gypsum blocks	low resolution slow reaction time not suited for clays need to resaturate in dry soils

Adapted from R. Munoz-Carpena, *Field Devices for Monitoring Soil Water Content* (University of Florida Extension Service Bulletin 343, 2004), http://edis.ifas.ufl.edu/ae266.

TABLE 5.4. APPROXIMATE SOIL WATER CHARACTERISTICS FOR TYPICAL SOIL CLASSES

Characteristic	Sandy Soil	Loamy Soil	Clay Soil
Dry weight 1 cu ft	90 lb	80 lb	75 lb
Field capacity—% of dry weight	10%	20%	35%
Permanent wilting percentage	5%	10%	19%
Percent available water	5%	10%	16%
Water available to plants			
lb/cu ft	4 lb	8 lb	12 lb
in./ft depth	¾ in.	1½ in.	2¼ in.
gal/cu ft	½ gal	1 gal	1½ gal
Approximate depth of soil that will be wetted by each 1 in. water applied if half the available water has been used	24 in.	16 in.	11 in.
Suggested lengths of irrigation runs	330 ft	660 ft	1,320 ft

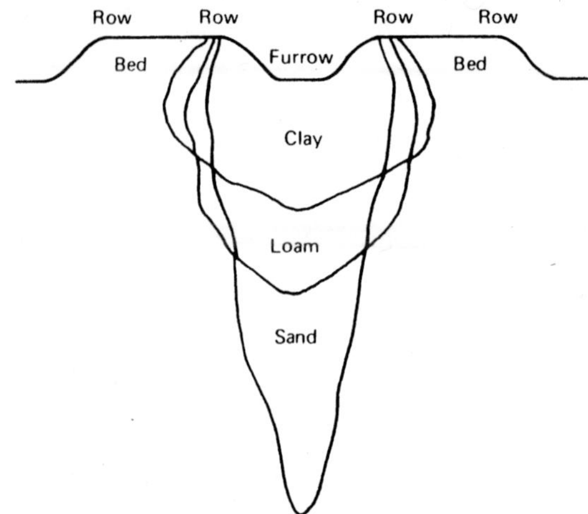

Figure 5.1. Arrangement of beds for furrow irrigation. Beds intended for two rows are usually on 36-, 40-, or 42-in. centers, with the surface 4–6 in. above the bottom of the furrow. The depth of penetration of an equal quantity of water varies with the class of soil as indicated.

RATES OF WATER APPLICATION FOR VARIOUS IRRIGATION METHODS

The infiltration rate has an important bearing on the intensity and frequency with which water should be applied by any method of irrigation.

Normally, sandy soils have a high infiltration rate and clay soils have a low one. The rate is affected by soil texture, structure, dispersion, and the depth of the water table. The longer the water is allowed to run, the more the infiltration rate decreases.

With furrows, use a flow of water initially 2–3 times that indicated to fill the run as quickly as possible. Then cut back the flow to the indicated amount. This prevents excessive penetration at the head and equalizes the application of water throughout the whole furrow.

TABLE 5.5. APPROXIMATE FLOW OF WATER PER FURROW AFTER WATER REACHES THE END OF THE FURROW

		Slope of Land (%)		
		0–0.2	0.2–0.5	0.5–1
Infiltration Rate of Soil (in./hr)	Length of Furrow (ft)	Flow of Water per Furrow (gal/min)		
High (1.5 or more)	330	9	4	3
	660	20	9	7
	1,320	45	20	15
Medium (0.5–1.5)	330	4	3	1.5
	660	10	7	3.5
	1,320	25	15	7.5
Low (0.1–0.5)	330	2	1.5	1
	660	4	3.5	2
	1,320	9	7.5	4

TABLE 5.6. APPROXIMATE MAXIMUM WATER INFILTRATION RATES FOR VARIOUS SOIL TYPES

Soil Type	Infiltration Rate[1] (in./hr)
Sand	2.0
Loamy sand	1.8
Sandy loam	1.5
Loam	1.0
Silt and clay loam	0.5
Clay	0.2

[1] Assumes a full crop cover. For bare soil, reduce the rate by half.

TABLE 5.7. PERCENT OF AVAILABLE WATER DEPLETED FROM SOILS AT VARIOUS TENSIONS

Tension—less than—(bars)[1]	Loamy Sand	Sandy Loam	Loam	Clay
0.3	55	35	15	7
0.5	70	55	30	13
0.8	77	63	45	20
1.0	82	68	55	27
2.0	90	78	72	45
5.0	95	88	80	75
15.0	100	100	100	100

Adapted from Cooperative Extension, University of California Soil and Water Newsletter No. 26 (1975).

[1] 1 bar = 100 kilopascals

TABLE 5.8. SPRINKLER IRRIGATION: APPROXIMATE APPLICATION OF WATER

	Slope of Land (%)	
	0–5	5–12
Infiltration Rate of Soil (in./hr)	Approximate Application (in./hr)	
High (1.5 or more)	1.0	0.75
Medium (0.5–1.5)	0.5	0.40
Low (0.1–0.5)	0.2	0.15

TABLE 5.9. BASIN IRRIGATION: APPROXIMATE AREA

	Quantity of Water to be Supplied	
	450 gal/min or 1 cu ft/sec	900 gal/min or 2 cu ft/sec
Infiltration Rate of Soil (in./hr)	Approximate Area (acre/basin)	
High (1.5 or more)	0.1	0.2
Medium (0.5–1.5)	0.2	0.4
Low (0.1–0.5)	0.5	1.0

TABLE 5.10. VOLUME OF WATER APPLIED FOR VARIOUS FLOW RATES AND TIME PERIODS

	Volume (acre-in.) Applied			
Flow Rate (gpm)	1 hr	8 hr	12 hr	24 hr
25	0.06	0.44	0.66	1.33
50	0.11	0.88	1.33	2.65
100	0.22	1.77	2.65	5.30
200	0.44	3.54	5.30	10.60
300	0.66	5.30	7.96	15.90
400	0.88	7.07	10.60	21.20
500	1.10	8.84	13.30	26.50
1,000	2.21	17.70	26.50	53.00
1,500	3.32	26.50	39.80	79.60
2,000	4.42	35.40	53.00	106.00

Adapted from A. Smajstrla and D. S. Harrison, Florida Cooperative Extension, Agricultural Engineering Fact Sheet AE18 (1982).

TABLE 5.11. APPROXIMATE TIME REQUIRED TO APPLY VARIOUS DEPTHS OF WATER PER ACRE WITH DIFFERENT FLOWS[1]

Flow of Water			Approximate Time Required per Acre for a Depth of:							
			1 in.		2 in.		3 in.		4 in.	
gpm	sec-ft	Approximate acre-in./hr	hr	min	hr	min	hr	min	hr	min
50	0.11	1/8	9	03	18	06	27	09	36	12
100	0.22	1/4	4	32	9	03	13	35	18	06
150	0.33	5/16	3	01	6	02	9	03	12	04
200	0.45	7/16	2	16	4	32	6	47	9	03
250	0.56	9/16	1	49	3	37	5	26	7	14
300	0.67	11/16	1	31	3	01	4	32	6	02
350	0.78	3/4	1	18	2	35	3	53	5	10
400	0.89	7/8	1	08	2	16	3	24	4	32
450	1.00	1	1	00	2	01	3	01	4	01
500	1.11	1 1/8		54	1	49	2	43	3	37
550	1.23	1 3/16		49	1	39	2	28	3	18
600	1.34	1 5/16		45	1	31	2	16	3	01
650	1.45	1 7/16		42	1	24	2	05	2	48
700	1.56	1 9/16		39	1	18	1	56	2	35
750	1.67	1 21/32		36	1	12	1	49	2	24

TABLE 5.11. **APPROXIMATE TIME REQUIRED TO APPLY VARIOUS DEPTHS OF WATER PER ACRE WITH DIFFERENT FLOWS[1] (*Continued*)**

Flow of Water			Approximate Time Required per Acre for a Depth of:							
			1 in.		2 in.		3 in.		4 in.	
gpm	sec-ft	Approximate acre-in./hr	hr	min	hr	min	hr	min	hr	min
800	1.78	1¾		34	1	08	1	42	2	16
850	1.89	1⅞		32	1	04	1	36	2	08
900	2.01	2		30	1	00	1	31	2	01
950	2.12	2 3/32		29		57	1	26	1	54
1,000	2.23	2 3/16		27		54	1	21	1	49
1,050	2.34	2 5/16		26		52	1	18	1	44
1,100	2.45	2 7/16		25		49	1	14	1	38
1,150	2.56	2½		24		47	1	11	1	34
1,200	2.67	2⅝		23		45	1	08	1	31
1,300	2.90	2⅞		21		42	1	03	1	24
1,400	3.12	3 1/16		20		39		58	1	18
1,500	3.34	3 5/16		18		36		54	1	12

[1]If a sprinkler system is used, the time required should be increased by 2–10% to compensate for the water that will evaporate before reaching the soil.

TO DETERMINE THE WATER NEEDED TO WET VARIOUS DEPTHS OF SOIL

Example: You wish to wet a loam soil to a 12-in. depth when half the available water in that zone is gone. Move across the chart from the left on the 12-in. line. Stop when you reach the diagonal line marked "loams." Move upward from that point to the scale at the top of the chart. You will see that about ¾ in. water is needed.

Depth of water required, inches, based on depletion of about half the available water in the effective root zone.

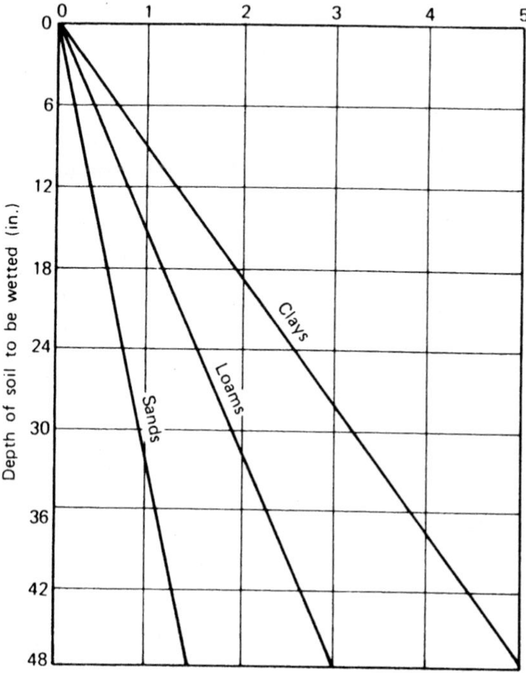

Figure 5.2. Chart for determining the amount of water needed to wet various depths of soil.

USE OF SIPHONS

Siphons of metal, plastic, or rubber can be used to carry water from a ditch to the area or furrow to be irrigated.

The inside diameter of the pipe and the head—the vertical distance from the surface of the water in the ditch to the surface of the water on the outlet side—determine the rate of flow.

When the outlet is not submerged, the head is measured to the center of the siphon outlet. You can determine how many gallons per minute are flowing through each siphon from the chart below.

Example: You have a head of 4 in. and are using 2-in. siphons. Follow the 4-in. line across the chart until you reach the curve for 2-in. siphons. Move straight down to the scale at the bottom. You will find that you are putting on about 28 gal/min.

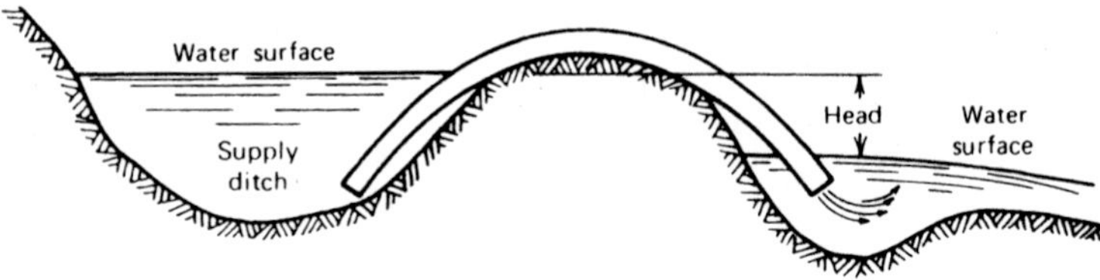

Figure 5.3. Method of measuring the head for water carried from a supply ditch to a furrow by means of a siphon. Adapted from University of California Division of Agricultural Science Leaflet 2956 (1977).

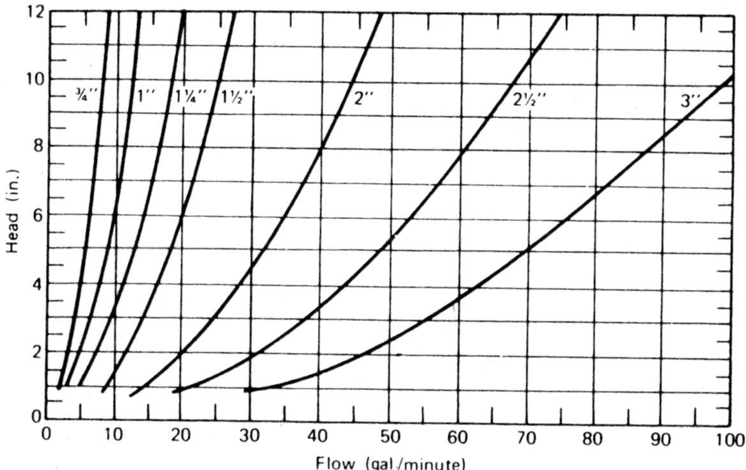

Figure 5.4. Chart for determining the flow of water through small siphons. Adapted from University of California Division of Agricultural Science Leaflet 2956 (1977). Also: E. C. Martin, *Measuring Water Flow in Surface Irrigation and Gated Pipe* (University of Arizona College Agriculture and Life Sciences, Arizona Water Series 31, 2004), http://cals.arizona.edu/pubs/water/az1329.pdf.

APPLICATION OF FERTILIZER IN WATER FOR FURROW IRRIGATION

There are certain limitations to the method of applying fertilizer solutions or soluble fertilizers in water supplied by furrow irrigation. You do not get uniform distribution of the fertilizer over the whole irrigated area. More of the dissolved material may enter the soil near the head than at the end of the furrow. You must know how long it will be necessary to run water in order to irrigate a certain area so as to meter the fertilizer solution properly. Soils vary considerably in their ability to absorb water.

Fertilizer solutions can be dripped from containers into the water. Devices are available that meter dry fertilizer materials into the irrigation water where they dissolve.

The rate of flow of dry soluble fertilizer or of fertilizer solutions into an irrigation head ditch can be calculated as follows:

$$\frac{\text{area to be irrigated (acres/hr)} \times \text{amount of nutrient wanted (lb/acre)}}{\text{nutrients in solution (lb/gal)}} = \text{flow rate of fertilizer solution (gal/hr)}$$

$$\frac{\text{area to be irrigated (acres)} \times \text{amount of soluble fertilizer (lb/acre or gal/acre)}}{\text{time of irrigation (hr)}} = \text{flow rate of fertilizer solution (lb/hr or gal/hr)}$$

Knowing the gallons of solution per hour to be added to the irrigation water, you can adjust the flow from the tank as directed by the following table.

TABLE 5.12. RATE OF FLOW OF FERTILIZER SOLUTIONS

Amount of Solution Desired (gal/hr)	Approximate Time (sec) to Fill a 4-oz Container	Approximate Time (sec) to Fill an 8-oz Container
½	225	450
1	112	224
2	56	112
3	38	76
4	28	56
5	22	44
6	18	36
7	16	32
8	14	28
9	12	24
10	11	22
12	9	18
14	8	16
16	7	14
18	6	12
20	5.5	11

05

OVERHEAD IRRIGATION

LAYOUT OF A SPRINKLER SYSTEM

Each irrigation system presents a separate engineering problem. The advice of a competent engineer is essential. Many factors must be taken into consideration in developing a plan for the equipment:

Water supply available at period of greatest use
Distance from source of water to field to be irrigated
Height of field above water source and topography of the land
Type of soil (rate at which it absorbs water and its water-holding capacity)
Area to be irrigated
Desired frequency of irrigation
Quantity of water to be applied
Time on which application is to be made
Type of power available
Normal wind velocity and direction
Possible future expansion of the installation

Specific details of the plan must then include the following:

Size of power unit and pump to do the particular job
Pipe sizes and lengths for mains and laterals
Operating pressures of sprinklers
Size and spacing of sprinklers
Friction losses in the system

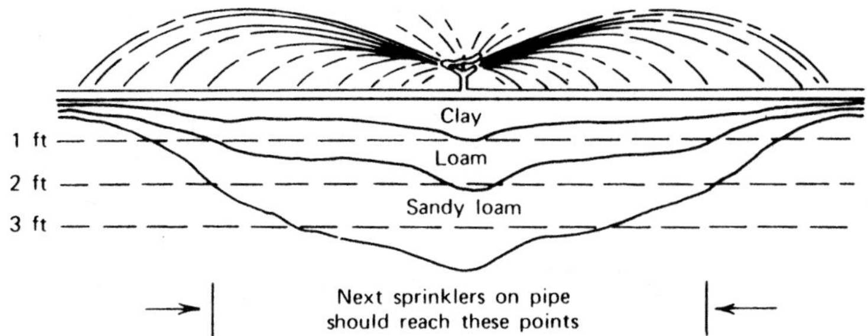

Figure 5.5. The diagram shows the approximate depth of penetration of available water from a 3-in. irrigation on various classes of soil. To avoid uneven water distribution, there should be enough distance between sprinklers to allow a 40% overlap in diameter of the area they are to cover.

TABLE 5.13. ACREAGE COVERED BY MOVES OF PIPE OF VARIOUS LENGTHS

Lateral Move of Pipe (ft)	Length of Sprinkler Pipe (ft)	Area Covered per Move (acres)
20	2,640	1.21
20	1,320	0.61
20	660	0.30
20	330	0.15
30	2,640	1.82
30	1,320	0.91
30	660	0.46
30	330	0.23
40	2,640	2.42
40	1,320	1.21
40	660	0.61
40	330	0.30
50	2,640	3.03
50	1,320	1.52
50	660	0.76
50	330	0.38
60	2,640	3.64
60	1,320	1.82
60	660	0.91
60	330	0.46
80	2,640	4.85
80	1,320	2.42
80	660	1.21
80	330	0.61
100	2,640	6.06
100	1,320	3.03
100	660	1.52
100	330	0.76

CALCULATION OF RATES OF SPRINKLER APPLICATIONS

To determine the output per sprinkler needed to put on the desired rate of application:

$$\frac{\begin{array}{c}\text{distance between}\\ \text{sprinklers (ft)}\end{array} \times \begin{array}{c}\text{distance between}\\ \text{line settings (ft)}\end{array} \times \begin{array}{c}\text{precipitation}\\ \text{rate (in./hr)}\end{array}}{96.3}$$

$$= \text{sprinkler rate (gal/minute)}$$

Example: $\dfrac{30 \times 50 \times 0.4}{96.3} = 6.23$ gal/minute per sprinkler

To determine the rate at which water is being applied:

$$\frac{\begin{array}{c}\text{sprinkler rate}\\ \text{(gal/minute)}\end{array} \times 96.3}{\begin{array}{c}\text{distance between}\\ \text{sprinklers (ft)}\end{array} \times \begin{array}{c}\text{distance between}\\ \text{line settings (ft)}\end{array}} = \text{precipitation rate (in./hr)}$$

Manufacturer's specifications give the gallons per minute for each type of sprinkler at various pressures.

Example: $\dfrac{10 \times 96.3}{40 \times 50} = 0.481$ in./hr

TABLE 5.14. PRECIPITATION RATES FOR VARIOUS NOZZLE SIZES, PRESSURE, AND SPACINGS

Nozzle Size (in.)	Pressure (psi)	Discharge [1] (gpm)	Diameter of Spray [2] (ft)	Precipitation Rate at Spacings (in./hr)[1]		
				30 × 40 ft	30 × 45 ft	40 × 40 ft
1/16	45	0.76	60–72	0.061		
1/16	50	0.80	61–73	0.064		
1/16	55	0.85	62–74	0.068		
1/16	60	0.88	63–75	0.071		
1/16	65	0.93	64–76	0.075		
5/64	45	1.19	59–73	0.095	0.085	
5/64	50	1.25	62–72	0.100	0.089	
5/64	55	1.30	64–74	0.104	0.094	0.079
5/64	60	1.36	67–76	0.110	0.097	0.082
5/64	65	1.45	68–77	0.116	0.103	0.087
3/32	45	1.72	68–76	0.138	0.123	0.103
3/32	50	1.80	69–77	0.145	0.128	0.108
3/32	55	1.88	70–78	0.151	0.134	0.113
3/32	60	1.98	71–79	0.159	0.141	0.119
3/32	65	2.08	72–80	0.167	0.148	0.125
7/64	45	2.32	71–78	0.186	0.165	0.140
7/64	50	2.44	72–80	0.196	0.174	0.147
7/64	55	2.56	74–81	0.205	0.182	0.154

7/64	60	2.69	76–82	0.216	0.192	0.161
7/64	65	2.79	77–83	0.224	0.199	0.168
1/8	45	3.04	76–82	0.244	0.217	0.183
1/8	50	3.22	78–82		0.230	0.193
1/8	55	3.39	79–83		0.242	0.204
1/8	60	3.55	80–84		0.253	0.213
1/8	65	3.70	81–85			0.222

Adapted from A. W. Marsh et al., "Solid Set Sprinklers for Starting Vegetable Crops," University of California Division of Agricultural Science Leaflet 2265 (1977). Also, H. W. Otto and J. Meyer, "Tips on Irrigating Vegetables," Family Farm Series Publications: *Vegetable Crop Production* (University of California), http://www.sfc.ucdavis.edu/pubs/family_farm_series/veg/irrigating/irrigating.html.

[1] Three-digit numbers are shown here only to indicate the progression as nozzle size and pressure increase.

[2] Range of diameters of spray for different makes and models of sprinklers.

TABLE 5.15. GUIDE FOR SELECTING SIZE OF ALUMINUM PIPE FOR SPRINKLER LATERAL LINES

Sprinkler Discharge (gpm)	Maximum Number of Sprinklers to Use on Single Lateral Line					
	30-ft Sprinkler Spacing for Pipe Diameter (in.):			40-ft Sprinkler Spacing for Pipe Diameter (in.):		
	2	3	4	2	3	4
0.75	47	95	200	43	85	180
1.00	40	80	150	36	72	125
1.25	34	69	118	31	62	104
1.50	30	62	100	28	56	92
1.75	27	56	92	25	50	83
2.00	25	51	84	23	46	76
2.25	23	47	78	21	43	71
2.50	21	44	73	19	40	66
2.75	20	42	68	18	38	62
3.00	19	40	65	17	36	58
3.25	18	38	62	16	34	56
3.50	17	36	59	15	32	53
3.75	16	34	56	14	31	51
4.00	16	33	54	14	30	48

Adapted from A. W. Marsh et al., "Solid Set Sprinklers for Starting Vegetable Crops," University of California Division of Agricultural Science Leaflet 2265 (1977). Also, H. W. Otto and J. Meyer, "Tips on Irrigating Vegetables," Family Farm Series Publications: *Vegetable Crop Production* (University of California), http://www.sfc.ucdavis.edu/pubs/family_farm_series/veg/irrigating/irrigating.html.

TABLE 5.16. GUIDE TO MAIN-LINE PIPE SIZES[1]

Distance (ft)	Water Flow (gpm) for Pipe Diameter (in.):								
	200	400	600	800	1,000	1,200	1,400	1,600	1,800
200	3	4	5	5	6	6	6	7	7
400	4	5	5	6	6	7	7	8	8
600	4	5	6	7	7	7	8	8	8
800	4	5	6	7	7	8	8	8	10
1,000	5	6	6	7	8	8	8	10	10
1,200	5	6	7	7	8	8	10	10	10

Adapted from A. W. Marsh et al., "Solid Set Sprinklers for Starting Vegetable Crops," University of California Division of Agricultural Science Leaflet 2265 (1977). Also, H. W. Otto and J. Meyer, "Tips on Irrigating Vegetables," Family Farm Series Publications: *Vegetable Crop Production* (University of California), http://www.sfc.ucdavis.edu/pubs/family_farm_series/veg/irrigating/irrigating.html.

[1] Using aluminum pipe (C = 120) with pressure losses ranging from 5 to 15 psi, average about 10.

TABLE 5.17. CONTINUOUS POWER OUTPUT REQUIRED AT TRACTOR POWER TAKEOFF TO PUMP WATER

		Flow (gpm)								
		100	200	300	400	500	600	700	800	1,000
Pressure[1] (psi)	Head[1] (ft)	Horsepower Required[2]								
50	116	3.9	7.8	11.7	16	20	23	27	31	39
55	128	4.3	8.7	13	17	22	26	30	35	43
60	140	4.7	9.5	14	19	24	28	33	38	47
65	151	5.1	10	15	20	25	30	36	41	51
70	162	5.5	11	16	22	27	33	38	44	55
75	173	5.8	12	17	23	29	35	41	47	58
80	185	6.2	12	19	25	31	37	44	50	62

Adapted from A. W. Marsh et al., "Solid Set Sprinklers for Starting Vegetable Crops," University of California Division of Agricultural Science Leaflet 2265 (1977). Also, H. W. Otto and J. Meyer, "Tips on Irrigating Vegetables," Family Farm Series Publications: *Vegetable Crop Production* (University of California), http://www.sfc.ucdavis.edu/pubs/family_farm_series/veg/irrigating/irrigating.html.

[1] Including nozzle pressure, friction loss, and elevation lift.

[2] Pump assumed to operate at 75% efficiency.

TABLE 5.18. FLOW OF WATER REQUIRED TO OPERATE SOLID SET SPRINKLER SYSTEMS

Irrigation rate (in./hr)	Area Irrigated per Set (acres) 4 gpm[1]	cfs[2]	8 gpm	cfs	12 gpm	cfs	16 gpm	cfs	20 gpm	cfs
0.06	108	0.5	217	0.5	326	1.0	435	1.0	543	1.5
0.08	145	0.5	290	1.0	435	1.0	580	1.5	725	2.0
0.10	181	0.5	362	1.0	543	1.5	724	2.0	905	2.5
0.12	217	0.5	435	1.0	652	1.5	870	2.0	1,086	2.5
0.15	271	1.0	543	1.5	815	2.0	1,086	2.5	1,360	3.5
0.20	362	1.0	724	2.0	1,086	2.5	1,448	2.5	1,810	4.5

Adapted from A. W. Marsh et al., "Solid Set Sprinklers for Starting Vegetable Crops," University of California Division of Agricultural Science Leaflet 2265 (1977). Also, H. W. Otto and J. Meyer, "Tips on Irrigating Vegetables," Family Farm Series Publications: *Vegetable Crop Production* (University of California), http://www.sfc.ucdavis.edu/pubs/family_farm_series/veg/irrigating/irrigating.html.

[1] Gallons per minute pumped into the sprinkler system to provide an average precipitation rate as shown. Pump must have this much or slightly greater capacity.

[2] Cubic feet per second—the flow of water to the next larger ½ cfs that must be ordered from the water district, assuming that the district accepts orders only in increments of ½ cfs. Actually, ½ cfs = 225 gpm.

APPLYING FERTILIZER THROUGH A SPRINKLER SYSTEM

Anhydrous ammonia, aqua ammonia, and nitrogen solutions containing free ammonia should not be applied by sprinkler irrigation because of the excessive loss of the volatile ammonia. Ammonium nitrate, ammonium sulfate, calcium nitrate, sodium nitrate, and urea are all suitable materials for use through a sprinkler system. The water containing the ammonia salts should not have a reaction on the alkaline side of neutrality, or the loss of ammonia will be considerable.

It is best to put phosphorus fertilizers directly in the soil by a band application. Potash fertilizers can be used in sprinkler lines. However, a soil application ahead of or at planting time usually proves adequate and can be made efficiently at that time.

Manganese, boron, and copper can be applied through the sprinkler system. See pages 242–243 for possible rates of application.

The fertilizing material is dissolved in a tank of water. Calcium nitrate, ammonium sulfate, and ammonium nitrate dissolve completely. The solution can then be introduced into the water line, either by suction or by pressure from a pump. See page 172 for relative solubility of fertilizer materials.

Introduce the fertilizer into the line slowly, taking 10–20 min to complete the operation.

After enough of the fertilizer solution has passed into the pipelines, shut the valve if suction by pump is used. This prevents unpriming the pump. Then run the system for 10–15 min to wash the fertilizer off the leaves. This also flushes out the lines, valves, and pump, if one has been used to force or suck the solution into the main line.

TABLE 5.19. AMOUNT OF FERTILIZER TO USE FOR EACH SETTING OF THE SPRINKLER LINE

Length of Line (ft)	Lateral Move of Line (ft)	Nutrient per Setting of Sprinkler Line (lb):									
		10	20	30	40	50	60	70	80	90	100
		Nutrient Application Desired (lb/acre)									
330	40	3	6	9	12	15	18	21	24	27	30
	60	4	9	12	18	22	27	31	36	40	45
	80	6	12	18	24	30	36	42	48	54	60
660	40	6	12	18	24	30	36	42	48	54	60
	60	9	18	24	36	45	54	63	72	81	90
	80	12	24	36	48	60	72	84	96	108	120
990	40	9	18	24	36	45	54	63	72	81	90
	60	13	27	40	54	67	81	94	108	121	135
	80	18	36	54	72	90	108	126	144	162	180
1,320	40	12	24	36	48	60	72	84	96	108	120
	60	18	36	54	72	90	108	126	144	162	180
	80	24	48	72	96	120	144	168	192	216	240

It is necessary to calculate the actual pounds of a fertilizing material that must be dissolved in the mixing tank in order to supply a certain number of pounds of the nutrient to the acre at each setting of the sprinkler line. This is done as follows. To apply 40 lb nitrogen to the acre when the sprinkler line is 660 ft long and will be moved 80 ft, if sodium nitrate is used, divide 48 (as shown in the table) by 0.16 (the percentage of nitrogen in sodium nitrate). This equals 300 lb, which must be dissolved in the tank and applied at each setting of the pipe. Do the same with ammonium nitrate: Divide 48 by 0.33, which equals about 145 lbs.

SPRINKLER IRRIGATION FOR COLD PROTECTION

Sprinklers are often used to protect vegetables from freezing. Sprinkling provides cold protection because the latent heat of fusion is released when water changes from liquid to ice. When water is freezing, its temperature is near 32°F. The heat liberated as the water freezes maintains the temperature of the vegetable near 32°F even though the surroundings may be colder. As long as there is a mixture of both water and ice present, the temperature remains near 32°F. For all of the plant to be protected, it must be covered or encased in the freezing ice-water mixture. Enough water must be applied so that the latent heat released compensates for the heat losses.

References

- R. Evans and R. Sneed, *Selection and Management of Efficient Hand-move, Solid-set, and Permanent Sprinkler Irrigation Systems* (North Carolina State University. Publication EBAE 91-152, 1996), http://www.bae.ncsu.edu/programs/extension/evans/ebae-91-152.html.
- R. Snyder, *Principles of Frost Protection* (University of California FP005, 2001), http://biomet.ucdavis.edu/frostprotection/Principles%20of%20Frost%20Protection/FP005.html.

TABLE 5.20. APPLICATION RATE RECOMMENDED FOR COLD PROTECTION UNDER DIFFERENT WIND AND TEMPERATURE CONDITIONS

	Wind Speed (mph)		
	0–1	2–4	5–8
Minimum Temperature Expected (°F)		(in./hr)	
27	0.10	0.10	0.10
26	0.10	0.10	0.14
24	0.10	0.16	0.30
22	0.12	0.24	0.50
20	0.16	0.30	0.60

Adapted from D. S. Harrison, J. F. Gerber, and R. E. Choate, *Sprinkler Irrigation for Cold Protection,* Florida Cooperative Extension Circular 348 (1974).

06
DRIP OR TRICKLE IRRIGATION

Drip or *trickle irrigation* refers to the frequent slow application of water directly to the base of plants. Vegetables are usually irrigated by double-wall, thin-wall, or heavy-wall tubing to supply a uniform rate along the entire row.

Pressure in the drip lines typically varies from 8 to 10 psi and about 12 psi in the submains. Length of the drip lines may be as long as 600 ft, but 200–250 ft is more common. Rate of water application is about ¼–½ gpm/100 ft of row. One acre of plants in rows 100 ft long and 4 ft apart use about 30 gpm water. Unless clear, sediment-free water is available, it is necessary to install a filter in the main line in order to prevent clogging of the small pores in the drip lines.

Drip irrigation provides for considerable saving in water application, particularly during early plant growth. The aisles between rows remain dry because water is applied only next to plants in the row.

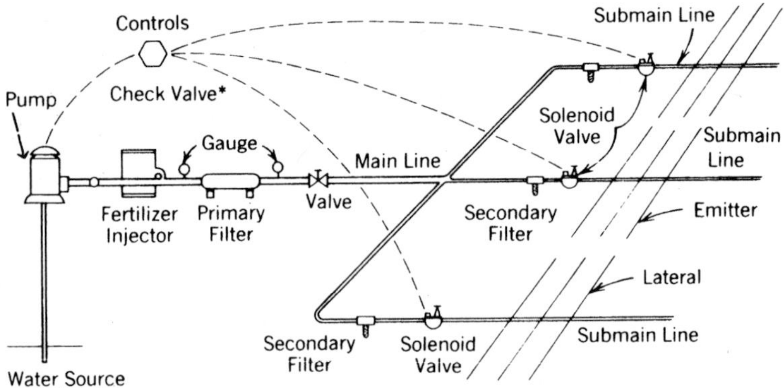

Figure 5.6. Drip or trickle irrigation system components.

TABLE 5.21. VOLUME OF WATER TO APPLY (GAL) BY DRIP IRRIGATION PER 100 LINEAR FT BED FOR A GIVEN WETTED SOIL VOLUME, AVAILABLE WATER-HOLDING CAPACITY, AND AN ALLOWABLE DEPLETION OF 1/2[1]

Wetted Soil Volume per 100 ft (cubic ft)	Available Water-holding Capacity (in. water per ft soil)							
	0.25	0.50	0.75	1.00	1.25	1.50	1.75	2.00
	(gal per 100 linear bed feet)							
25	2.2	4.3	6.5	8.7	10.8	13.0	15.2	17.3
50	4.3	8.7	13.0	17.3	21.6	26.0	30.3	34.6
75	6.5	13.0	19.5	26.0	32.5	39.0	45.5	51.9
100	8.7	17.3	26.0	34.6	43.3	51.9	60.6	69.3
125	10.8	21.6	32.5	43.3	54.1	64.9	75.8	86.6
150	13.0	26.0	39.0	51.9	64.9	77.9	90.9	103.9
175	15.2	30.3	45.5	60.6	75.8	90.9	106.1	121.2
200	17.3	34.6	51.9	69.3	86.6	103.9	121.2	138.5
225	19.5	39.0	58.4	77.9	97.4	116.9	136.4	155.8
250	21.6	43.3	64.9	86.6	108.2	129.9	151.5	173.1
275	23.8	47.6	71.4	95.2	119.0	142.8	166.7	190.5
300	26.0	51.9	77.9	103.9	129.9	155.8	181.8	207.8
350	30.3	60.6	90.9	121.2	151.5	181.8	212.1	242.4
400	34.6	69.3	103.9	138.5	173.1	207.8	242.4	277.0
450	39.0	77.9	116.9	155.8	194.8	233.8	272.7	311.7
500	43.3	86.6	129.9	173.1	216.4	259.7	303.0	346.3
550	47.6	95.2	142.8	190.5	238.1	285.7	333.3	380.9
600	51.9	103.9	155.8	207.8	259.7	311.7	363.6	415.6
700	60.6	121.2	181.8	242.4	303.0	363.6	424.2	484.8
800	69.3	138.5	207.8	277.0	346.3	415.6	484.8	554.1
900	77.9	155.8	233.8	311.7	389.6	467.5	545.4	623.3

Adapted from G. A. Clark, C. D. Stanley, and A. G. Smajstrla, *Micro-irrigation on Mulched Bed Systems: Components, System Capacities, and Management* (Florida Cooperative Extension Service Bulletin 245, 2002), http://edis.ifas.ufl.edu/ae042.

[1] An irrigation application efficiency of 90% is assumed.

TABLE 5.22. DISCHARGE PER GROSS ACRE (GPM/ACRE) FOR DRIP IRRIGATION BASED ON IRRIGATED LINEAR BED FEET AND EMITTER DISCHARGE

Linear Bed Feet per Acre	Emitter Discharge (gpm/100 ft)						
	0.25	0.30	0.40	0.50	0.75	1.00	1.50
	(gal per min/acre)						
3,000	7.5	9.0	12.0	15.0	22.5	30.0	45.0
3,500	8.8	10.5	14.0	17.5	26.3	35.0	52.5
4,000	10.0	12.0	16.0	20.0	30.0	40.0	60.0
4,500	11.3	13.5	18.0	22.5	33.8	45.0	67.5
5,000	12.5	15.0	20.0	25.0	37.5	50.0	75.0
5,500	13.8	16.5	22.0	27.5	41.3	55.0	82.5
6,000	15.0	18.0	24.0	30.0	45.0	60.0	90.0
6,500	16.3	19.5	26.0	32.5	48.8	65.0	97.5
7,000	17.5	21.0	28.0	35.0	52.5	70.0	105.0
7,500	18.8	22.5	30.0	37.5	56.3	75.0	112.5
8,000	20.0	24.0	32.0	40.0	60.0	80.0	120.0
8,500	21.3	25.5	34.0	42.5	63.8	85.0	127.5
9,000	22.5	27.0	36.0	45.0	67.5	90.0	135.0
9,500	23.8	28.5	38.0	47.5	71.3	95.0	142.5
10,000	25.0	30.0	40.0	50.0	75.0	100.0	150.0

Adapted from G. A. Clark, C. D. Stanley, and A. G. Smajstrla, *Micro-irrigation on Mulched Bed Systems: Components, System Capacities, and Management* (Florida Cooperative Extension Service Bulletin 245, 2002), http://edis.ifas.ufl.edu/ae042.

TABLE 5.23. VOLUME OF WATER (GAL WATER PER ACRE PER MINUTE) DELIVERED UNDER VARIOUS BED SPACINGS WITH ONE TAPE LATERAL PER BED AND FOR SEVERAL EMITTER FLOW RATES

Bed Spacing (in.)	Drip Tape per Acre (ft)	Emitter Flow Rate (gal per min per 100 ft)			
		0.50	0.40	0.30	0.25
		(gal per acre/min)			
24	21,780	108.9	87.1	65.3	54.5
30	17,420	87.1	69.7	52.3	43.6
36	14,520	72.6	58.1	43.6	36.6
42	12,450	62.2	49.8	37.3	31.1
48	10,890	54.5	43.6	32.7	27.2
54	9,680	48.4	38.7	29.0	24.2
60	8,710	43.6	34.9	26.1	21.8
72	7,260	36.3	29.0	21.8	18.2
84	6,220	31.1	24.9	18.7	15.6
96	5,450	27.2	21.8	16.3	13.6
108	4,840	24.2	19.4	14.5	12.1
120	4,360	21.8	17.4	13.1	10.0

TABLE 5.24. VOLUME OF AVAILABLE WATER IN THE WETTED CYLINDRICAL DISTRIBUTION PATTERN UNDER A DRIP IRRIGATION LINE BASED ON THE AVAILABLE WATER-HOLDING CAPACITY OF THE SOIL

Available Water (%)	Wetted Radius (in.)[1] 6	9	12	15	18
	(gal available water per 100 emitters)				
3	9	20	35	55	79
4	12	26	47	74	106
5	15	33	59	92	132
6	18	40	71	110	159
7	21	46	82	129	185
8	24	53	94	147	212
9	26	60	106	165	238
10	29	66	118	184	265
11	32	73	129	202	291
12	35	79	141	221	318
13	38	86	153	239	344
14	41	93	165	257	371
15	44	99	176	276	397

Adapted from G. A. Clark and A. G. Smajstrla, *Application Volumes and Wetted Patterns for Scheduling Drip Irrigation in Florida Vegetable Production*, Florida Cooperative Extension Service Circular 1041 (1993).

[1] For a 1-ft depth of wetting.

TABLE 5.25. MAXIMUM APPLICATION TIMES FOR DRIP-IRRIGATED VEGETABLE PRODUCTION ON SANDY SOILS WITH VARIOUS WATER-HOLDING CAPACITIES

Available Water-holding Capacity (in. water per in. soil)	Tubing Flow Rate (gpm per 100 ft)				
	0.2	0.3	0.4	0.5	0.6
	(maximum min per application)[1]				
0.02	41	27	20	16	14
0.03	61	41	31	24	20
0.04	82	54	41	33	27
0.05	102	68	51	41	34
0.06	122	82	61	49	41
0.07	143	95	71	57	48
0.08	163	109	82	65	54
0.09	184	122	92	73	61
0.10	204	136	102	82	68
0.11	224	150	112	90	75
0.12	245	163	122	98	82

Adapted from C. D. Stanley and G. A. Clark, "Maximum Application Times for Drip-irrigated Vegetable Production as Influenced by Soil Type or Tubing Emission Characteristics," Florida Cooperative Extension Service Drip Tip No. 9305 (1993).

[1] Assumes 10-in.-deep root zone and irrigation at 50% soil moisture depletion.

TREATING IRRIGATION SYSTEMS WITH CHLORINE

Chlorine can be used in irrigation systems to control the growth of algae and other microorganisms such as bacteria and fungi. These organisms are found in surface and ground water and can proliferate with the nutrients present in the water inside the drip tube. Filtration alone cannot remove all of these contaminants. Hypochlorous acid, the agent responsible for controlling microorganisms in drip tubes, is more active under slightly acidic water conditions. Chlorine gas, solid (calcium hypochlorite), or liquid (sodium hypochlorite) are sources of chlorine; however, all forms might not be legal for injecting into irrigation systems. For example, only sodium hypochlorite is legal for use in Florida.

When sodium hypochlorite is injected, the pH of the water rises. The resulting chloride and sodium ions are not detrimental to crops at typical injection rates. Chlorine materials should be injected at a rate to provide 1–2 ppm free residual chlorine at the most distant part of the irrigation system.

In addition to controlling microorganisms, hypochlorous acid also reacts with iron in solution to oxidize the ferrous form to the ferric form, which precipitates as ferric hydroxide. If irrigation water contains iron, this reaction with injected chlorine should occur before the filter system so the precipitate can be removed.

Adapted from G. A. Clark and A. G. Smajstrla, *Treating Irrigation Systems with Chlorine* (Florida Cooperative Extension Service Circular 1039, 2002), http://edis.ifas.ufl.edu/ae080.

TABLE 5.26. LIQUID CHLORINE (SODIUM HYPOCHLORITE) INJECTION

	% Concentration of chlorine in stock			
Treatment level (ppm)	1	3	5.25	10
	gal/hr injection per 100 gal/min irrigation flow rate			
1	0.54	0.18	0.10	0.05
2	1.1	0.36	0.21	0.14
3	1.6	0.54	0.31	0.16
4	2.2	0.72	0.41	0.22
5	2.7	0.90	0.51	0.27
6	3.3	1.1	0.62	0.32
8	4.4	1.5	0.82	0.43
10	5.5	1.8	1.0	0.54
15	8.3	2.8	1.5	0.81
20	11.0	3.7	2.1	1.1
30	16.5	5.5	3.1	1.6

Adapted from G. Clark and A. Smajstrla, *Treating Irrigation Systems with Chlorine* (Florida Cooperative Extension Service Circular 1039, 2002), http://edis.ifas.ufl.edu/ae080.

METHODS OF INJECTING FERTILIZER AND OTHER CHEMICAL SOLUTIONS INTO IRRIGATION PIPELINE

Four principal methods are used to inject fertilizers and other solutions into drip irrigation systems: (1) pressure differential; (2) the venturi (vacuum); (3) centrifugal pumps; and (4) positive displacement pumps. It is essential that irrigation systems equipped with a chemical injection system have a vacuum breaker (anti-siphon device) and a backflow preventer (check valve) installed upstream from the injection point. The vacuum-breaking valve and backflow preventer prevent chemical contamination of the water source in case of a water pressure loss or power failure. Operators may need a license to chemigate in some states. Local backflow regulations should be consulted prior to chemigation to insure compliance.

TABLE 5.27. REQUIRED VOLUME (GAL) OF CHEMICAL MIXTURE TO PROVIDE A DESIRED LEVEL OF AN ACTIVE CHEMICAL FOR DIFFERENT CONCENTRATIONS (LB/GAL) OF THE CHEMICAL IN THE STOCK SOLUTION

	S_{mx}							
	Mass of Chemical (lb) Per Gal Stock Solution							
Weight of Chemical Desired (lb)	0.2	0.4	0.6	0.8	1.0	2.0	3.0	4.0
	(gal stock solution)							
20	100	50	33	25	20	10	7	5
40	200	100	67	50	40	20	13	10
60	300	150	100	75	60	30	20	15
80	400	200	133	100	80	40	27	20
100	500	250	167	125	100	50	33	25
150	750	375	250	188	150	75	50	38
200	1,000	500	333	250	200	100	67	50
250	1,250	625	417	313	250	125	83	63
300	1,500	750	500	375	300	150	100	75
350	1,750	875	583	438	350	175	117	88
400	2,000	1,000	667	500	400	200	133	100
450	2,250	1,125	750	563	450	225	150	113
500	2,500	1,250	833	625	500	250	167	125

Adapted from G. A. Clark, D. Z. Haman, and F. S. Zazueta, *Injection of Chemicals into Irrigation Systems: Rates, Volumes, and Injection Periods* (Florida Cooperative Extension Service Bulletin 250, 2002), http://edis.ifas.ufl.edu/ae116.

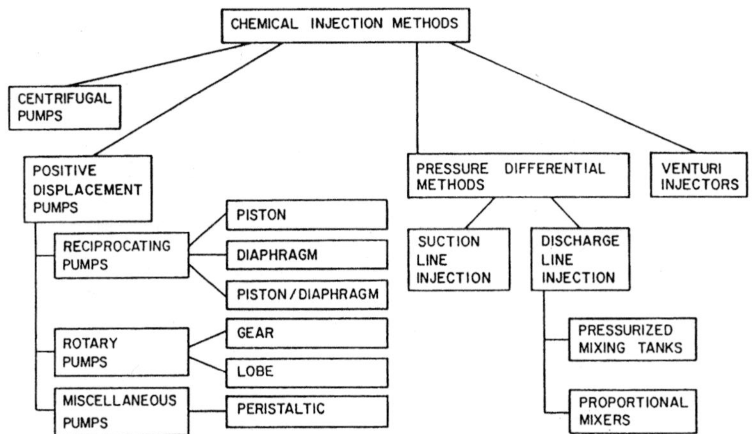

Figure 5.7. Classification of chemical injection methods for irrigation systems.

Adapted from D. Z. Haman, A. G. Smajstrla, and F. S. Zazueta, *Chemical Injection Methods for Irrigation* (Florida Cooperative Extension Service Circular 864, 2003). http://edis.ifas.ufl.edu/wi004.

TABLE 5.28. COMPARISON OF VARIOUS CHEMICAL INJECTION METHODS

Injector	Advantages	Disadvantages
Centrifugal Pumps		
Centrifugal pump injector	Low cost. Can be adjusted while running.	Calibration depends on system pressure. Cannot accurately control low injection rates.
Positive Displacement Pumps		
Piston pumps	High precision. Linear calibration. Very high pressure. Calibration independent of pressure.	High cost. May need to stop to adjust calibration. Chemical flow not continuous.
Diaphragm pumps	Adjust calibration while injecting. High chemical resistance.	Nonlinear calibration. Calibration depends on system pressure. Medium to high cost. Chemical flow not continuous.
Piston/diaphragm	High precision. Linear calibration. High chemical resistance. Very high pressure. Calibration independent of pressure.	High cost. May need to stop to adjust calibration.

Rotary Pumps		
Gear pumps Lobe pumps	Injection rate can be adjusted when running.	Fluid pumped cannot be abrasive. Injection rate is dependent on system pressure. Continuity of chemical flow depends on number of lobes in a lobe pump.
Miscellaneous		
Peristaltic pumps	High chemical resistance. Major adjustment can be made by changing tubing size. Injection rate can be adjusted when running.	Short tubing life expectancy. Injection rate dependent on system pressure. Low to medium injection pressure.
Pressure Differential Methods		
Suction line injection	Very low cost. Injection can be adjusted while running.	Permitted only for surface water source and injection of fertilizer. Injection rate depends on main pump operation.
Discharge Line Injection		
Pressurized mixing tanks	Low to medium cost. Easy operation. Total chemical volume controlled.	Variable chemical concentration. Cannot be calibrated accurately for constant injection rate.
Proportional mixers	Low to medium cost. Calibrate while operating. Injection rates accurately controlled.	Pressure differential required. Volume to be injected is limited by the size of the injector. Frequent refills required.

TABLE 5.28. COMPARISON OF VARIOUS CHEMICAL INJECTION METHODS (*Continued*)

Injector	Advantages	Disadvantages
Venturi Injectors		
Venturi injector	Low cost. Water powered. Simple to use. Calibrate while operating. No moving parts.	Pressure drop created in the system. Calibration depends on chemical level in the tank.
Combination Methods		
Proportional mixers/ venturi	Greater precision than proportional mixer or venturi alone.	Higher cost than proportional mixer or venturi alone.

Adapted from D. Haman, A. Smajstrla, and F. Zazueta, *Chemical Injection Methods for Irrigation* (Florida Cooperative Extension Service Circular 864, 2003), http://edis.ifas.ufl.edu/wi004.

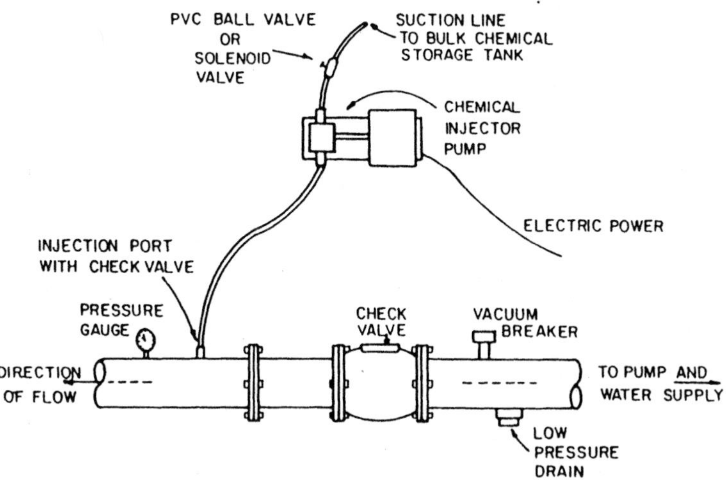

Figure 5.8. Single antisyphon device assembly.

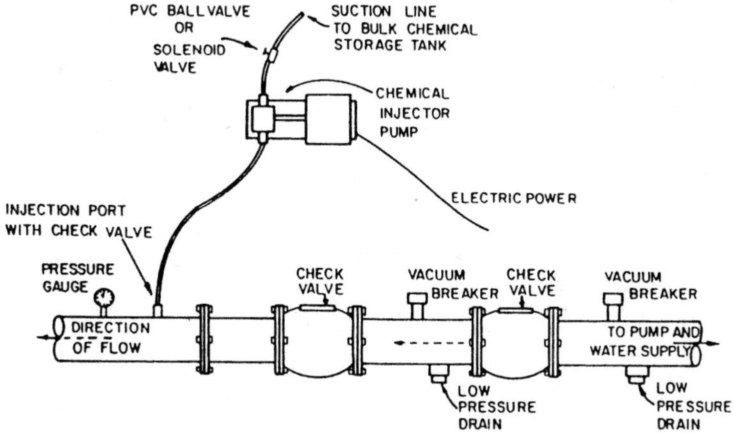

Figure 5.9. Double antisyphon device assembly.

Adapted from A. G. Smajstrla, D. S. Harrison, W. J. Becker, F. S. Zazueta, and D. Z. Haman, *Backflow Prevention Requirements for Florida Irrigation Systems,* Florida Cooperative Extension Service Bulletin 217 (1985).

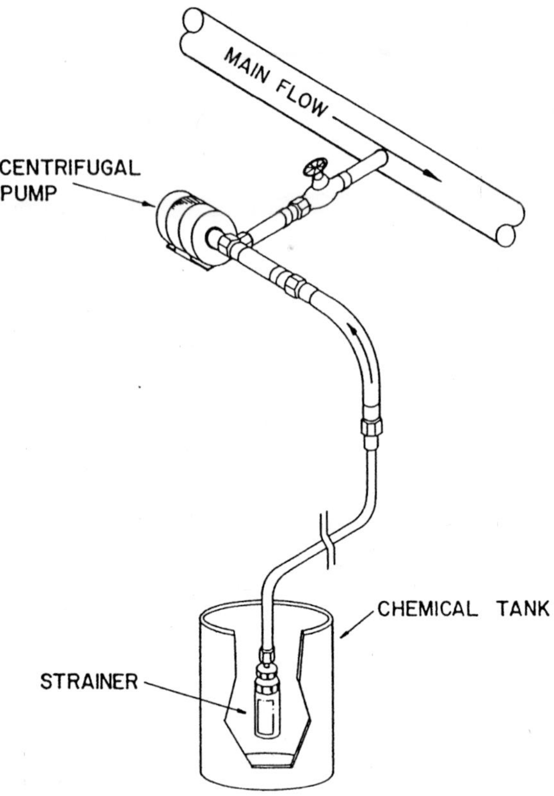

Figure 5.10. Centrifugal pump chemical injector.

Adapted from D. Z. Haman, A. G. Smajstrla, and F. S. Zazueta, *Chemical Injection Methods for Irrigation* (Florida Cooperative Extension Service Circular 864, 2003), http://edis.ifas.ufl.edu/wi004.

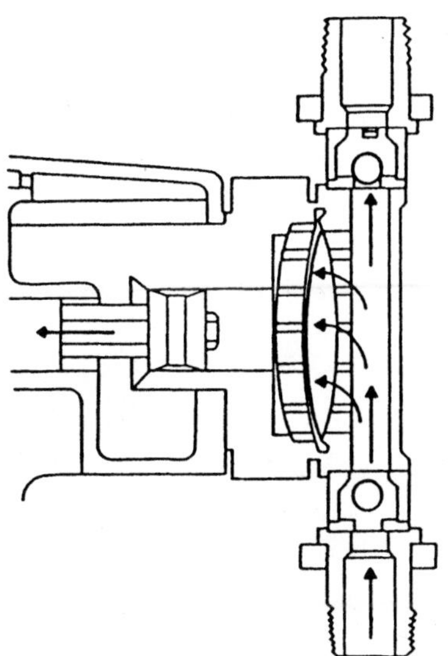

Figure 5.11. Diaphragm pump—suction stroke.

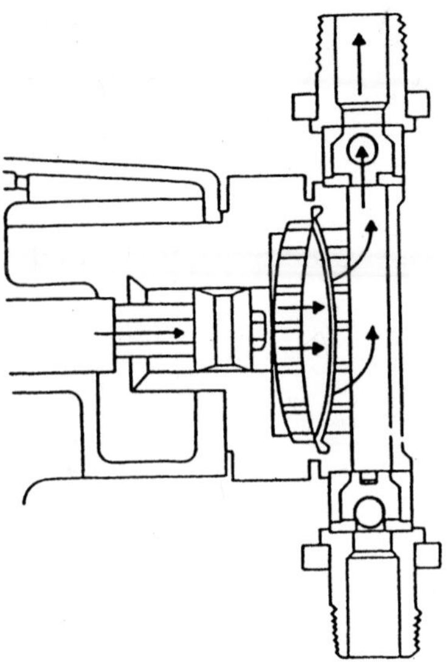

Figure 5.12. Diaphragm pump—discharge stroke.

Adapted from D. Z. Haman, A. G. Smajstrla, and F. S. Zazueta, *Chemical Injection Methods for Irrigation* (Florida Cooperative Extension Service Circular 864, 2003), http://edis.ifas.ufl.edu/wi004.

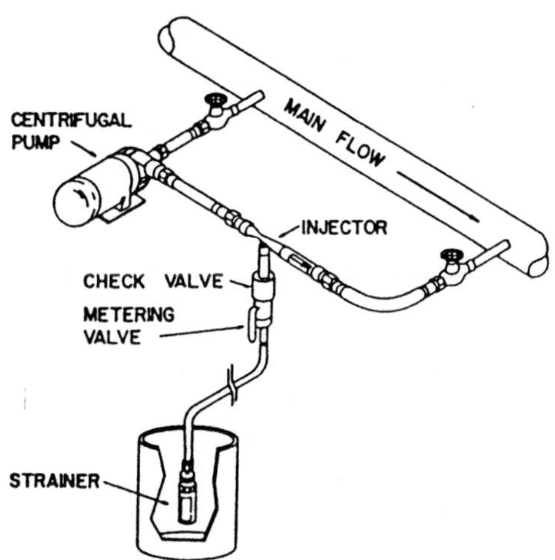

Figure 5.13. Venturi injector.

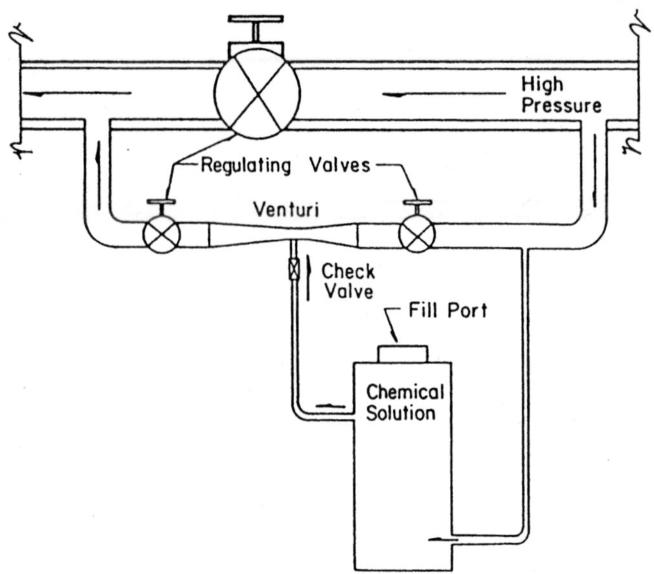

Figure 5.14. Proportional mixer/venturi.

Adapted from D. Z. Haman, A. G. Smajstrla, and F. S. Zazueta, *Chemical Injection Methods for Irrigation* (Florida Cooperative Extension Service Bulletin 864, 2003), http://edis.ifas.ufl.edu/wi004.

TABLE 5.29. WATER QUALITY GUIDELINES FOR IRRIGATION[1]

	Degree of Problem		
Situation	None	Increasing	Severe
Salinity			
EC (dS/m) or	Less than 0.75	0.75–3.0	More than 3.0
TDS (mg/L)	Less than 480	480–1,920	More than 1920
Permeability			
Low EC (dS/m) or	More than 0.5	0.5–0	—
Low TDS (mg/L)	More than 320	320–0	—
SAR	Less than 6.0	6.0–9.0	More than 9.0
Toxicity of Specific Ions to Sensitive Crops			
ROOT ABSORPTION			
Sodium (evaluated by SAR)	SAR less than 3	3–9	More than 9
Chloride			
meq/L	Less than 2	2–10	More than 10.0
mg/L	Less than 70	70–345	More than 345
Boron (mg/L)	1.0	1.0–2.0	2.0–10.0

TABLE 5.29. WATER QUALITY GUIDELINES FOR IRRIGATION[1] (*Continued*)

Situation	Degree of Problem		
	None	Increasing	Severe
RELATED TO FOLIAR ABSORPTION (SPRINKLER IRRIGATED)			
Sodium			
meq/L	Less than 3.0	More than 3	—
mg/L	Less than 70	70	—
Chloride			
meq/L	Less than 3.0	More than 3	—
mg/L	Less than 100	100	—
MISCELLANEOUS			
NH_4 and NO_3-N (mg/L)	Less than 5	5–30	More than 30
HCO_3			—
meq/L	Less than 1.0	1.5–8.5	More than 8.5
mg/L	Less than 40	40–520	More than 520
pH	Normal range: 6.5–8.3	More than 8.3	—

Adapted from D. S. Farnham, R. F. Hasek, and J. L. Paul, "Water Quality," University of California Division of Agricultural Science Leaflet 2995 (1985); and B. Hanson, S. Gratham, and A. Fulton, Agricultural Salinity and Drainage, University of California Division of Agriculture and Natural Resources Publication 3375 (1999).

[1] Interpretation is related to type of problem and its severity but is modified by circumstances of soil, crop, and locality.

TABLE 5.30. MAXIMUM CONCENTRATIONS OF TRACE ELEMENTS IN IRRIGATION WATERS

Element	For Waters Used Continuously on All Soils (mg/L)	For Use Up to 20 Years on Fine-textured Soils of pH 6.0–8.5 (mg/L)
Aluminum	5.0	20.0
Arsenic	0.10	2.0
Beryllium	0.10	0.50
Boron	0.75	2.0–10.0
Cadmium	0.01	0.05
Chromium	0.10	1.0
Cobalt	0.05	5.0
Copper	0.20	5.0
Fluoride	1.0	15.0
Iron	5.0	20.0
Lead	5.0	10.0
Lithium	2.5	2.5
Manganese	0.20	10.0
Molybdenum	0.01	0.05[1]
Nickel	0.20	2.0
Selenium	0.02	0.02
Vanadium	0.10	1.0
Zinc	2.00	10.0

Adapted from D. S. Farnham, R. F. Hasek, and J. L. Paul, "Water Quality," University of California Division of Agricultural Science Leaflet 2995 (1985).

[1] Only for acid, fine-textured soils or acid soils with relatively high iron oxide contents.

TABLE 5.31. ESTIMATED YIELD LOSS TO SALINITY OF IRRIGATION WATER

Crop	Electrical Conductivity of Water (mmho/cm or dS/m) for Following % Yield Loss			
	0	10	25	50
Bean	0.7	1.0	1.5	2.4
Carrot	0.7	1.1	1.9	3.1
Strawberry	0.7	0.9	1.2	1.7
Onion	0.8	1.2	1.8	2.9
Radish	0.8	1.3	2.1	3.4
Lettuce	0.9	1.4	2.1	3.4
Pepper	1.0	1.5	2.2	3.4
Sweet potato	1.0	1.6	2.5	4.0
Sweet corn	1.1	1.7	2.5	3.9
Potato	1.1	1.7	2.5	3.9
Cabbage	1.2	1.9	2.9	4.6
Spinach	1.3	2.2	3.5	5.7
Cantaloupe	1.5	2.4	3.8	6.1
Cucumber	1.7	2.2	2.9	4.2
Tomato	1.7	2.3	3.4	5.0
Broccoli	1.9	2.6	3.7	5.5
Beet	2.7	3.4	4.5	6.4

Adapted from R. S. Ayers, *Journal of the Irrigation and Drainage Division* 103 (1977): 135–154.

TABLE 5.32. RELATIVE TOLERANCE OF VEGETABLE CROPS TO BORON IN IRRIGATION WATER[1]

10–15 ppm	4–6 ppm	2–4 ppm	1–2 ppm	0.5–1 ppm
Asparagus	Beet	Artichoke	Broccoli	Bean
	Parsley	Cabbage	Carrot	Garlic
	Tomato	Cantaloupe	Cucumber	Lima bean
		Cauliflower	Pea	Onion
		Celery	Pepper	
		Corn	Potato	
		Lettuce	Radish	
		Turnip		

Adapted from E. V. Mass, "Salt Tolerance of Plants," *Applied Agricultural Research* 1(1):12–26 (1986).

[1] Maximum concentrations of boron in soil water without yield reduction

PART 6

VEGETABLE PESTS AND PROBLEMS

01
AIR POLLUTION

AIR POLLUTION DAMAGE TO VEGETABLE CROPS

Plant damage by pollutants depends on meteorological factors leading to air stagnation, the presence of a pollution source, and the susceptibility of the plants. Among the pollutants that affect vegetable crops are sulfur dioxide (SO_2), ozone (O_3), peroxyacetyl nitrate (PAN), chlorine (Cl_2), and ammonia (NH_3). The following symptoms are most likely to be observed on vegetables produced near heavily urbanized areas or industrialized areas, particularly where weather conditions frequently lead to air stagnation.

Sulfur dioxide: SO_2 causes acute and chronic plant injury. Acute injury is characterized by dead tissue between the veins or on leaf margins. The dead tissue may be bleached, ivory, tan, orange, red, reddish brown, or brown, depending on the plant species, time of year, and weather. Chronic injury is marked by brownish red, turgid, or bleached white areas on the leaf blade. Young leaves rarely display damage, whereas fully expanded leaves are very sensitive.

Ozone: Common symptoms of O_3 injury are very small, irregularly shaped spots that are dark brown to black (stipple-like) or light tan to white (fleck-like) on the upper leaf surface. Very young and old leaves are normally resistant to ozone. Recently matured leaves are most susceptible. Injury is usually more pronounced at the leaf tip and along the margins. With severe damage, symptoms may extend to the lower leaf surface.

Peroxyacetyl nitrate: Typically, PAN affects the under-leaf surface of newly matured leaves and causes bronzing, glazing, or silvering on the lower surface of sensitive leaf areas. The leaf apex of broad-leaved plants becomes sensitive to PAN approximately five days after leaf emergence. About four leaves on a shoot are sensitive at any one time. PAN toxicity is specific for tissue in a particular stage of development. Only with successive exposure to PAN will the entire leaf develop injury. Injury may consist of bronzing or glazing with little or no tissue collapse on the upper leaf surface. Pale green to white stipple-like areas may appear on upper and lower leaf surfaces. Complete tissue collapse in a diffuse band across the leaf is helpful in identifying PAN injury.

Chlorine: Injury from chlorine is usually of an acute type and is similar in pattern to sulfur dioxide injury. Foliar necrosis and bleaching are common. Necrosis is marginal in some species but scattered in others, either between or along veins. Lettuce plants exhibit necrotic injury on the margins of outer leaves, which often extends in solid areas toward the center and base of the leaf. Inner leaves remain unmarked.

Ammonia: Field injury from NH_3 is primarily due to accidental spillage. Slight amounts of the gas produce color changes in the pigments of vegetable skin. The dry outer scales of red onion may become greenish or black, whereas scales of yellow or brown onion may turn dark brown.

Hydrochloric acid gas: HCl causes an acid-type burn. The usual acute response is a bleaching of tissue. Leaves of lettuce, endive, and escarole exhibit a tip burn that progresses toward the center of the leaf and soon dries out. Tomato plants develop interveinal bronzing.

Original material from Commercial Vegetable Production Recommendations, Maryland Agricultural Extension Service EB-236 (1986) and Bulletin 137 (2005).

Additional Resource H. Griffiths, *Effects of Air Pollution on Agricultural Crops* (Ontario Ministry of Agriculture, Food, and Rural Affairs, 2003), http://www.omafra.gov.on.ca/english/crops/facts/01-015.htm.

REACTION OF VEGETABLE CROPS TO AIR POLLUTANTS

Vegetable crops may be injured following exposure to high concentrations of atmospheric pollutants. Prolonged exposure to lower concentrations may also result in plant damage. Injury appears progressively as leaf chlorosis (yellowing), necrosis (death), and perhaps restricted growth and yields. On occasion, plants may be killed, but usually not until they have suffered persistent injury. Symptoms of air pollution damage vary with the individual crops and plant age, specific pollutant, concentration, duration of exposure, and environmental conditions.

RELATIVE SENSITIVTY OF VEGETABLE CROPS TO AIR POLLUTANTS

TABLE 6.1. SENSITIVITY OF VEGETABLES TO SELECTED AIR POLLUTANTS

Pollutant	Sensitive	Intermediate	Tolerant
Ozone	Bean	Carrot	Beet
	Broccoli	Endive	Cucumber
	Onion	Parsley	Lettuce
	Potato	Parsnip	
	Radish	Turnip	
	Spinach		
	Sweet corn		
	Tomato		
Sulfur dioxide	Bean	Cabbage	Cucumber
	Beet	Pea	Onion
	Broccoli	Tomato	Sweet corn
	Brussels sprouts		
	Carrot		
	Endive		
	Lettuce		
	Okra		
	Pepper		
	Pumpkin		
	Radish		
	Rhubarb		
	Spinach		
	Squash		
	Sweet potato		
	Swiss chard		
	Turnip		
Fluoride	Sweet corn		Asparagus
			Squash
			Tomato
Nitrogen dioxide	Lettuce		Asparagus
			Bean
PAN	Bean	Carrot	Broccoli
	Beet		Cabbage

TABLE 6.1. SENSITIVITY OF VEGETABLES TO SELECTED AIR POLLUTANTS (*Continued*)

Pollutant	Sensitive	Intermediate	Tolerant
	Celery		Cauliflower
	Endive		Cucumber
	Lettuce		Onion
	Mustard		Radish
	Pepper		Squash
	Spinach		
	Sweet corn		
	Swiss chard		
	Tomato		
Ethylene	Bean	Carrot	Beet
	Cucumber	Squash	Cabbage
	Pea		Endive
	Southern pea		Onion
	Sweet potato		Radish
	Tomato		
2,4-D	Tomato	Potato	Bean
			Cabbage
			Eggplant
			Rhubarb
Chlorine	Mustard	Bean	Eggplant
	Onion	Cucumber	Pepper
	Radish	Southern pea	
	Sweet corn	Squash	
		Tomato	
Ammonia	Mustard	Tomato	
Mercury vapor	Bean	Tomato	
Hydrogen sulfide	Cucumber	Pepper	Mustard
	Radish		
	Tomato		

Adapted from J. S. Jacobson and A. C. Hill (eds.), *Recognition of Air Pollution Injury to Vegetation* (Pittsburgh: Air Pollution Control Association, 1970); M. Treshow, *Environment and Plant Response* (New York: McGraw-Hill, 1970); R.G. Pearson et al., *Air Pollution and Horticultural Crops,* Ontario Ministry of Agriculture and Food AGDEX 200/691 (1973); and H. Griffiths, *Effects of Air Pollution on Agricultural Crops* (Ontario Ministry of Agriculture, Food, and Rural Affairs, 2003), http://www.omafra.gov.on.ca/english/crops/facts/01-015.htm.

02
INTEGRATED PEST MANAGEMENT

BASICS OF INTEGRATED PEST MANAGEMENT

Integrated Pest Management (IPM) attempts to make the most efficient use of the strategies available to control pest populations by taking action to prevent problems, suppress damage levels, and use chemical pesticides only where needed. Rather than seeking to eradicate all pests entirely, IPM strives to prevent their development or to suppress their population numbers below levels that would be economically damaging.

Integrated means that a broad, interdisciplinary approach is taken using scientific principles of crop protection in order to fuse into a single system a variety of methods and tactics.

Pest includes insects, mites, nematodes, plant pathogens, weeds, and vertebrates that adversely affect crop quality and yield.

Management refers to the attempt to control pest populations in a planned, systematic way by keeping their numbers or damage within acceptable levels.

Effective IPM consists of four basic principles:

1. *Exclusion* seeks to prevent pests from entering the field in the first place.
2. *Suppression* refers to the attempt to suppress pests below the level at which they would be economically damaging.
3. *Eradication* strives to eliminate entirely certain pests.
4. *Plant resistance* stresses the effort to develop healthy, vigorous varieties that are resistant to certain pests.

In order to carry out these four basic principles, the following steps are recommended:

1. *The identification of key pests and beneficial organisms* is a necessary first step.
2. *Preventive cultural practices* are selected to minimize pest population development.
3. *Pest populations must be monitored* by trained scouts who routinely sample fields.

4. *A prediction of loss and risks* involved is made by setting an economic threshold. Pests are controlled only when the pest population threatens acceptable levels of quality and yield. The level at which the pest population or its damage endangers quality and yield is often called the *economic threshold*. The economic threshold is set by predicting potential loss and risks at a given pest population density.
5. *An action decision must be made*. In some cases, pesticide application is necessary to reduce the crop threat, whereas in other cases, a decision is made to wait and rely on closer monitoring.
6. *Evaluation and follow-up* must occur throughout all stages in order to make corrections, assess levels of success, and project future possibilities for improvement.

To be effective, IPM must make use of the following tools:

- *Pesticides*. Some pesticides are applied preventively—for example, herbicides, fungicides, and nematicides. In an effective IPM program, pesticides are applied on a prescription basis tailored to the particular pest and chosen for minimum impact on people and the environment. Pesticides are applied only when a pest population is diagnosed as large enough to threaten acceptable levels of yield and quality and no other economic control measure is available.
- *Resistant crop varieties* are bred and selected when available in order to protect against key pests.
- *Natural enemies* are used to regulate the pest population whenever possible.
- *Pheromone* (sex lure) traps are used to lure and destroy male insects to help monitor populations.
- *Preventive measures* such as soil fumigation for nematodes and assurance of good soil fertility help provide a healthy, vigorous plant.
- *Avoidance* of peak pest populations can be brought about by a change in planting times or pest-controlling crop rotation.
- *Improved application* is achieved by keeping equipment up to date and in excellent shape.
- *Other assorted cultural practices* such as flooding and row and plant spacing can influence pest populations.

Resources for Integrated Pest Management:

- Pest Management (IPM), http://attra.ncat.org/attra-pub/ipm.html (2001)

- UC IPM Online—Statewide Integrated Pest Management Program, http://www.ipm.ucdavis.edu/
- The Pennsylvania Integrated Pest Management Program, http://paipm.cas.psu.edu/
- North Central Region—National IPM Network, http://www.ipm.iastate.edu/ipm/nipmn/
- Northeastern IPM Center, http://northeastipm.org/
- Integrated Pest Management in the Southern Region, http://ipm-www.ento.vt.edu/nipmn/
- New York State Integrated Pest Management Program at Cornell University, http://www.nysipm.cornell.edu/
- IPM Florida, http://ipm.ifas.ufl.edu/

03
SOIL SOLARIZATION

BASICS OF SOIL SOLARIZATION

Soil solarization is a nonchemical pest control method that is particularly effective in areas that have high temperatures and long days for the required 4–6 weeks. In the northern hemisphere, this generally means that solarization is done during the summer months in preparation for a fall crop or for a crop in the following spring.

Soil solarization captures radiant heat energy from the sun, thereby causing physical, chemical, and biological changes in the soil. Transparent polyethylene plastic placed on moist soil during the hot summer months increases soil temperatures to levels lethal to many soil borne plant pathogens, weed seeds, and seedlings (including parasitic seed plants), nematodes, and some soil-residing mites. Soil solarization also improves plant nutrition by increasing the availability of nitrogen and other essential nutrients.

Time of Year

Highest soil temperatures are obtained when the days are long, air temperatures are high, the sky is clear, and there is no wind.

Plastic Color

Clear polyethylene should be used, not black plastic. Transparent plastic results in greater transmission of solar energy to the soil. Polyethylene should be UV stabilized so the tarp does not degrade during the solarization period.

Plastic Thickness

Polyethylene 1 mil thick is the most efficient and economical for soil heating. However, it is easier to rip or puncture and is less able to withstand high winds than thicker plastic. Users in windy areas may prefer to use plastic 1½–2 mil thick. Thick transparent plastic (4–6 mil) reflects more solar energy than does thinner plastic (1–2 mil) and results in slightly lower temperatures.

Preparation of the Soil

It is important that the area to be treated is level and free of weeds, plants, debris, and large clods that would raise the plastic off the ground. Maximum soil heating occurs when the plastic is close to the soil; therefore, air pockets caused by large clods or deep furrows should be avoided. Soil should be tilled as deep as possible and have moisture at field capacity.

Partial Versus Complete Soil Coverage

Polyethylene tarps may be applied in strips (a minimum of 2–3 ft wide) over the planting bed or as continuous sheeting glued, heat-fused, or held in place by soil. In some cases, strip coverage may be more practical and economical than full soil coverage, because less plastic is needed and plastic connection costs are avoided.

Soil Moisture

Soil must be moist (field capacity) for maximum effect because moisture not only makes organisms more sensitive to heat but also conducts heat faster and deeper into the soil.

Duration of Soil Coverage

Killing of pathogens and pests is related to time and temperature exposure. The longer the soil is heated, the deeper the control. Although some pest organisms are killed within days, 4–6 weeks of treatment in full sun during the summer is usually best. The upper limit for temperature is about 115°F for vascular plants, 130°F for nematodes, 140°F for fungi, and 160–212°F for bacteria.

Original material adapted from G. S. Pullman, J. E. DeVay, C. L Elmore, and W. H. Hart, "Soil Solarization," California Cooperative Extension Leaflet 21377 (1984), and from D. O. Chellemi, "Soil Solarization for Management of Soilborne Pests," Florida Cooperative Extension Fact Sheet PPP51 (1995).

Additional Resources on Soil Solarization:

- International Workgroup on Soil Solarization and Integrated Management of Soilborne Pests, http://www.uckac.edu/iwgss//
- C. Strausbaugh *Soil Solarization for Control of Soilborne Pest Problems* (University of Idaho, 1996), http://www.uidaho.edu/ag/plantdisease/soilsol.htm

- The Soil Solarization Home (Hebrew University of Jerusalem, 1998), http://agri3.huji.ac.il/~katan/
- Soil Solarization (University of California), http://ucce.ucdavis.edu/files/filelibrary/40/942.pdf
- A. Hagan and W. Gazaway, *Soil Solarization for the Control of Nematodes and Soilborne Diseases* (Auburn University, 2002), http://www.aces.edu/pubs/docs/A/ANR-0713/

04
PESTICIDE USE PRECAUTIONS

GENERAL SUGGESTIONS FOR PESTICIDE SAFETY

All chemicals are potentially hazardous and should be used carefully. Follow exactly the directions, precautions, and limitations given on the container label. Store all chemicals in a safe place where children, pets, and livestock cannot reach them. Do not reuse pesticide containers. Avoid inhaling fumes and dust from pesticides. Avoid spilling chemicals; if they are accidentally spilled, remove contaminated clothing and thoroughly wash the skin with soap and water immediately.

Observe the following rules:

1. Avoid drift from the application area to adjacent areas occupied by other crops, humans, livestock, or bodies of water.
2. Wear goggles, an approved respirator, and neoprene gloves when loading or mixing pesticides. Aerial applicators should be loaded by a ground crew.
3. Pour chemicals at a level well below the face to avoid splashing or spilling onto the face or eyes.
4. Have plenty of soap and water on hand to wash contaminated skin in the event of spilling.
5. Change clothing and bathe after the job is completed.
6. Know the insecticide, the symptoms of overexposure to it, and a physician who can be called quickly. In case symptoms appear (contracted pupils, blurred vision, nausea, severe headache, dizziness), stop operations at once and contact a physician.

THE WORKER PROTECTION STANDARD (WPS)

Glossary of terms for WPS:

Agricultural employer: Any person who:

- *employs or contracts for the services of agricultural workers* (including themselves and family members) for any type of compensation to perform tasks relating to the production of agricultural plants;
- *owns or operates an agricultural establishment* that uses such workers;

- *employs pesticide handlers* (including family members) for any type of compensation; or
- *is self-employed as a pesticide handler.*

Agricultural establishment: Any farm, greenhouse, nursery, or forest that produces agricultural plants.

Agricultural establishment owner: Any person who owns, leases, or rents an agricultural establishment covered by the Worker Protection Standard (WPS).

Agricultural worker: Any person employed by an agricultural employer to do tasks such as harvesting, weeding, or watering related to the production of agricultural plants on a farm, forest, nursery, or greenhouse. By definition, *agricultural workers* do *not* apply pesticides or handle pesticide containers or equipment. Any employee can be an agricultural worker while performing one task and a pesticide handler while performing a different task.

Immediate family: The spouse, children, stepchildren, foster children, parents, stepparents, foster parents, brothers and sisters of an agricultural employer.

Personal protective equipment (PPE): Clothing and equipment, such as goggles, gloves, boots, aprons, coveralls and respirators, that provide protection from exposure to pesticides.

Pesticide handler: Any person employed by an agricultural establishment to mix, load, transfer, or apply pesticides or do other tasks that involve direct contact with pesticides.

Restricted entry interval (REI): The time immediately after a pesticide application when entry into the treated area is limited. Lengths of restricted entry intervals range between 4 and 72 hours. The exact amount of time is product specific and indicated on the pesticide label.

Adapted from O. N. Nesheim and T. W. Dean, "The Worker Protection Standard," in S. M. Olson and E. H. Simonne (eds.), *Vegetable Production Handbook for Florida* (Florida Cooperative Extension Service, 2005–2006), http://edis.ifas.ufl.edu/CV138.

DESCRIPTION OF THE WORKER PROTECTION STANDARD

The worker protection standard (WPS) is a set of regulations issued by the U.S. Environmental Protection Agency (EPA) designed to protect agricultural workers and pesticide handlers from exposure to pesticides.

http://www.epa.gov/pesticides/safety/workers/PART170.htm

The WPS applies to all agricultural employers whose employees are involved in the production of agricultural plants on a farm, forest, nursery, or greenhouse. The employers include owners or managers of farms, forests, nurseries, or greenhouses as well as commercial (custom) applicators and crop advisors who provide services for the production of agricultural plants on these sites. The WPS requires that specific protections be provided to workers and pesticide handlers to prevent pesticide exposure. The agricultural employer is responsible for providing those protections to employees.

Information at a Central Location

An information display at a central location must be provided and contain:

1. An approved EPA safety poster showing how to keep pesticides from a person's body and how to clean up any contact with a pesticide
2. Name, address, and telephone number of the nearest emergency medical facility
3. Information about each pesticide application, including description of treated area, product name, EPA registration number of product, active ingredient of pesticide, time and date of application, and the restricted entry interval (REI) for the pesticide

Pesticide Safety Training

Employers must provide pesticide safety training for pesticide handlers and agricultural workers unless the handler or worker is a certified pesticide applicator. Workers must receive training within 5 days of beginning work.

Decontamination Areas Must Provide:

1. Water for washing and eye flushing
2. Soap
3. Single-use towels
4. Water for whole-body wash (pesticide handler sites only)
5. Clean coveralls (pesticide handler sites only)
6. Provide each handler at least 1 pt clean water for flushing eyes

Restricted Entry Interval

Pesticides have restricted entry intervals, a period after a pesticide application during which employers must keep everyone, except appropriately trained and equipped handlers, out of treated areas. Employers must orally inform all of their agricultural workers who will be within a quarter-mile of a treated area. For certain pesticides, treated fields must also be posted with a WPS poster.

Personal Protective Equipment

Personal protective equipment (PPE) must be provided by the employer to all pesticide handlers.

1. All PPE must be clean, in operating condition, used correctly, inspected each day before use, and repaired or replaced as needed.
2. Respirators must fit correctly and filters or canisters replaced at recommended intervals.
3. Handlers must be warned about symptoms of heat illness when wearing PPE.
4. Handlers must be provided clean, pesticide-free areas to store PPE.
5. Contaminated PPE must not be worn or taken home and must be cleaned separate from other laundry.
6. Employers are responsible for providing clean PPE for each day.
7. Coveralls contaminated with undiluted Danger or Warning category pesticide must be discarded.

Adapted from O. N. Nesheim and T. W. Dean, "The Worker Protection Standard," http://edis.ifas.ufl.edu/CV138, and O. N. Nesheim, "Interpreting PPE Statements on Pesticide Labels," in S. M. Olson and E. H. Simonne (eds.), *Vegetable Production Handbook for Florida* (Florida Cooperative Extension Service, 2005–2006), http://edis.ifas.ufl.edu/CV285.

Additional resource on the Worker Protection Standard:

http://www.epa.gov/pesticides/safety/workers/PART170.htm

ADDITIONAL INFORMATION ON SAFETY AND RULES AFFECTING PESTICIDE USE

Recordkeeping

Growers are required to keep records on applications of restricted-use pesticides.

http://www.environment.nsw.gov.au/envirom/recordkeeping.htm

Sample recordkeeping forms can be found at:

http://www.epa.nsw.gov.au/resources/pesticidesrkform.pdf

SARA Title III Emergency Planning and Community Right-to-Know Act

This law sets rules for farmers to inform the proper authorities about storage of extremely hazardous materials. Each state has related SARA Title III statutes.

Lists of chemicals are available at:

http://www.epa.gov/ceppo/pubs/title3.pdf

Right-to-Know

Employees have a right to know about chemical use on the farm. Right-to-know training is typically provided through the County Extension Office.

Endangered Species Act

The EPA has determined threshold pesticide application rates that may affect listed species. This information is presented on the pesticide label.

http://www.fws.gov/endangered/esa.html
http://www.epa.gov/region5/defs/html/esa.htm

Invasive Species

An *invasive species* is defined as a species that is non-native (or alien) to the ecosystem under consideration and whose introduction causes or is likely to cause economic or environmental harm, or harm to human health. Some websites:

http://www.fws.gov/contaminants/Issues/InvasiveSpecies.cfm
http://invasivespeciesinfo.gov
http://aquat1.ifas.ufl.edu
http://plants.nrcs.usda.gov/cgi_bin/topics.cgi?earl=noxious.cgi
http://www.invasive.org/

PESTICIDE HAZARDS TO HONEYBEES

Honeybees and other bees are necessary for pollination of vegetables in the gourd family—cantaloupe and other melons, cucumber, pumpkin, squash, and watermelon. Bees and other pollinating insects are necessary for all the insect-pollinated vegetables grown for seed production. Some pesticides are extremely toxic to bees and other pollinating insects, so certain precautions are necessary to avoid injury to them.

Recommendations for Vegetable Growers and Pesticide Applicators to Protect Bees

1. Participate actively in areawide integrated crop management programs.
2. Follow pesticide label directions and recommendations.
3. Apply hazardous chemicals in late afternoon, night, or early morning (generally 6 P.M. to 7 A.M.) when honeybees are not actively foraging. Evening applications are generally somewhat safer than morning applications.
4. Use pesticides that are relatively nonhazardous to bees whenever this is consistent with other pest-management strategies. Choose the least hazardous pesticide formulations or tank mixes.
5. Become familiar with bee foraging behavior and types of pesticide applications that are hazardous to bees.
6. Know the location of all apiaries in the vicinity of fields to be sprayed.
7. Avoid drift, overspray, and dumping of toxic materials in noncultivated areas.
8. Survey pest populations and be aware of current treatment thresholds in order to avoid unnecessary pesticide use.
9. Determine if bees are foraging in target areas so protective measures can be taken.

TOXICITY OF CHEMICALS USED IN PEST CONTROL

The danger in handling pesticides does not depend exclusively on toxicity values. Hazard is a function of both toxicity and the amount and type of exposure. Some chemicals are very hazardous from dermal (skin) exposure as well as oral (ingestion). Although inhalation values are not given, this type of exposure is similar to ingestion. A compound may be highly toxic but present little hazard to the applicator if the precautions are followed carefully.

Toxicity values are expressed as acute oral LD_{50} in terms of milligrams of the substance per kilogram (mg/kg) of test animal body weight required to kill 50% of the population. The acute dermal LD_{50} is also expressed in mg/kg. These acute values are for a single exposure and not for repeated exposures such as may occur in the field. Rats are used to obtain the oral LD_{50}, and the test animals used to obtain the dermal values are usually rabbits.

TABLE 6.2. CATEGORIES OF PESTICIDE TOXICITY[1]

		LD_{50} Value (mg/kg)	
Categories	Signal Word	Oral	Dermal
I	Danger–Poison	0–50	0–200
II	Warning	50–500	200–2,000
III	Caution	500–5,000	2,000–20,000
IV	None[2]	5,000	20,000

[1] EPA-accepted categories.
[2] No signal word required based on acute toxicity; however, products in this category usually display "Caution."

Resources on Toxicity of Pesticides

- *Commercial Vegetable Production Recommendations,* Maryland Agricultural Extension Service Bulletin 137 (2005).
- L. Moses, R. Meister, and C. Sine, *Farm Chemicals Handbook '99* (Willoughby, Ohio: Meister, 1999).

PESTICIDE FORMULATIONS

Several formulations of many pesticides are available commercially. Some are emulsifiable concentrates, flowables, wettable powders, dusts, and granules. After any pesticide recommendation, one of these formulations is suggested; however, unless stated to the contrary, equivalent rates of another formulation or concentration of that pesticide can be used. In most cases, pesticide experts suggest that sprays rather than dusts be applied to control pests of vegetables. This is because sprays allow better control and result in less drift.

PREVENTING SPRAY DRIFT

Pesticides sprayed onto soil or crops may be subject to movement or drift away from the target, due mostly to wind. Drift may lead to risks to nearby people and wildlife, damage to nontarget plants, and pollution of surface water or groundwater.

Factors That Affect Drift

1. *Droplet size.* Smaller droplets can drift longer distances. Droplet size can be managed by selecting the proper nozzle type and size, adjusting the spray pressure, and increasing the viscosity of the spray mixture.
2. *Equipment adjustments.* Routine calibration of spraying equipment and general maintenance should be practiced. Equipment can be fitted with hoods or shields to reduce drift away from the sprayed area.
3. *Weather conditions.* Spray applicators must be aware of wind speed and direction, relative humidity, temperature, and atmospheric stability at time of spraying.

Spraying Checklist to Minimize Drift

1. Do not spray on windy days ($>$12 mph).
2. Avoid spraying on extremely hot and dry days.
3. Use minimum required pressure.
4. Select correct nozzle size and spray pattern.
5. Keep the boom as close as possible to the target.
6. Install hoods or shields on the spray boom.
7. Leave an unsprayed border of 50–100 ft near water supplies, wetlands, neighbors, and non-target crops.
8. Spray when wind direction is favorable for keeping drift off of non-target areas.

ESTIMATION OF CROP AREA

To calculate approximately the acreage of a crop in the field, multiply the length of the field by the number of rows or beds. Divide by the factor for spacing of beds.

Examples: Field 726 ft long with 75 rows 48 in. apart.

$$\frac{726 \times 75}{10890} = 5 \text{ acres}$$

Field 500 ft long with 150 beds on 40-in. centers.

$$\frac{500 \times 150}{13068} = 5.74 \text{ acres}$$

TABLE 6.3. FACTORS USED IN CALCULATING TREATED CROP AREA

Row or Bed Spacing (in.)	Factor
12	43,560
18	29,040
24	21,780
30	17,424
36	14,520
40	13,068
42	12,445
48	10,890
60	8,712
72	7,260
84	6,223

TABLE 6.4. DISTANCE TRAVELED AT VARIOUS TRACTOR SPEEDS

mph	ft/min	mph	ft/min
1.0	88	3.1	273
1.1	97	3.2	282
1.2	106	3.3	291
1.3	114	3.4	299
1.4	123	3.5	308
1.5	132	3.6	317
1.6	141	3.7	325
1.7	150	3.8	334
1.8	158	3.9	343
1.9	167	4.0	352
2.0	176	4.1	361
2.1	185	4.2	370
2.2	194	4.3	378
2.3	202	4.4	387
2.4	211	4.5	396
2.5	220	4.6	405
2.6	229	4.7	414
2.7	237	4.8	422
2.8	246	4.9	431
2.9	255	5.0	440
3.0	264		

CALCULATIONS OF SPEED OF EQUIPMENT AND AREA WORKED

To review the actual performance of a tractor, determine the number of seconds required to travel a certain distance. Then use the formula

$$\text{speed (mph)} = \frac{\text{distance traveled (ft)} \times 0.682}{\text{time to cover distance (sec)}}$$

or the formula

$$\text{speed (mph)} = \frac{\text{distance traveled (ft)}}{\text{time to cover distance (sec)} \times 1.47}$$

Another method is to walk beside the machine counting the number of normal paces (2.93 ft) covered in 20 seconds. Move decimal point 1 place. Result equals tractor speed (mph).

Example: 15 paces/20 sec = 1.5 mph.

The working width of an implement multiplied by mph equals the number of acres covered in 10 hr. This includes an allowance of 17.5% for turning at the ends of the field. By moving the decimal point 1 place, which is equivalent to dividing by 10, the result is the acreage covered in 1 hr.

Example: A sprayer with a 20-ft boom is operating at 3.5 mph. Thus, 20 × 3.5 = 70 acres/10 hr or 7 acres/hr.

TABLE 6.5. APPROXIMATE TIME REQUIRED TO WORK AN ACRE[1]

Rate (mph):	1	2	3	4	5	10
Rate (ft/min):	88	176	264	352	440	880
Effective Working Width of Equipment (in.)	Approximate Time Required (min/acre)					
18	440	220	147	110	88	44
24	330	165	110	83	66	33
36	220	110	73	55	44	22
40	198	99	66	50	40	20
42	189	95	63	47	38	19
48	165	83	55	41	33	17
60	132	66	44	33	26	13
72	110	55	37	28	22	11
80	99	50	33	25	20	10
84	94	47	31	24	19	9
96	83	42	28	21	17	8
108	73	37	24	19	15	7
120	66	33	22	17	13	6
240	33	17	11	8	7	3
360	22	11	7	6	4	2

[1] These figures are calculated on the basis of 75% field efficiency to allow for turning and other lost time.

USE OF PESTICIDE APPLICATION EQUIPMENT

Ground Application

Boom-type Sprayers

High-pressure, high-volume sprayers have been used for row-crop pest control for many years. Now a trend exists toward the use of sprayers that utilize lower volumes and pressures.

Airblast-type Sprayers

Airblast sprayers are used in the vegetable industry to control insects and diseases. Correct operation of an airblast sprayer is more critical than for a boom-type sprayer. Do not operate an airblast sprayer under high wind conditions. Wind speed below 5 mph is preferable unless it becomes necessary to apply the pesticide for timely control measures, but drift and nearby crops must be considered. Do not overextend the coverage of the machine.

Considerable visible mist from the machine moves into the atmosphere and does not deposit on the plant. If in doubt, use black plastic indicator sheets in the rows to determine deposit and coverage before a pest problem appears as evidence. Use correct gallonage and pressures to obtain proper droplet size and ensure uniform coverage across the effective swath width. Adjust the vanes and nozzles on the sprayer unit to give best coverage. Vane adjustment must occur in the field, depending on terrain, wind, and crop. Cross-drives in the field allow the material to be blown down the rows instead of across them and help give better coverage in some crops, such as staked or trellised tomato.

Air-boom Sprayers

These sprayers combine the airblast spray with the boom spray delivery characteristics. Air-boom sprayers are becoming popular with vegetable producers in an effort to achieve high levels of spray coverage with minimal quantities of pesticide.

Electrostatic Sprayers

These sprayers create an electrical field through which the spray droplets move. Charged spray droplets are deposited more effectively onto plant surfaces, and less drift results.

Aerial Application

Spraying should not be done when wind is more than 6 mph. A slight crosswind during spraying is advantageous in equalizing the distribution of the spray within the swath and between swaths. Proper nozzle angle and arrangements along the boom are critical and necessary to obtain proper distribution at ground level. Use black plastic indicator sheets in the rows to determine deposit and coverage patterns. Cover a swath no wider than is reasonable for the aircraft and boom being used. Fields of irregular shape or topography and ones bounded by woods, power lines, or other flight hazards should not be sprayed by aircraft.

CALIBRATION OF FIELD SPRAYERS

Width of Boom

The boom coverage is equal to the number of nozzles multiplied by the space between two nozzles.

Ground Speed (mph)

Careful control of ground speed is important for accurate spray application. Select a gear and throttle setting to maintain constant speed. A speed of 2–3 mph is desirable. From a running start, mark off the beginning and ending of a 30-sec run. The distance traveled in this 30-sec period divided by 44 equals the speed in mph.

Sprayer Discharge (gpm)

Run the sprayer at a certain pressure, and catch the discharge from each nozzle for a known length of time. Collect all the discharge and measure the total volume. Divide this volume by the time in minutes to determine discharge in gallons per minute.

Before Calibrating

1. Thoroughly clean all nozzles, screens, etc., to ensure proper operation.
2. Check that all nozzles are the same.
3. Check the spray patterns of all nozzles for uniformity. Check the volume of delivery by placing similar containers under each nozzle. Replace nozzles that do not have uniform patterns or do not fill containers at the same rate.
4. Select an operating speed. Note the tachometer reading or mark the throttle setting. When spraying, be sure to use the same speed as used for calibrating.
5. Select an operating pressure. Adjust to desired pressure (pounds per square in. [psi]) while the pump is operating at normal speed and water is actually flowing through the nozzles. This pressure should be the same during calibration and field spraying.

Calibration (Jar Method)

Either a special calibration jar or a homemade one can be used. If you buy one, carefully follow the manufacturer's instructions.

Make accurate speed and pressure readings and jar measurements. Make several checks.

Any 1-qt or larger container, such as a jar or measuring cup, if calibrated in fl oz, can easily be used in the following manner:

1. Measure a course on the same type of surface (sod, plowed, etc.) and same type of terrain (hilly, level, etc.) as that to be sprayed, according to nozzle spacing as follows:

TABLE 6.6. COURSE LENGTH SELECTED BASED ON NOZZLE SPACING

Nozzle spacing (in.)	16	20	24	28	32	36	40
Course length (ft)	255	204	170	146	127	113	102

2. Time the seconds it takes the sprayer to cover the measured distance at the desired speed.
3. With the sprayer standing still, operate at selected pressure and pump speed. Catch the water from several nozzles for the number of seconds measured in step 2.
4. Determine the average output per nozzle in ounces. The ounces per nozzle equal the gal/acre applied for one nozzle per spacing.

Calibration (Boom or Airblast Sprayer)

1. Fill sprayer with water.
2. Spray a measured area (width of area covered × distance traveled) at constant speed and pressure selected from manufacturer's information.
3. Measure amount of water necessary to refill tank (gal used).
4. Multiply gallons used by 43,560 and divide by the number of sq ft in area sprayed. This gives gal/acre.

$$\text{gal/acre} = \frac{\text{gal used} \times 43{,}560}{\text{area sprayed (sq ft)}}$$

5. Add correct amount of spray material to tank to give the recommended rate per acre.

EXAMPLE

Assume: 10 gal water used to spray an area 660 ft long and 20 ft wide
Tank size: 100 gal
Spray material: 2 lb (actual)/acre
Calculation:

$$\frac{\text{gal used} \times 43{,}560}{\text{area sprayed (sq ft)}} = \frac{10 \times 43{,}560}{660 \times 20} = 33 \text{ gal/acre}$$

$$\frac{\text{tank capacity}}{\text{gal/acre}} \quad \frac{100 \text{ (tank size)}}{33} = 3.03 \text{ acres sprayed per tank}$$

3.03×2 (lb/acre) $= 6.06$ lb material per tank

If 80% material is used:

$$\frac{6.06}{0.8} = 7.57 \text{ lb material needed per tank to give 2 lb/acre rate}$$

Adapted from *Commercial Vegetable Production Recommendations* (Maryland Agricultural Extension Service Bulletin 137, 2005), http:///hortweb.cas.psu.edu/extension/images/PA%2005%20commercial%20veg%20Recommends.pdf.

CALIBRATION OF GRANULAR APPLICATORS

Sales of granular fertilizer, herbicides, insecticides, etc., for application through granular application equipment have been on the increase.

Application rates of granular application equipment are affected by several factors: gate openings or settings, ground speed of the applicator, shape and size of granular material, and roughness of the ground.

Broadcast Application

1. From the label, determine the application rate.
2. From the operator's manual, set the dial or feed gate to apply desired rate.
3. On a level surface, fill the hopper to a given level and mark this level.
4. Measure the test area—length of run depends on size of equipment. It need not be one long run but can be multiple runs at shorter distances.

5. Apply material to measured area, operating at the speed applicator will travel during application.
6. Weigh the amount of material required to refill the hopper to the marked level.
7. Determine the application rate:

$$\text{area covered} = \frac{\text{number of runs} \times \text{length of run} \times \text{width of application}}{43{,}560}$$

$$\text{application rate} = \frac{\text{amount applied (pounds to refill hopper)}}{\text{area covered}}$$

Note: Width of application is width of the spreader for drop or gravity spreaders. For spinner applicators, it is the working width (distance between runs). Check operator's manual for recommendations, generally one-half to three-fourths of overall width spread.

Example:
Assume: 50 lb/acre rate
Test run: 200 ft
Four runs made
Application width: 12 ft
11.5 lb to refill hopper

Calculation:

$$\text{area covered} = \frac{4 \times 200 \times 12}{43{,}560} = 0.22 \text{ acre}$$

$$\text{application rate} = \frac{11.5}{0.22} = 52.27 \text{ lb}$$

8. If application rate is not correct, adjust feed gate opening and recheck.

Band Application

1. From the label, determine the application rate.
2. From the operator's manual, determine the applicator setting and adjust accordingly.

3. Fill the hopper half full.
4. Operate the applicator until all units are feeding.
5. Stop the applicator; remove the feed tubes at the hopper.
6. Attach paper or plastic bags over the hopper openings.
7. Operate the applicator over a measured distance at the speed equipment will be operated.
8. Weigh and record the amount delivered from each hopper. (Compare to check that all hoppers deliver the same amount.)
9. Calculate the application rate:

$$\text{area covered in bands} = \frac{\text{length of run} \times \text{band width} \times \text{number of bands}}{43{,}560}$$

Application Rate

$$\frac{\text{amount applied in bands} = \text{total amount collected}}{\text{area covered in bands}}$$

Changing from Broadcast to Band Application

$$\frac{\text{band width in in.}}{\text{row spacing in in.}} \times \begin{array}{l}\text{broadcast}\\ \text{rate}\\ \text{per acre}\end{array} = \text{amount needed per acre}$$

Adapted from *Commercial Vegetable Production Recommendations* (Maryland Agricultural Extension Service Bulletin 137, 2005), http:///hortweb.cas.psu.edu/extension/images/PA%2005%20commercial%20veg%20Recommends.pdf.

CALIBRATION OF AERIAL SPRAY EQUIPMENT

Calibration

$$\text{acres covered} = \frac{\text{length of swath (miles)} \times \text{width (ft)}}{8.25}$$

$$\text{acres/min} = \frac{2 \times \text{swath width} \times \text{mph}}{1000}$$

$$\text{GPM} = \frac{2 \times \text{swath width} \times \text{mph} \times \text{gal/acre}}{1000}$$

Adapted from O. C. Turnquist et al., *Weed, Insect, and Disease Control Guide for Commercial Vegetable Growers,* Minnesota Agricultural Extension Service Special Report 5 (1978).

Resources on Calibration of Aerial Sprayers

- D. Overhults, *Applicator Training Manual for Aerial Application of Pesticides* (University of Kentucky), http://www.uky.edu/Agriculture/PAT/Cat11/Cat11.htm.
- *Agricultural Aircraft Calibration and Setup for Spraying* (Kansas State University Publication MF-1059, 1992), http://www.oznet.ksu.edu/library/ageng2/mf1059.pdf.

CALIBRATION OF DUSTERS

Select a convenient distance that multiplied by the width covered by the duster, both expressed in feet, equals a convenient fraction of an acre. With the hopper filled to a marked level, operate the duster at this distance. Take a known weight of dust in a bag or other container and refill hopper to the marked level. Weigh the dust remaining in the container. The difference is the quantity of dust applied to the fraction of an acre covered.

Example:

Distance duster is operated × width covered by duster = area dusted

$$= 108.9 \text{ ft} \times 10 \text{ ft} = 1089 \text{ sq ft} \frac{1089 \text{ sq ft}}{43{,}560} = \frac{1}{40} \text{ acre}$$

If it takes 1 lb dust to refill the hopper, the rate of application is 40 lb/acre.

MORE INFORMATION ON CALIBRATION OF SPRAYERS

Florida

http://edis.ifas.ufl.edu/WG013
http://edis.ifas.ufl.edu/TOPIC_Herbicide_Calibration_and_Application
http://edis.ifas.ufl.edu/TOPIC_Calibration

Minnesota

http://www.extension.umn.edu/distribution/cropsystems/DC3885.html

North Dakota

http://www.ext.nodak.edu/extpubs/ageng/machine/ae73-5.htm

Other

http://www.dupont.com/ag/vm/literature/K-04271.pdf

HAND SPRAYER CALIBRATION

Ohio

http://ohioline.osu.edu/for-fact/0020.html

Colorado

http://www.co.larimer.co.us/publicworks/weeds/calibration.htm

SPRAY EQUIVALENTS AND CONVERSIONS

Pesticide containers give directions usually in terms of pounds or gallons of material in 100 gal water. The following tables make the conversion for smaller quantities of spray solution easy.

TABLE 6.7. SOLID EQUIVALENT TABLE

100 gal	25 gal	5 gal	1 gal
4 oz	1 oz	3⁄16 oz	½ oz
8 oz	2 oz	⅜ oz	1 tsp
1 lb	4 oz	⅞ oz	2 tsp
2 lb	8 oz	1¾ oz	4 tsp
3 lb	12 oz	2⅜ oz	2 tbsp
4 lb	1 lb	3¼ oz	2 tbsp + 2 tsp

TABLE 6.8. LIQUID EQUIVALENT TABLE

100 gal	25 gal	5 gal	1 gal
1 gal	1 qt	6½ oz	1¼ oz
2 qt	1 pt	3¼ oz	⅝ oz
1 qt	½ pt	1 9⁄16 oz	5⁄16 oz
1½ pt	6 oz	1¼ oz	¼ oz
1 pt	4 oz	⅞ oz	3⁄16 oz
8 oz	2 oz	7⁄16 oz	½ tsp
4 oz	1 oz	¼ oz	¼ tsp

TABLE 6.9. DILUTION OF LIQUID PESTICIDES TO VARIOUS CONCENTRATIONS

Dilution	1 gal	3 gal	5 gal
1:100	2 tbsp + 2 tsp	½ cup	¾ cup + 5 tsp
1:200	4 tsp	¼ cup	6½ tbsp
1:800	1 tsp	1 tbsp	1 tbsp + 2 tsp
1:1000	¾ tsp	2½ tsp	1 tbsp + 1 tsp

TABLE 6.10. PESTICIDE DILUTION CHART

Amount of formulation necessary to obtain various amounts of active ingredients.

Insecticide Formulation	Amount of formulation (at left) needed to obtain the following amounts of active ingredients			
	¼ lb	½ lb	¾ lb	1 lb
1% dust	25	50	75	100
2% dust	12½	25	37½	50
5% dust	5	10	15	20
10% dust	2½	5	7½	10
15% wettable powder	1⅔ lb	3⅓ lb	5 lb	6⅔ lb
25% wettable powder	1 lb	2 lb	3 lb	4 lb
40% wettable powder	⅝ lb	1¼ lb	1⅞ lb	2½ lb
50% wettable powder	½ lb	1 lb	1½ lb	2 lb
73% wettable powder	⅓ lb	⅔ lb	1 lb	1⅓ lb
23–25% liquid concentrate (2 lbs active ingredient/gal)	1 pt	1 qt	3 pt	2 qt
42–46% liquid concentrate (4 lbs active ingredient/gal)	½ pt	1 pt	1½ pt	1 qt
60–65% liquid concentrate (6 lbs active ingredient/gal)	⅓ pt	⅔ pt	1 pt	1⅓ pt
72–78% liquid concentrate (8 lbs active ingredient/gal)	¼ pt	½ pt	¾ pt	1 pt

TABLE 6.11. PESTICIDE APPLICATION RATES FOR SMALL CROP PLANTINGS

Distance Between Rows (ft)	Amount (gal/acre)	Amount (per 100 ft of row)	Length of Row Covered (ft/gal)
1	75	22 oz	581
	100	30 oz	435
	125	1 qt 5 oz	348
	150	1 qt 12 oz	290
	175	1 qt 20 oz	249
	200	1 qt 27 oz	218
2	75	1 qt 12 oz	290
	100	1 qt 27 oz	218
	125	2 qt 10 oz	174
	150	2 qt 24 oz	145
	175	3 qt 7 oz	124
	200	3 qt 21 oz	109
3	75	2 qt 2 oz	194
	100	2 qt 24 oz	145
	125	3 qt 14 oz	116
	150	4 qt 4 oz	97
	175	4 qt 26 oz	83
	200	5 qt 16 oz	73

GUIDELINES FOR EFFECTIVE PEST CONTROL

Often, failure to control a pest is blamed on the pesticide when the cause lies elsewhere. Some common reasons for failure follow:

1. Delaying applications until pests are already well established.
2. Making applications with insufficient gallonage or clogged or poorly arranged nozzles.
3. Selecting the wrong pesticide.

The following points are suggested for more effective pest control:

1. *Inspect fields regularly.* Frequent examinations (at least twice per week) help determine the proper timing of the next pesticide application.
2. *Control insects and mites according to economic thresholds or schedule.* Economic thresholds assist in determining whether pesticide applications or other management actions are needed to avoid economic loss from pest damage. Thresholds for insect pests are generally expressed as a numerical count of a given life stage or as a damage level based on certain sampling techniques. They are intended to reflect the population size that would cause economic damage and thus warrant the cost of treatment. Guidelines for other pests are usually based on the field history, crop development, variety, weather conditions, and other factors.

   Rather than using economic thresholds, many pest problems can be predicted to occur at approximately the same time year after year. One application before buildup often eliminates the need for several applications later in the season. Often, less toxic and safer-to-handle chemicals are effective when pests are small in size and population.
3. *Take weather conditions into account.* Spray only when wind velocity is less than 10 mph. Dust only when it is perfectly calm. Do not spray when sensitive plants are wilted during the heat of the day. If possible, make applications when ideal weather conditions prevail.

   Biological insecticides are ineffective in cool weather. Some pyrethroid insecticides (permethrin) do not perform well when field temperatures reach 85°F and above. Best control results from these insecticides are achieved when the temperature is in the 70s or low 80s (evening or early morning).

   Sprinkler irrigation washes pesticide deposits from foliage. Wait at least 48 hr after pesticide application before sprinkler irrigating. More frequent pesticide applications may be needed during and after periods of heavy rainfall.
4. *Strive for adequate coverage of plants.* The principal reason aphids, mites, cabbage loopers, and diseases are serious pests is that they occur beneath leaves, where they are protected from spray deposit or dust particles. Improved control can be achieved by adding and arranging nozzles so the application is directed toward the plants from the sides as well as from the top. In some cases, nozzles should be arranged so the application is directed beneath the leaves. As the season progresses, plant size increases, and so does the need for increased spray gallonage to ensure adequate coverage. Applying sprays with sufficient spray volume and pressure is important. Sprays from high-volume, high-pressure rigs (airblast) should be applied at

40–100 gal/acre at approximately 400 psi pressure. Sprays from low-volume, low-pressure rigs (boom type) should be applied at 50–100 gal/acre at approximately 100–300 psi pressure.

5. *Select the proper pesticide.* Know the pests to be controlled and choose the recommended pesticide and rate of application.

   For certain pests that are extremely difficult to control or are resistant, it may be important to alternate labeled insecticides, especially with different classes of insecticides; for example, alternate a pyrethroid insecticide with either a carbamate or an organophosphate insecticide.

6. *Assess pesticide compatibility.* To determine if two pesticides are compatible, use the following jar test before tank-mixing pesticides or pesticides and fluid fertilizers:
   a. Add 1 pt water or fertilizer solution to a clean quart jar. Then add the pesticide to the water or fertilizer solution in the same proportion as used in the field.
   b. To a second clean quart jar, add 1 pt water or fertilizer solution. Then add ½ tsp of an adjuvant to keep the mixture emulsified. Finally, add the pesticide to the water-adjuvant or fertilizer-adjuvant in the same proportion as to be used in the field.
   c. Close both jars tightly and mix thoroughly by inverting 10 times. Inspect the mixtures immediately and again after standing for 30 min. If the mix in either jar remains uniform for 30 min, the combination can be used. If either mixture separates but readily remixes, constant agitation is required. If nondispersible oil, sludge, or clumps of solids form, do not use the mixture.

7. *Calibrate application equipment.* Periodic calibration of sprayers, dusters, and granule distributors is necessary to ensure accurate delivery rates of pesticides per acre. See pages 333–338.

8. *Select correct sprayer tips.* The selection of proper sprayer tips for use with various pesticides is very important. Flat fan-spray tips are designed for preemergence and postemergence application of herbicides. They can also be used with insecticides, fertilizers, and other pesticides. Flat fan-spray tips produce a tapered-edge spray pattern for uniform coverage where patterns overlap. Some flat fan-spray tips (SP) are designed to operate at low pressure (15–40 psi) and are usually used for preemergence herbicide applications. These lower pressures result in large spray particles than those from standard flat tips operating at higher pressures (30–60 psi). Spray nozzles with even flat-spray tips (often designated E) are designed for band spraying where uniform distribution is desired over a zone 8–14

in. wide; they are generally used for herbicides.

Flood-type nozzle tips are generally used for complete fertilizer, liquid nitrogen, and so on, and sometimes for spraying herbicides onto the soil surface prior to incorporation. They are less suited for spraying postemergence herbicides or for applying fungicides or insecticides to plant foliage. Coverage of the target is often less uniform and complete when flood-type nozzles are used, compared with the coverage obtained with other types of nozzles. Place flood-type nozzles a maximum of 20 in. apart, rather than the standard 40-in. spacing, for better coverage. This results in an overlapping spray pattern. Spray at the maximum pressure recommended for the nozzle.

Wide-spray angle tips with full or hollow cone patterns are usually used for fungicides and insecticides. They are used at higher water volume and spray pressures than are commonly recommended for herbicide application with flat fan or flood-type nozzle tips.

9. *Consider the relationship of pH and pesticides.* Unsatisfactory results with some pesticides may be related to the pH of the mixing water. Some materials carry a label cautioning the user against mixing the pesticide with alkaline materials because they undergo a chemical reaction known as *alkaline hydrolysis.* This reaction occurs when the pesticide is mixed in water with a pH greater than 7. Many manufacturers provide information on the rate at which their product hydrolyzes. The rate is expressed as *half-life,* meaning the time it takes for 50% hydrolysis or breakdown to occur. Check the pH of the water. If acidification is necessary, use one of the several commercial nutrient buffer materials available on the market.

Adapted from *Commercial Vegetable Production Recommendations, Pennsylvania, Delaware, Maryland, Virginia, and New Jersey* (2005), http://hortweb.cas.psu.edu/extension/images/PA%2005%20Commercial%20Veg%20Recommends.pdf.

SPRAY ADJUVANTS OR ADDITIVES

Adjuvants are chemicals that, when added to a liquid spray, make it mix, wet, spread, stick, or penetrate better. Water is almost a universal diluent for pesticide sprays. However, water is not compatible with oily pesticides, and an *emulsifier* may be needed in order to obtain good mixing. Furthermore, water from sprays often remains as large droplets on leaf surfaces. A *wetting agent* lowers the interfacial tension between the spray droplet and the leaf surface and thus moistens the leaf. *Spreaders* are closely related to wetters and help build a deposit on the leaf and improve weatherability. *Stickers* cause pesticides to adhere to the sprayed surface and are often called spray-stickers. They are oily and serve to increase the amounts of suspended solids held on the leaves or fruits by holding the particles in a resin-like film. *Extenders* form a sticky, elastic film that holds the pesticide on the leaves and thus reduces the rate of loss caused by sunlight and rainfall.

There are a number of adjuvants on the market. Read the label not only for dosages but also for crop uses and compatibilities, because some adjuvants must not be used with certain pesticides. Although many formulations of pesticides contain adequate adjuvants, some do require additions on certain crops, especially cabbage, cauliflower, onion, and pepper.

Spray adjuvants for use with herbicides often serve a function distinct from that of adjuvants used with insecticides and fungicides. For example, adjuvants such as oils used with atrazine greatly improve penetration of the chemical into crop and weed leaves, rather than just give more uniform coverage. Do not use any adjuvant with herbicides unless there are specific recommendations for its use. Plant damage or even crop residues can result from using an adjuvant that is not recommended.

Resources

- J. Witt, *Agricultural Spray Adjuvants* (Cornell University Pest Management Educational Program), http://pmep.cce.cornell.edu/facts-slides-self/facts/gen-peapp-adjuvants.html
- B. Young, *Compendium of Herbicide Adjuvants,* 7th ed. (Southern Illinois University, 2004), http://www.herbicide-adjuvants.com/index-7th-edition.html.

06
VEGETABLE SEED TREATMENTS

Various vegetable seed treatments prevent early infection by seedborne diseases, protect the seed from infection by soil microorganisms, and guard against a poor crop stand or crop failure caused by attacks on seeds by soil insects. Commercial seed is often supplied with the appropriate treatment.

Two general categories of vegetable seed treatments are used. *Eradication treatments* kill disease-causing agents on or within the seed, whereas *protective treatments* are applied to the surface of the seed to protect against seed decay, damping off, and soil insects. Hot-water treatment is the principal means of eradication, and chemical treatments usually serve as protectants. Follow time-temperature directions precisely for hot-water treatment and label directions for chemical treatment. When insecticides are used, seeds should also be treated with a fungicide.

HOT-WATER TREATMENT

To treat seeds with hot water, fill cheesecloth bags half full, wet seed and bag with warm water, and treat at exact time and temperature while stirring to maintain a uniform temperature. Use an accurate thermometer.

TABLE 6.12. HOT-WATER TREATMENT OF SEEDS

Seed	Temperature (°F)	Time (min)	Diseases Controlled
Broccoli, cauliflower, collards, kale, kohlrabi, turnip	122	20	Alternaria, blackleg, black rot
Brussels sprouts, cabbage	122	25	Alternaria, blackleg, black rot
Celery	118	30	Early blight, late blight
Eggplant	122	30	Phomopsis blight, anthracnose
Pepper	122	25	Bacterial spot, rhizoctonia
Tomato	122	25	Bacterial canker, bacterial spot, bacterial speck
	132	30	Anthracnose

CHEMICAL SEED TREATMENTS

Several fungicides are commonly used for treating vegetable seed. They protect against fungal attack during the germination process, resulting in more uniform plant stands. Seed protectants are most effective under cool germination conditions in a greenhouse or field, when germination is likely to be slow, and where germinating seeds might be exposed to disease-causing organisms. Seed treatments are applied as a dust or slurry and, when dry, can be dusty. To reduce dust, the fungicide-treated seed can be covered by a thin polymer film; many seed companies offer film-coated seeds. Large-seeded vegetables may require treatment with a labeled insecticide as well as a fungicide. Always follow label directions when pesticides are used.

Certain bacterial diseases on the seed surface can be controlled by other chemical treatments:

1. *Tomato bacterial canker*. Soak seeds in 1.05% sodium hypochlorite solution for 20–40 min or 5% hydrochloric acid for 5–10 hr. Rinse and dry.
2. *Tomato bacterial spot*. Soak seeds in 1.3% sodium hypochlorite for 1 min. Rinse and dry.
3. *Pepper bacterial spot*. Soak seeds in 1.3% sodium hypochlorite for 1 min. Rinse and dry.

DO NOT USE CHEMICALLY TREATED SEED FOR FOOD OR FEED.

Adapted from A. F. Sherf and A. A. MacNab, *Vegetable Diseases and Their Control*, Wiley, New York (1986).

Additional References for Seed Treatment

- M. McGrath, *Treatments for Managing Bacterial Pathogens in Vegetable Seed* (2005), http://vegetablemdonline.ppath.cornell.edu/NewsArticles/All_BactSeed.htm
- J. Boucher and J. Nixon, *Preventing Bacterial Diseases of Vegetables with Hot-water Seed Treatment,* University of Connecticut Cooperative Extension Service; and R. Hazard and R. Wick (University of Massachusetts Cooperative Extension Service), http://www.hort.uconn.edu/ipm/homegrnd/htms/54sedtrt.htm
- S. Miller and M. Lewis Ivey, *Hot Water and Chlorine Treatment of Vegetable Seeds to Eradicate Bacterial Plant Pathogens* (Cooperative Extension Service, Ohio State University), http://ohioline.osu.edu/hyg-fact/3000/3085.html

ORGANIC SEED TREATMENTS

With the increasing interest in organic vegetables, the need for information about organic seed production, sources, and treatment is greater. Seed quality and vigor are important aspects for high-quality organic seeds. The seed industry is working to provide vegetable seeds that meet the requirements for organic crop production. Likewise, the seed treatment technology for organic seeds is increasing in scope—for example, seed coating and treatment to increase germination uniformity and adaptation to mechanized seeding, among other needs. The following lists a few publications that address organic seed quality and seed treatment:

- J. Bonina and D. J. Cantliffe, *Seed Production and Seed Sources of Organic Vegetables* (University of Florida Cooperative Extension Service), http://edis.ifas.ufl.edu/hs227
- S. Koike, R. Smith, and E. Brennan, *Investigation of Organic Seed Treatments for Spinach Disease Control,* http://vric.ucdavis.edu/scrp/sum-koike.html

07
NEMATODES

PLANT PARASITIC NEMATODES

Nematodes are unsegmented round worms that range in size from microscopic to many inches long. Some nematodes, usually those that are microscopic or barely visible without magnification, attack vegetable crops and cause maladies, restrict yields, or, in severe cases, lead to total crop failure. Many kinds of nematode are known to infest the roots and aboveground plant parts of vegetable crops. Their common names are usually descriptive of the affected plant part and the resulting injury.

TABLE 6.13. COMMONLY OBSERVED NEMATODES IN VEGETABLE CROPS

Common Name	Scientific Name
Awl nematode	*Dolichodorus* spp.
Bud and leaf nematode	*Aphelenchoides* spp.
Cyst nematode	*Heterodera* spp.
Dagger nematode	*Xiphinema* spp.
Lance nematode	*Hopolaimus* spp.
Root-lesion nematode	*Pratylenchus* spp.
Root-knot nematode	*Meloidogyne* spp.
Spiral nematode	*Helicotylenchus* spp. and *Scutellonema* spp.
Sting nematode	*Belonolaimus* spp.
Stubby-root nematode	*Trichodorus* spp.
Stunt nematode	*Tylenchorhynchus* spp.

Nematodes are most troublesome in areas with mild winters where soils are not subject to freezing and thawing. Management practices and chemical control are both required to keep nematode numbers low enough to permit normal plant growth where populations are not kept in check naturally by severe winters.

The first and most obvious control for nematodes is avoiding their introduction into uninfected fields or areas. This may be done by quarantine over large geographical areas or by means of good sanitation in smaller areas. A soil sample for a nematode assay through the County Extension Service can provide information on which nematodes are present and their population levels. This information is valuable for planning a nematode management program.

Once nematodes have been introduced into a field, several management practices help control them: rotating with crops that a particular species of nematode does not attack, frequent disking during hot weather, and alternating flooding and drying cycles.

If soil management practices are not possible or are ineffective, chemicals (nematicides) may have to be used to control nematodes. Some fumigants are effective against soilborne disease, insects, and weed seeds; these are termed *multipurpose soil fumigants.* Growers should select a chemical for use against the primary problem to be controlled and use it according to label directions.

TABLE 6.14. PLANT PARASITIC NEMATODES KNOWN TO BE OF ECONOMIC IMPORTANCE TO VEGETABLES

Nematode	Bean and Peas	Carrot	Celery	Crucifers	Cucurbits	Leaf Crops	Okra	Onion	Potato	Sweet Corn	Sweet Potato	Tomato	Pepper	Eggplant
Root knot	X	X	X	X	X	X	X	X	X		X	X	X	X
Sting	X	X	X	X	X	X	X	X	X	X		X	X	X
Stubby root	X		X	X		X		X	X	X		X	X	X
Root lesion										X				
Cyst				X										
Awl	X		X							X				
Stunt										X				
Lance										X				
Spiral														
Ring														
Dagger														
Bud and leaf														
Reniform	X										X			

From J. Noling, "Nematodes and Their Management," in S. M. Olson and E. H. Simonne (eds.), *Vegetable Production Handbook for Florida* (Florida Cooperative Extension Service, 2005–2006), http://edis.ifas.ufl.edu/CV112.

MANAGEMENT TECHNIQUES FOR CONTROLLING NEMATODES

1. *Crop rotation.* Exposing a nematode population to an unsuitable host crop is an effective means of reducing nematodes in a field. Cover crops should be established rapidly and kept weed free. There are many cover crop options, but growers must consider the species of nematode in question and how long the alternate crop is needed. Also, certain cover crops may be effective as non-hosts for nematodes but may not fit into a particular cropping sequence.
2. *Fallowing.* Practicing clean fallowing in the intercropping season is probably the single most effective nonchemical means to reducing nematode populations. Clean disking of the field must be practiced frequently to keep weeds controlled because certain nematodes can survive on weeds.
3. *Plant resistance.* Certain varieties have genetic resistance to nematode damage. Where possible, these varieties should be selected when their other horticultural traits are acceptable. There are nematode-resistant varieties in tomato and pepper.
4. *Soil amendments.* Certain soil amendments, such as compost, manures, cover crops, chitin, and other materials, have been shown to reduce nematode populations. The age of the material, amount applied, and level of incorporation affect performance.
5. *Flooding.* Extended periods of flooding can reduce nematode populations. This technique should be practiced only where flooding is approved by environmental agencies. Alternating periods of flooding and drying seems to be more effective than a single flooding event.
6. *Soil solarization.* Nematodes can be killed by elevated heat created in the soil by covering it with clear polyethylene film for extended periods. Solarization works best in clear, hot, and dry climates. More information on solarization can be found on pages 317–319.
7. *Crop management.* Growers should rapidly destroy and till crops that are infested with nematodes. Irrigation water should not drain from infested fields to noninfested fields, and equipment should be cleaned between infested and clean fields.
8. *Chemical control.* Some fumigant and nonfumigant nematicides are approved for use against nematodes, but not all vegetable crops have nematicides recommended for use. Effectiveness of the treatment depends on many factors, including timing, placement in the soil, soil moisture, soil temperature, soil type, and presence of plastic mulch. Growers should refer to their state Extension recommendations for the approved chemicals for particular crops.

Adapted from J. Noling, "Nematodes and Their Management," in S. M. Olson and E. H. Simonne (eds.), *Vegetable Production Handbook for Florida* (Florida Cooperative Extension Service, 2005–2006), http://edis.ifas.ufl.edu/CV112.

08
DISEASES

GENERAL DISEASE CONTROL PROGRAM

Diseases of vegetable crops are caused by fungi, bacteria, viruses, and mycoplasms. For a disease to occur, organisms must be transported to a susceptible host plant. This may be done by infected seeds or plant material, contaminated soil, wind, water, animals (including humans), or insects. Suitable environmental conditions must be present for the organism to infect and thrive on the crop plant. Effective disease control requires knowledge of the disease life cycle, time of likely infection, agent of distribution, plant part affected, and the symptoms produced by the disease.

Crop rotation: Root-infecting diseases are the usual targets of crop rotation, although rotation can help reduce innocula of leaf- and stem-infecting organisms. Land availability and costs are making rotation challenging, but a well-planned rotation program is still an important part of an effective disease control program.

Site selection: Consider using fields that are free of volunteer crops and perimeter weeds that may harbor disease organisms. If aerial applications of fungicides are to be used, try to select fields that are geometrically adapted to serial spraying (long and wide), are away from homes, and have no bordering trees or power lines.

Deep plowing: Use tillage equipment such as plows to completely bury plant debris in order to fully decompose plant material and kill disease organisms.

Weed control: Certain weeds, particularly those botanically related to the crop, may harbor disease agents, especially viruses, that could move to the crop. Also, weeds within the crop could harbor diseases and by their physical presence interfere with deposition of fungicides on the crop. Volunteer plants from previous crops should be carefully controlled in nearby fallow fields.

Resistant varieties: Where possible, growers should choose varieties that carry genetic resistance to disease. Varieties with disease resistance require less pesticide application.

Seed protection: Seeds can be treated with fungicides to offer some degree of protection of the young seedling against disease attack. Seeds planted in warm soil germinate fast and possibly can outgrow disease development.

Healthy transplants: Growers should always purchase or grow disease-free transplants. Growers should contract with good transplant growers and should inspect their transplants before having them shipped to the farm. Paying a little extra to a reputable transplant producer is good insurance.

Field observation: Check fields periodically for disease development by walking the field and inspecting the plants up close, not from behind the windshield. Have any suspicious situations diagnosed by a competent disease diagnosis laboratory.

Foliar fungicides: Plant disease outbreaks sometimes can be prevented or minimized by timely use of fungicides. For some diseases, it is essential to have a preventative protectant fungicide program in place. For successful fungicide control, growers should consider proper chemical selection, use well-calibrated sprayers, use correct application rate, and follow all safety recommendations for spray application.

Integrated approach: Successful vegetable growers use a combination of these strategies or the entire set of strategies listed above. Routine implementation of combinations of strategies is needed where a grower desires to reduce the use of fungicides on vegetable crops.

COMMON VEGETABLE DISEASES

Some of the more common vegetable diseases are described below. Consult your local County Extension Service for recommendations for specific fungicide information, as products and use recommendations change frequently. When using fungicides, read the label and carefully follow the instructions. Do not exceed maximum rates given, observe the interval between application and harvest, and apply only to those crops for which use is approved. Make a record of the product used, trade name, concentration of the fungicide, dilution, rate applied per acre, and dates of application. Follow local recommendations for efficacy and read the directions on the label for proper use.

TABLE 6.15. DISEASE CONTROL FOR VEGETABLES

Crop	Disease	Description	Control
Asparagus	Fusarium root rot	Damping off of seedlings. Yellowing, stunting, or wilting of the growing stalks; vascular bundle discoloration. Crown death gives fields a spotty appearance.	Use disease-free crowns. Select fields where asparagus has not grown for 8 years. Use a fungicide crown dip before planting.
	Rust	Reddish or black pustules on stems and foliage.	Cut and burn diseased tops. Use resistant varieties. Use approved fungicides.
Bean	Anthracnose	Brown or black sunken spots with pink centers on pods, dark red or black cankers on stems and leaf veins.	Use disease-free seed and rotate crops every 2 years. Plow stubble. Do not cultivate when plants are wet. Use approved fungicides.
	Bacterial blight	Large, dry, brown spots on leaves, often encircled by yellow border; water-soaked spots on pods; reddish cankers on stems. Plants may be girdled.	Use disease-free seed. Do not cultivate when plants are wet. Use 3-year rotation. Use approved fungicides.
	Mosaic (several)	Mottled (light and dark green) and curled leaves; stunting, reduced yields.	Use mosaic-resistant varieties. Control weeds in areas adjacent to field. Control insect (aphid, white fly) carrier with insecticides.
	Powdery mildew	Faint, slightly discolored spots appear first on leaves, later on stems and pods, from which white powdery spots develop and may cover the entire plant.	Use approved fungicides.

	Rust	Red to black pustules on leaves; leaves yellow and drop.	Use approved fungicides.
	Seed rot	Seed or seedling decay, which results in poor stands. Occurs most commonly in cold, wet soils.	Crop rotation. Treat seed with approved fungicides.
	White mold	Water-soaked spots on plants. White, cottony masses on pods.	Use approved fungicides.
Beet	Cercospora	Numerous light tan to brown spots with reddish to dark brown borders on leaves.	Long rotation. Use approved fungicides.
	Damping off	Seed decay in soil; young seedlings collapse and die.	Avoid wet soils, rotate crops. Treat seed with approved fungicides.
	Downy mildew	Lighter than normal leaf spots on upper surface and white mildew areas on lower side. Roots, leaves, flowers, and seed balls distorted on stecklings.	Use approved fungicides.
Broccoli, Brussels sprouts, cabbage, cauliflower, kale, kohlrabi	Alternaria leaf spot	Damping off of seedlings. Small, circular yellow areas that enlarge in concentric circles and become black and sooty.	Use approved fungicides.
	Black leg	Sunken areas on stem near ground line resulting in girdling; gray spots speckled with black dots on leaves and stems.	Use hot-water-treated seed and long rotation. Sanitation.

TABLE 6.15. DISEASE CONTROL FOR VEGETABLES (*Continued*)

Crop	Disease	Description	Control
	Black rot	Yellowing and browning of the foliage; blackened veins; stems show blackened ring when cross-sectioned.	Use hot-water-treated seed and long rotation. Do not work wet fields. Sanitation.
	Club root	Yellow leaves or green leaves that wilt on hot days; large, irregular swellings or clubs on roots.	Start plants in new, steamed, or fumigated plant beds. Adjust soil pH to 7.2 with hydrated lime before planting. Use approved fungicides.
	Downy mildew	Begins as slight yellowing on upper side of leaves; white mildew on lower side; spots enlarge until plant dies.	Use approved fungicides.
	Fusarium yellows	Yellowish green leaves; stunted plants; lower leaves drop.	Use yellows-resistant varieties.
Cantaloupe (*See* Vine Crops)			
Carrot	Alternaria leaf blight	Small, brown to black, irregular spots with yellow margins may enlarge to infect the entire top.	Use approved fungicides.
	Cercospora leaf blight	Small, necrotic spots that may enlarge and infect the entire top.	Use approved fungicides.
	Aster Yellows	Purpling of tops; yellowed young leaves at center of crown followed by bushiness due to excessive petiole formation. Roots become woody and form numerous adventitious roots.	Control leafhopper carrier with insecticides.

Celery	Aster yellows	Yellowed leaves; stunting; tissues brittle and bitter in taste.	Use resistant varieties. Control leafhopper carrier with insecticides. Control weeds in adjacent areas.
	Bacterial blight	Bright yellow leaf spots, center turns brown, and a yellow halo appears with enlargement.	Seedbed sanitation. Copper compounds.
	Early blight	Dead, ash gray, velvety areas on leaves.	Use approved fungicides.
	Late blight	Yellow spots on old leaves and stalks that turn dark gray speckled with black dots.	Use approved fungicides.
	Mosaic	Dwarfed plants with narrow, gray, or mottled leaves.	Control weeds in adjacent areas. Control aphid carrier with insecticides.
	Pink rot	Water-soaked spots; white- to pink-colored cottony growth at base of stalk leads to rotting.	Crop rotation. Flooding for 4–8 weeks. Use approved fungicides.
Cucumber (*See* Vine Crops)			
Eggplant	Anthracnose	Sunken, tan fruit lesions.	Use disease-free seed. Use approved fungicides.
	Phomopsis blight	Young plants blacken and die; older plants have brown spots on leaves and fruit covered with brownish black pustules.	Use resistant varieties. Use approved fungicides.
	Verticillium wilt	Slow wilting; browning between leaf veins; stunting.	Fumigate soil with approved fumigants. Use verticillium-tolerant varieties. Use long rotation.
Endive, escarole, lettuce	Aster yellows	Center leaves bleached, dwarfed, curled, or twisted. Heads do not form; young plants particularly affected.	Control leafhopper carrier with insecticides.

TABLE 6.15. DISEASE CONTROL FOR VEGETABLES (*Continued*)

Crop	Disease	Description	Control
	Big vein	Leaves with light green, enlarged veins developing into yellow, crinkled leaves; stunting; delayed maturity.	Avoid cold, wet soils. Use tolerant varieties. Crop rotation.
	Bottom rot	Damage begins at base of plants; blades of leaves rot first, then the midrib, but the main stem is hardly affected.	Avoid wet, poorly drained areas. Plant on raised beds. Practice 3-year rotation. Use approved fungicides.
	Downy mildew	Light green spots on upper side of leaves; lesions enlarge and white mycelium appears on opposite side of spots; browning and dwarfing of plant.	Use approved fungicides.
	Drop	Wilting of outer leaves; watery decay on stems and old leaves.	Crop rotation. Deep plowing. Raised beds. Use approved fungicides.
	Mosaic	Mottling (yellow and green), ruffling, or distortion of leaves; plants have unthrifty appearance.	Use virus-free MTO seed. Plant away from old lettuce beds. Control weeds. Control aphid carrier with insecticide.
	Tipburn	Edges of tender leaves brown and die; may interfere with growth; most severe on head lettuce.	Use tolerant varieties. Prevent stress by providing good growing conditions.
Lima bean	Downy mildew	Purpling and distortion of leaf veins; white downy mold on pods; blackened beans.	Use resistant varieties and disease-free seed. Use approved fungicides.
Okra	Southern blight	Mass of pinkish fungus bodies around base of plant; sudden loss of leaves.	Crop rotation. Deep plowing of plant stubble.
	Verticillium wilt	Stunting; chlorosis; shedding of leaves.	Crop rotation. Avoid planting where disease was previously present.

Onion	Blight (blast)	Papery spots on leaves; browning and death of upper portion of leaves; delayed maturity.	Use approved fungicides.
	Downy mildew	Begins as pale green spot near tip of leaf; purple mold found when moisture present; infected leaves olive-green to black.	Use approved fungicides.
	Neck rot	Soft, brownish tissue around neck; scales around neck are dry, and black sclerotia may form. Essentially a dry rot if soft rot bacteria not present.	Undercut and windrow plants until inside neck tissues are dry before storage. Cure at 93–95°F for 5 days.
	Pink rot	Plants are affected from seedling stage onward throughout life cycle. Affected roots turn pink, shrivel, and die.	Avoid infected soils. Use tolerant varieties.
	Purple blotch	Small, white sunken lesions with purple centers enlarge to girdle leaf or seed stem. Leaves and stems fall over 3–4 weeks after infection in severe cases. Bulb rot at and after harvest.	Use approved fungicides.
	Smut	Black spots on leaves; cracks develop on side of spot revealing black, sooty powder within.	Crop rotation. Use approved fungicides.
Parsnip	Canker	Brown discoloration near shoulder or crown of root.	Ridge soil over shoulders.
	Leaf blight	Leaves and petioles turn yellow and then brown. Entire plant may be killed.	Practice 2-year rotation; use well-drained soil with pH 7.0.
Pea	Powdery mildew	White, powdery mold on leaves, stems, and pods; affected areas become brown and necrotic.	Use disease-free seed and resistant varieties. Use approved fungicides.

TABLE 6.15. DISEASE CONTROL FOR VEGETABLES (*Continued*)

Crop	Disease	Description	Control
	Root rot	Rotted and yellowish brown or black stems (below ground) and roots; outer layers of root slough off, leaving a central core.	Early planting and 3-year rotation. Do not double-crop with bean. Seed treatment.
	Virus	Several viruses affect pea, causing mottling, distortion of leaves, rosetting, chlorosis, or necrosis.	Use resistant varieties. Control aphid carrier with insecticides.
	Wilt	Yellowing leaves; dwarfing, browning of xylem; wilting.	Early planting and 3-year rotation. Use resistant varieties.
Pepper	Anthracnose	Dark, round spots with black specks on fruits.	Use approved fungicides.
	Bacterial leaf spot	Yellowish green spots on young leaves; raised, brown spots on undersides of older leaves; brown, cracked, rough spots on fruit; old leaves turn yellow.	Use disease-free seed, hot-water-treated seed. Use approved bactericides. Use resistant varieties.
	Mosaic	Mottled (yellowed and green) and curled leaves; fruits yellow or show green ring spots; stunted; reduced yields.	Use resistant varieties. Control insect carriers (particularly aphids) and weed hosts. Use stylet oil.
Potato	Early blight	Dark brown spots on leaves; foliage injured; reduced yields.	Bury all cull potatoes. Use approved fungicides.
	Late blight	Dark, then necrotic area on leaves and stem; infected tubers rot in storage. Disease is favored by moist conditions.	Bury cull piles. Use approved fungicides.

	Rhizoctonia	Necrotic spots, girdling and death of sprouts before or shortly after emergence. Brown to black raised spots on mature tubers.	Avoid deep planting to encourage early emergence. Use disease-free seed. Use approved fungicides.
	Scab	Rough, scabby, raised, or pitted lesions on tubers.	Crop rotation. Use resistant varieties. Maintain soil pH about 5.3.
	Virus	A large number of viruses infect potato, causing leaf mottling, distortion, and dwarfing. Some viruses cause irregularly shaped or necrotic area in tubers.	Use certified seed. Control aphid and leafhopper carriers with insecticides.
Radish	Downy mildew	Internal discoloration of root crown tissue. Outer surface may become dark and rough at the soil line.	Select clean, well-drained soils. Use approved fungicides.
	Fusarium wilt	Young plants yellow and die rapidly in warm weather. Stunting, unilateral leaf yellowing; vascular discoloration of fleshy roots.	Use tolerant varieties. Avoid infested soil.
Rhubarb	Crown rot	Wilting of leaf blades; browning at base of leaf stalk leading to decay.	Plant in well-drained soil.
	Leaf spot	Tiny, greenish yellow spots (resembling mosaic) on upper side of leaf, eventually browning and forming a white spot surrounded by a red band; these spots may drop out to give a shot-hole appearance.	Use approved fungicides.
Rutabaga, turnip	Alternaria	Small, circular, yellow areas that enlarge in concentric circles and become a black sooty color. Roots may become infested in storage.	Use hot-water-treated seed. Use approved fungicides.

TABLE 6.15. DISEASE CONTROL FOR VEGETABLES (*Continued*)

Crop	Disease	Description	Control
	Anthracnose	Small, water-soaked spots on all above-ground parts, which become light-colored and may drop out. Small, sunken, dry spots on turnip roots, which are subject to secondary decay.	Use approved fungicides.
	Club root	Tumor-like swellings on tap root. Main root may be distorted. Diseased roots decay prematurely.	Avoid soil previously infested with club root. Adjust acid soil to pH 7.3 by liming.
	Downy mildew	Small, purplish, irregular spots on leaves, stems, and seedpods that produce fluffy white growth. Desiccation of roots in storage.	Use approved fungicides.
	Mosaic virus	Stunted plants having ruffled leaves. Infected roots store poorly.	Destroy volunteer plants. Control aphid carrier with insecticides.
Southern pea	Fusarium wilt	Yellowed leaves; wilted plants; interior of stems lemon yellow.	Avoid infested soil.
Spinach	Blight (CMV)	Yellowed and curled leaves; stunted plants; reduced yields.	Use tolerant varieties. Control aphid carrier with insecticides.
	Downy mildew	Yellow spots on upper surface of leaves; downy or violet-gray mold on undersides.	Use resistant varieties. Use approved fungicides.
Squash (*See* Vine Crops)			

Strawberry	Anthracnose	Spotting and girdling of stolens and petioles, crown rot, fruit rot, and a black leaf spot; commonly occurs in southeastern United States.	Use disease-free plants and resistant varieties. Use approved fungicides.
	Gray mold	Rot on green or ripe fruit, beginning at calyx or contact with infected fruit; affected area supports white or gray mycelium.	Use less susceptible varieties. Use approved fungicides.
	Leaf scorch	Numerous irregular, purplish blotches with brown centers; entire leaves dry up and appear scorched.	Use disease-free plants and resistant varieties. Renew perennial plantings frequently. Use approved fungicides.
	Leaf spot	Indefinite-shaped spots with brown, gray, or white centers and purple borders.	Use disease-free plants and resistant varieties. Use approved fungicides.
	Powdery mildew	Characteristic white mycelium on leaves, flower, and fruit.	Use resistant varieties. Use approved fungicides.
	Red stele	Stunted plants having roots with red stele seen when root is cut lengthwise.	Improve drainage and avoid compaction of soil. Use disease-free plants and resistant varieties.
	Verticillium	Marginal and interveinal necrosis of outer leaves; inner leaves remain green.	Preplant soil fumigation. Use resistant varieties.
Sweet corn	Bacterial blight	Dwarfing; premature tassels die; yellow bacterial slime oozes from wet stalks; stem dries and dies.	Use resistant varieties. Control corn flea beetle with insecticides.
	Leaf blight	Canoe-shaped spots on leaves.	Use resistant varieties. Use approved fungicides.
	Maize dwarf mosaic	Stunting; mottling of new leaves in whorl and poor ear fill at the base.	Use tolerant varieties. Plant tolerant varieties around susceptible ones. Control aphid carrier with insecticides.
	Seed rot	Seed decays in soils.	Use seed treated with approved fungicides.

TABLE 6.15. DISEASE CONTROL FOR VEGETABLES (*Continued*)

Crop	Disease	Description	Control
	Smut	Large, smooth, white galls, or outgrowths on ears, tassels, and nodes; covering dries and breaks open to release black, powdery, or greasy spores.	Use tolerant varieties. Control corn borers with insecticides.
Sweet potato	Black rot	Black depressions on sweet potato; black cankers on underground stem parts.	Select disease-free potato seed. Rotate crops and planting beds. Use vine cuttings for propagation rather than slips.
	Internal cork	Dark brown to black, hard, corky lesions in flesh developing in storage at high temperature. Yellow spots with purple borders on new growth of leaves.	Select disease-free seed potatoes.
	Pox	Plants dwarfed; only one or two vines produced; leaves thin and pale green; soil rot pits on roots.	Use disease-free stock and clean planting beds. Sulfur to lower soil pH to 5.2.
	Scurf	Brown to black discoloration of root; uniform rusting of root surface.	Rotation of crops and beds. Use disease-free stock. Use vine cuttings rather than slips.
	Stem rot	Yellowing between veins; vines wilt; stems darken inside and may split.	Select disease-free seed potatoes. Rotate fields and plant beds.
Tomato	Anthracnose	Begins with circular, sunken spots on fruit; as spots enlarge, center becomes dark and fruit rots.	Use approved fungicides.

Bacterial canker	Wilting; rolling, and browning of leaves; pith may discolor or disappear; fruit displays bird's-eye spots.	Use hot-water-treated seed. Avoid planting in infested fields for 3 years.
Bacterial spot	Young lesions on fruit appear as dark, raised spots; older lesions blacken and appear sunken with brown centers; leaves brown and dry.	Use hot-water-treated seed. Use approved bactericides.
Early blight	Dark brown spots on leaves; brown cankers on stems; girdling; dark, leathery, decayed areas at stem end of fruit.	Use approved fungicides.
Late blight	Dark, water-soaked spots on leaves; white fungus on undersides of leaves; withering of leaves; water-soaked spots on fruit turn brown. Disease is favored by moist conditions.	Use approved fungicides.
Fusarium wilt	Yellowing and wilting of lower, older leaves; disease eventually affects whole plant.	Use resistant varieties.
Gray leaf spot	Symptoms appear first in seedlings. Small brown to black spots on leaves, which enlarge and have shiny gray centers. The centers may drop out to give shotgun appearance. Oldest leaves affected first.	Use resistant varieties. Use approved fungicides.
Leaf mold	Chlorotic spots on upper side of oldest leaves appear in humid weather. Underside of leaf spot may have green mold. Spots may merge until entire leaf is affected. Disease advances to younger leaves.	Use resistant varieties. Stake and prune to provide air movement. Use approved fungicides.

TABLE 6.15. DISEASE CONTROL FOR VEGETABLES (*Continued*)

Crop	Disease	Description	Control
	Mosaic	Mottling (yellow and green) and roughening of leaves; dwarfing; reduced yields; russeting of fruit.	Avoid contact by smokers. Control aphid carrier with insecticides. Stylet oil may be effective.
	Tomato spotted wilt virus	Brown spots, some circular on youngest leaves, stunted plant. Fruits misshapen often with circular brown rings, a diagnostic characteristic of this disease.	Use a combination of resistant varieties, highly reflective mulch to repel the silverleaf white fly, and plant activators.
	Verticillium wilt	Differs from fusarium wilt by appearance of disease on all branches at the same time; yellow areas on leaves become brown; midday wilting; dropping of leaves beginning at bottom.	Use resistant varieties.
Vine Crops: cantaloupe, cucumber, pumpkin, squash, watermelon	Alternaria leaf spot	Circular spots showing concentric rings as they enlarge, appear first on oldest leaves.	Field sanitation. Use disease-free seed. Use approved fungicides.
	Angular leaf spot	Irregular, angular, water-soaked spots on leaves that later turn gray and die. Dead tissue may tear away, leaving holes. Nearly circular fruit spots, which become white.	Use tolerant varieties. Use approved bactericides.

Anthracnose	Reddish black spots on leaves; elongated tan cankers on stems; fruits have sunken spots with flesh-colored ooze in center, later turning black.	Use tolerant varieties. Use approved fungicides.
Bacterial wilt	Vines wilt and die; stem sap produces strings; no yellowing occurs.	Control striped cucumber beetles with insecticides. Remove wilting plants from field.
Black rot (squash and pumpkin only)	Water-soaked areas appear on rinds of fruit in storage. Brown or black infected tissue rapidly invades entire plant.	Use disease-free seed. Crop rotation. Cure fruit for storage at 85°F for 2 weeks, store at 50–55°F. Use approved fungicides.
Downy mildew	Angular, yellow spots on older leaves; purple fungus on undersides of leaves when moisture present; leaves wither, die; fruit may be dwarfed, with poor flavor.	Use tolerant varieties. Use approved fungicides.
Fusarium wilt	Stunting and yellowing of vine; water-soaked streak on one side of vine eventually turns yellow, cracks, and oozes sap.	Use resistant varieties. Avoid infested soils.
Gummy stem blight	Lesions may occur on stems, leaves, and fruit from which a reddish gummy exudate may ooze.	Use disease-free seed. Rotate crops. Use approved fungicides.
Mosaic (several)	Mottling (yellow and green) and curling of leaves; mottled and warty fruit; reduced yields; burning and dwarfing of entire plant.	Control striped cucumber beetle or aphid with insecticides. Use resistant varieties. Destroy surrounding perennial weeds.

TABLE 6.15. DISEASE CONTROL FOR VEGETABLES (*Continued*)

Crop	Disease	Description	Control
	Powdery mildew	White, powdery growth on upper leaf surface and petioles; wilting of foliage.	Use tolerant varieties. Use approved fungicides.
	Cucumber scab	Water-soaked spots on leaves turning white; sunken cavity on fruit later covered by grayish olive fungus; fruit destroyed by soft rot.	Use resistant varieties. Use approved fungicides.
	Squash silverleaf	Silvering or white coloration to leaves. Associated with silverleaf whitefly feeding, disorder is worse in fall crops in southern United States.	Little control, except to avoid planting under high whitefly populations and use tolerant varieties.

DISEASE IDENTIFICATION WEBSITES WITH DISEASE PHOTOGRAPHS AND DIAGNOSTIC INFORMATION

Plant disease control begins with an accurate diagnosis and identification of the disease-causing organism or agent. Although photos are helpful in identifying plant diseases, we encourage the grower to consult a knowledgeable disease expert to provide confirmation of the identification before any control strategy is implemented. Here are a few websites containing photographs and helpful diagnostic information:

Arkansas, http://www.aragriculture.org/pestmanagement/diseases/image_library/default.htm

Florida, http://edis.ifas.ufl.edu/VH045

Maryland, http://www.agnr.umd.edu/users/hgic/diagn/home.html

Minnesota, http://www.extension.umn.edu/projects/yardandgarden/diagnostics/mainvegetables.html

New York, http://plantclinic.cornell.edu/vegetable/index.htm; http://vegetablemdonline.ppath.cornell.edu/PhotoPages/PhotoGallery.htm

Pennsylvania, http://vegdis.cas.psu.edu/Identification.html

Utah, http://extension.usu.edu/plantpath/vegetables/vegetables.htm

Washington, http://mtvernon.wsu.edu/path_team/diseasegallery.htm

Other, http://www.gardeners.com/gardening/content.asp?copy_id=5366

09
INSECTS

SOME INSECTS THAT ATTACK VEGETABLES

Most vegetable crops are attacked by insects at one time or another in the crop growth cycle. Growers should strive to minimize insect problems by employing cultural methods aimed at reducing insect populations. These tactics include using resistant varieties, reflective mulches, crop rotation for soilborne insects, destruction of weed hosts, stylet oils, insect repellants, trap crops, floating row covers, or other tactics. Sometimes insecticides must be used in an integrated approach to insect control when the economic threshold insect population is reached. When using insecticides, read the label and carefully follow the instructions. Do not exceed maximum rates given, observe the interval between application and harvest, and apply only to crops for which use is approved. Make a record of the product used, trade name, formulation, dilution, rate applied per acre, and dates of application. Read and follow all label precautions to protect the applicator and workers from insecticide injury and the environment from contamination. Follow local recommendations for efficacy and read the label for proper use.

TABLE 6.16. INSECTS THAT ATTACK VEGETABLES

Crop	Insect	Description
Artichoke	Aphid	Small, green, pink, or black soft-bodied insects that rapidly reproduce to large populations. Damage results from sucking plant sap; indirectly from virus transmission to crop plants.
	Plume moth	Small wormlike larvae blemish bracts and may destroy the base of the bract.
Asparagus	Beetle and twelve-spotted beetle	Metallic blue or black beetles (¼ in.) with yellowish wing markings and reddish, narrow head. Larvae are humpbacked, slate gray. Both feed on shoots and foliage.
	Cutworm	Dull-colored moths lay eggs in the soil. The produce dark-colored, smooth worms, 1–2 in. long, that characteristically curl up when disturbed. May feed below ground or above-ground at night.
Bean	Aphid	*See* Artichoke.
	Corn earworm	Gray-brown moth (1½ in.) with dark wing tips deposits eggs, especially on fresh corn silk. Brown, green, or pink larvae (2 in.) feed on silk, kernels, and foliage.

TABLE 6.16. INSECTS THAT ATTACK VEGETABLES (*Continued*)

Crop	Insect	Description
	Leafhopper	Green, wedge-shaped, soft bodies (1⁄8 in.). When present in large numbers, sucking of plant sap causes plant distortion or burned appearance. Secondary damage results from transmission of yellows disease.
	Mexican bean beetle	Copper-colored beetle (1⁄4 in.) with 16 black spots on its back. Orange to yellow spiny larva (1⁄3 in.). Beetle and larvae feeding on leaf undersides cause a lacework appearance.
	Seed corn maggot	Grayish brown flies (1⁄5 in.) deposit eggs in the soil near plants. Cream-colored, wedge-shaped maggots (1⁄4 in.) tunnel into seeds, potato seed pieces, and sprouts.
	Spider mite	Reddish, yellow, or greenish tiny eight-legged spiders suck plant sap from leaf undersides, causing distortion. Fine webs may be visible when mites are present in large numbers. Mites are not true insects.
	Spotted cucumber beetle	Yellowish, elongated beetle (1⁄4 in.) with 11 or 12 black spots on its back. Leaf-feeding may destroy young plants when present in large numbers. Transmits bacterial wilt of curcurbits.

TABLE 6.16. INSECTS THAT ATTACK VEGETABLES (*Continued*)

Crop	Insect	Description
	Striped cucumber beetle	Yellow (1⁄5 in.) with three black stripes on its back; feeds on leaves. White larvae (1⁄3 in.) feed on roots and stems. Transmits bacterial wilt of curcurbits.
	Tarnish plant bug	Brownish, flattened, oval bugs (1⁄4 in.) with a clear triangular marking at the rear. Bugs damage plants by sucking plant sap.
Beet	Aphid	*See* Artichoke.
	Flea beetle	Small (1⁄6 in.), variable-colored, usually dark beetles, often present in large numbers in the early part of the growing season. Feeding results in numerous small holes, giving a shotgun appearance. Indirect damage results from diseases transmitted.
	Leaf miner	Tiny black and yellow adults. Yellowish white maggot-like larvae tunnel within leaves and cause white or translucent irregularly damaged areas.
	Webworm	Yellow to green worm (1 1⁄4 in.) with a black stripe and numerous black spots on its back.
Broccoli, Brussels sprouts, cabbage, cauliflower, kale, kohlrabi	Aphid	*See* Artichoke.
	Flea beetle	*See* Beet.
	Harlequin cabbage bug	Black, shield-shaped bug (3⁄8 in.) with red or yellow markings.

TABLE 6.16. INSECTS THAT ATTACK VEGETABLES (*Continued*)

Crop	Insect	Description
	Cabbage maggot	Housefly-like adult lays eggs in the soil at the base of plants. Yellowish, legless maggot (¼–⅓ in.) tunnels into roots and lower stem.
	Cabbage looper	A brownish moth (1½ in.) that lays eggs on upper leaf surfaces. Resulting worms (1½ in.) are green with thin white lines. Easily identified by their looping movement.
	Diamondback moth	Small, slender gray or brown moths. The folded wings of male moths show three diamond markings. Small (⅓ in.) larvae with distinctive *V* at rear, wiggle when disturbed.
	Imported cabbage worm	White butterflies with black wing spots lay eggs on undersides of leaves. Resulting worms (1¼ in.) are sleek, velvety, green.
Cantaloupe (*See* Vine Crops)		
Carrot	Leafhopper	*See* Bean.
	Rust fly	Shiny, dark fly with a yellow head; lays eggs in the soil at the base of plants. Yellowish white, legless maggots tunnel into roots.
Celery	Aphid	*See* Beet.
	Leaf miner	Adults are small, shiny, black flies with a bright yellow spot on upper thorax. Eggs are laid within the leaf. Larvae mine between upper and lower leaf surfaces.

TABLE 6.16. INSECTS THAT ATTACK VEGETABLES (*Continued*)

Crop	Insect	Description
	Spider mite	*See* Bean.
	Tarnished plant bug	*See* Bean.
	Loopers and worms	*See* Broccoli, etc.
Cucumber (*See* Vine Crops)		
Eggplant	Aphid	*See* Artichoke.
	Colorado potato beetle	Oval beetle (3/8 in.) with 10 yellow and 10 black stripes, lays yellow eggs on undersides of leaves. Brick red, humpbacked larvae (1/2 in.) have black spots. Beetles and larvae are destructive leaf feeders.
	Flea beetle	*See* Beet.
	Leaf miner	*See* Beet.
	Spider mite	*See* Bean.
Endive, escarole, lettuce	Aphid	*See* Artichoke.
	Flea beetle	*See* Beet.
	Leafhopper	*See* Bean.
	Leaf miner	*See* Beet.
	Looper	*See* Broccoli.
Mustard greens	Aphid	*See* Artichoke.
	Worms	*See* Broccoli, etc.
Okra	Aphid	*See* Artichoke.
	Green stinkbug	Large, flattened, shield-shaped, bright green bugs; various-sized nymphs with reddish markings.
Onion	Maggot	Slender, gray flies (1/4 in.) lay eggs in soil. Small (1/3 in.) maggots bore into stems and bulbs.
	Thrips	Yellow or brown, winged or wingless, tiny (1/25 in.). Damages plant by sucking plant sap, causing white areas or brown leaf tips.

TABLE 6.16. INSECTS THAT ATTACK VEGETABLES (*Continued*)

Crop	Insect	Description
Parsnip	Carrot rust fly	*See* Carrot.
Pea	Aphid	*See* Artichoke.
	Seed maggot	Housefly-like gray adults lay eggs that develop into maggots (¼ in.) with sharply pointed heads.
	Weevil	Brown-colored adults, marked by white, black, or gray (⅕ in.), lay eggs on young pods. Larvae are small and whitish, with a brown head and mouth. Adults feed on blossoms. May infect seed before harvest and remain in hibernation during storage.
Pepper	Aphid	*See* Artichoke.
	Corn borer	*See* Sweet corn.
	Flea beetle	*See* Beet.
	Leaf miner	*See* Beet.
	Maggot	Housefly-sized adults have yellow stripes on body and brown stripes on wings. Larvae are typical maggots with pointed heads.
	Weevil	Black-colored, gray- or yellow-marked snout beetle, with the snout about half the length of the body. Grayish white larvae are legless and have a pale brown head. Both adults and larvae feed on buds and pods; adults also feed on foliage.
Potato	Aphid	*See* Artichoke.
	Colorado potato beetle	*See* Eggplant.

TABLE 6.16. INSECTS THAT ATTACK VEGETABLES (*Continued*)

Crop	Insect	Description
	Cutworm	*See* Asparagus.
	Flea beetle	*See* Beet.
	Leafhopper	*See* Bean.
	Leaf miner	*See* Beet.
	Tuberworm	Small, narrow-winged, grayish brown moths (½ in.) lay eggs on foliage and exposed tubers in evening. Purplish or green caterpillars (¾ in.) with brown heads burrow into exposed tubers in the field or in storage.
	Wireworm	Adults are dark-colored, elongated beetles (click beetles). Yellowish, tough-bodied, segmented larvae feed on roots and tunnel through fleshy roots and tubers.
Radish	Maggot	*See* Broccoli, etc.
Rhubarb	Curculio	Yellow-dusted snout beetle that damages plants by puncturing stems.
Rutabaga, turnip	Flea beetle	*See* Beet.
	Maggot	*See* Broccoli, etc.
Squash (*See* Vine Crops)		
Southern pea	Curculio	Black, humpbacked snout beetle. Eats small holes in pods and peas. Larvae are white with yellowish head and no legs.
	Leafhopper	*See* Bean.
	Leaf miner	*See* Beet.
Spinach	Aphid	*See* Artichoke.
	Leaf miner	*See* Beet.

TABLE 6.16. INSECTS THAT ATTACK VEGETABLES (*Continued*)

Crop	Insect	Description
Strawberry	Aphid	*See* Artichoke.
	Mites	Several mite species attack strawberry. *See* Bean.
	Tarnished plant bug	*See* Bean.
	Thrips	*See* Onion.
	Weevils	Several weevil species attack strawberry.
	Worms	Several worm species attack strawberry.
Sweet corn	Armyworms	Moths (1½ in.) with dark gray front wings and light-colored hind wings lay eggs on leaf undersides. Tan, green, or black worms (1¼ in.) feed on plant leaves and corn ears.
	Earworm	*See* Bean.
	European corn borer	Pale, yellowish moths (1 in.) with dark bands lay eggs on undersides of leaves. Caterpillars hatch, feed on leaves briefly, and tunnel into stalk and to the ear.
	Flea beetle	*See* Beet.
	Japanese beetle	Shiny, metallic green with coppery brown wing covers, oval beetles (½ in.). Severe leaf feeding results in a lacework appearance. Larvae are grubs that feed on grass roots.
	Seed corn maggot	*See* Bean.

TABLE 6.16. INSECTS THAT ATTACK VEGETABLES (*Continued*)

Crop	Insect	Description
	Stalk borer	Grayish moths (1 in.) lay eggs on weeds. Small, white, brown-striped caterpillars hatch and tunnel into weed and crop stalks. Most damage is usually at edges of fields.
Sweet potato	Flea beetle	*See* Beet.
	Weevil	Blue-black and red adult (¼ in.) feeds on leaves and stems; grub-like larva tunnels into roots in the field and storage.
	Wireworm	*See* Potato.
Tomato	Aphid	*See* Artichoke.
	Colorado potato beetle	*See* Eggplant.
	Corn earworm (tomato fruitworm)	*See* Bean.
	Flea beetle	*See* Beet.
	Fruit fly	Small, dark-colored flies usually associated with overripe or decaying vegetables.
	Hornworm	Large (4–5 in.) moths lay eggs that develop into large (3–4 in.) green fleshy worms with prominent white lines on sides and a distinct horn at the rear. Voracious leaf feeders.
	Leaf miner	*See* Beet.

TABLE 6.16. **INSECTS THAT ATTACK VEGETABLES** (*Continued*)

Crop	Insect	Description
	Pinworm	Tiny yellow, gray, or green purple-spotted, brown-headed caterpillars cause small fruit lesions, mostly near calyx. Presence detected by large white blotches near folded leaves.
	Mite	*See* Bean.
	Stink bug	*See* Okra.
	White fly	Small, white flies that move when disturbed.
Vine Crops: cantaloupe, cucumber, pumpkin squash, watermelon	Aphid	*See* Artichoke.
	Cucumber beetle (spotted or striped)	
	Leafhopper	*See* Bean.
	Leaf miner	*See* Beet.
	Mite	*See* Bean.
	Pickleworm	White moths (1 in.), later become greenish with black spots, with brown heads and brown-tipped wings with white centers and a conspicuous brush at the tip of the body, lay eggs on foliage. Brown-headed, white, later becoming greenish with black spots. Larvae (¾ in.) feed on blossoms, leaves, and fruit.
	Squash bug	Brownish, flat stinkbug (⅝ in.). Nymphs (⅜ in.) are gray to green. Plant damage is due to sucking of plant sap.

TABLE 6.16. INSECTS THAT ATTACK VEGETABLES (*Continued*)

Crop	Insect	Description
	Squash vine borer	Black, metallic moth ($1\frac{1}{2}$ in.) with transparent hind wings and abdomen ringed with red and black; lays eggs at the base of the plant. White caterpillars bore into the stem and tunnel throughout.
	White fly	*See* Tomato.

IDENTIFICATION OF VEGETABLE INSECTS

Effective insect management requires accurate identification and a thorough knowledge of the insect's habits and life cycle. Previous editions of *Handbook for Vegetable Growers* contained drawings of selected insect pests of vegetables. Today, many fine websites that contain photographs of insect pests are available. Some Extension Services also have available CD-ROMs containing insect photographs. We have chosen to direct the reader to some of these websites to assist in the identification of insect pests. Although the photos are helpful in identifying pests, we encourage the grower to consult a knowledgeable insect expert to confirm the identification before any control strategy is implemented.

SOME USEFUL WEBSITES FOR INSECT IDENTIFICATION

California, http://www.ipm.ucdavis.edu/PMG/crops-agriculture.html; http://www.ipm.ucdavis.edu/PCA/pcapath.html#SPECIFIC

Colorado, http://lamar.colostate.edu/~gec/vg.htm

Florida, http://pests.ifas.ufl.edu (for a listing of web sites on insects, mites, and other topics and information regarding vegetable pest images and CDs)

Georgia, http://www.ent.uga.edu/veg/veg_crops.htm

Indiana, http://*www.entm.purdue.edu/entomology/vegisite/*

Iowa, http://www.ent.iastate.edu/imagegallery/

Kentucky, http://www.uky.edu/Agriculture/Entomology/entfacts/efveg.htm

Mississippi, http://msucares.com/insects/vegetable/

North Carolina, http://www.ces.ncsu.edu/depts/hort/consumer/hortinternet/vegetable.html; http://www.ces.ncsu.edu/chatham/ag/SustAg/insectlinks.html

South Carolina, http://entweb.clemson.edu/cuentres/cesheets/veg/

Texas, http://vegipm.tamu.edu/imageindex.html; http://insects.tamu.edu/images/insects/color/veindex.html

10
PEST MANAGEMENT IN ORGANIC PRODUCTION SYSTEMS

Diseases, insects, and nematodes can be controlled in organic vegetable production systems by combinations of tactics, including certain approved control materials. Pest management practices useful in organic vegetable production include:

Understanding the biology and ecology of pests
Encouraging natural enemies, predators, and parasites
Crop rotation
Trap crops
Crop diversification
Resistant varieties
Scouting for early detection
Optimal timing of planting (avoidance)
Controlling weed hosts
Controlling alternate host plants
Sanitation of field
Pest-free transplants
Tilling crop refuse
Exclusion, e.g., row covers
Traps, sticky tape, pheromone traps, etc.
Maintaining healthy crops
Maintaining optimum plant nutrition
Optimal pH control
Mulches
Spatial separation of crop and pest
Avoiding splashing water (drip irrigation instead of sprinklers)
Destroying cull piles
Providing for good air movement (proper plant and row spacing)
Using raised beds for water drainage
Flaming for weeds and Colorado potato beetle
Trellising or staking for air movement and to keep fruits from contact with the ground
Hand removal
Compost use (may contain antagonistic organism)
Using approved control materials

SELECTED RESOURCES FOR PEST CONTROL IN ORGANIC FARMING SYSTEMS

We located a variety of websites with information on organic pest management, many of which also contain links to other helpful sources of information. Some of these websites are listed below:

- B. Caldwell, E. Rosen, E. Sideman, A. Shelton, and C. Smart, *Resource Guide for Organic Insect and Disease Management* (Cornell University, 2005), http://www.nysaes.cornell.edu/pp/resourceguide/index.php.
- C. Weeden, A. Shelton, Y. Li, and M. Hoffman, *Biological Control: A Guide to Natural Enemies in North America* (Cornell University, 2005), http://www.nysaes.cornell.edu/ent/biocontrol/.
- R. Hazzard and P. Westgate, *Organic Insect Management in Sweet Corn* (University of Massachusetts, 2004), http://www.umassvegetable.org/soil_crop_pest_mgt/pdf_files/organic_insect_management_in_sweet_corn.pdf.
- Organic Farming—National Sustainable Agriculture Information Service, http://attra.ncat.org/organic.html 2005). This site has many links to other organic farming publications by NCAT (National Center for Appropriate Technology).
- S. Koike, M. Gaskell, C. Fouche, R. Smith, and J. Mitchell, *Plant Disease Management for Organic Crops* (University of California—Davis, 2000), http://anrcatalog.ucdavis.edu/pdf7252.pdf.
- *Insect and Disease Management in Organic Crop Systems* (Manitoba Agriculture, Food, and Rural Initiatives, 2004), http://www.gov.mb.ca/agriculture/crops/insects/fad64s00.html.
- *Alternative Disease, Pest, and Weed Control* (Alternative Farming Systems Information Center, 2003), http://www.nal.usda.gov/afsic/sbjdpwc.htm.
- Sustainable Agriculture Research and Education, http://www.sare.org/publications/organic/organic03.htm.
- *Organic Gardening: A Guide to Resources, 1989–September 2003,* http://www.nal.usda.gov/afsic/AFSIC_pubs/org_gar.htm#toc2c.
- Organic Trade Association, http://www.ota.com/index.html.
- *Organic/Sustainable Farming: Idaho OnePlan* (The Idaho Association of Soil Conservation Districts, 2004), http://www.oneplan.org/index.shtml.

11
WILDLIFE CONTROL

DEER

Repellants. May be effective for low-density deer populations. Apply before damage is expected, when no precipitation is expected, and when temperatures are 40–80°F.

Fencing. Woven wire fences are the most effective and should be 8–10 ft tall. Electric fences may act as a deterrent. Some growers have success with 5- or 6-ft high-tensile electric fences, even though deer may be able to jump them.

RACCOONS

Many states have laws controlling the manner in which raccoons can be removed. Usually trapping is the only means of ridding a field of raccoons. Crops can be protected with a double-strand electric fence with wires at 5 and 10 in. above the ground.

BIRDS

Exclusion. Bird proof netting can be used to protect vegetables of high value.

Sound devices. Some success has been reported with recorded distress calls. Other sound devices such as propane guns may be effective for short periods. Use of these devices should be random and with a range of sound frequency and intervals.

Visual devices. Eye-spot balloons have been used with some success against grackles, blue jays, crows, and starlings and might be the control method of choice for urban farms. Reflective tape has been used with variable success and is labor intensive to install.

MICE

Habitat control. Remove any possible hiding or nesting sites near the field. Sometimes mice nest underneath polyethylene mulch not applied tightly to the ground, or in thick windbreaks.

Traps and baits. Strategically placed traps and bait stations can be used to reduce mouse populations.

Transplanting. The seed of some particularly attractive crops, such as curcurbits, is a favorite mouse food, and the seed is often removed from the ground soon after planting. One option to reduce stand losses is transplanting instead of direct seeding.

PART 7

WEED MANAGEMENT

01
WEED MANAGEMENT STRATEGIES

Weeds reduce yield and quality of vegetables through direct competition for light, moisture, and nutrients as well as by interference with harvest operations. Early season competition is most critical, and a major emphasis on control should be made during this period. Common amaranth reduces yields of lettuce, watermelon, and muskmelon at least 20% if allowed to compete with these crops for only the first 3 weeks of growth. Weeds can be controlled, but this requires good management practices in all phases of production. Because there are many kinds of weeds, with much variation in growth habit, they obviously cannot be managed by a single method. The incorporation of several of the following management practices into vegetable strategies increases the effectiveness for controlling weeds.

Crop Competition

An often overlooked tool in reducing weed competition is to establish a good crop stand in which plants emerge and rapidly shade the ground. The plant that emerges first and grows the most rapidly is the plant with the competitive advantage. Utilization of good production management practices such as fertility, well-adapted varieties, proper water control (irrigation and drainage), and establishment of adequate plant populations is very helpful in reducing weed competition. Everything possible should be done to ensure that vegetables, not weeds, have the competitive advantage.

Crop Rotation

If the same crop is planted in the same field year after year, usually some weed or weeds are favored by the cultural practices and herbicides used on that crop. By rotating to other crops, the cultural practices and herbicide program are changed. This often reduces the population of specific weeds tolerant in the previous cropping rotation. Care should be taken, however, to not replant vegetables in soil treated with a nonregistered herbicide. Crop injury as well as vegetables containing illegal residues may result. Check the labels for plant-back limitations before application and planting rotational crops.

Mechanical Control

Mechanical control includes field preparation by plowing or disking, cultivation, mowing, hoeing, and hand pulling of weeds. Mechanical control practices are among the oldest weed management techniques. Weed control is a primary reason for preparing land for crops planted in rows. Seedbed preparation by plowing or disking exposes many weed seeds to variations in light, temperature, and moisture. For some weeds, this process breaks weed seed dormancy, leading to early season control with herbicides or additional cultivation. Cultivate only deep enough in the row to achieve weed control; deep cultivation may prune roots, bring weed seeds to the surface, and disturb the soil previously treated with a herbicide. Follow the same precautions between rows. When weeds can be controlled without cultivation, there is no advantage to the practice. In fact, there may be disadvantages, such as drying out the soil surface, bringing weed seeds to the surface, and disturbing the root system of the crop.

Mulching

The use of polyethylene mulch increases yield and earliness of vegetables. The proper injection of fumigants under the mulch controls nematodes, soil insects, soil borne diseases, and weed seeds. Mulches act as a barrier to the growth of many weeds. Nutsedge, however, is one weed that can and will grow through the mulch.

Prevention

Preventing weeds from infesting or reinfesting a field should always be considered. Weed seed may enter a field in a number of ways. It may be distributed by wind, water, machinery, in cover crop seed, and other means. Fence rows and ditch banks are often neglected when controlling weeds in the crop. Seed produced in these areas may move into the field. Weeds in these areas can also harbor insects and diseases (especially viruses) that may move onto the crop. It is also important to clean equipment before entering fields or when moving from a field with a high weed infestation to a relatively clean field. Nutsedge tubers especially are moved easily on disks, cultivators, and other equipment.

Herbicides

Properly selected herbicides are effective tools for weed control. Herbicides may be classified several ways depending on how they are applied and their mode of action in or on the plant. Generally, herbicides are either soil applied or foliage applied. They may be selective or nonselective, and they may be either contact or translocated through the plant. For example, paraquat is a foliage-applied, contact, nonselective herbicide, whereas atrazine usually is described as a soil-applied, translocated, selective herbicide.

Foliage-applied herbicides may be applied to leaves, stems, and shoots of plants. Herbicides that kill only those parts of the plants they touch are *contact herbicides.* Those herbicides that are taken into the plant and moved throughout it are *translocated herbicides.* Paraquat is a contact herbicide, whereas glyphosate (Roundup) or Sethoxydim (Poast) are translocated herbicides. For foliage-applied herbicides to be effective, they must enter the plant. Good coverage is very important. Most foliage-applied herbicides require either the addition of a specified surfactant or a specified formulation to be used for best control.

Soil-applied herbicides are either applied to the surface or incorporated. Surface-applied herbicides require rainfall or irrigation shortly after application for best results. Lack of moisture often results in poor weed control. Incorporated herbicides are not dependent on rainfall or irrigation and generally give more consistent and wider-spectrum control. They do, however, require more time and equipment for incorporation. Herbicides that specify incorporation into the soil improve the contact of the herbicide with the weed seed and/or minimize the loss of the herbicide by volatilization or photodecomposition. Some herbicides, if not incorporated, may be lost from the soil surface. Although most soil-applied herbicides must be moved into the soil to be effective, the depth of incorporation into the soil can be used to achieve selectivity. For example, if a crop seed is planted 2 in. deep in the soil and the herbicide is incorporated by irrigation only in the top 1 in., where most of the problem weed seeds are found, the crop roots will not come in contact with the herbicide. If too much irrigation or rain moves the herbicide down into the crop seed zone or if the herbicide is incorporated mechanically too deep, crop injury may result.

Adapted from W. M. Stall, "Weed Management," in S. M. Olson and E. H. Simone (eds.), *Commercial Vegetable Production Handbook for Florida* (Florida Cooperative Extension Service, Serv. SP-170, 2004), http://edis.ifas.ufl.edu/cv113.

02
WEED IDENTIFICATION

Accurate identification of the particular weed species is the first step to controlling the problem. Several university and cooperative extension websites offer information about and assistance with identifying weeds. Some sources offer information across crops and commodities, and many have very good photos of weed species. As good as these websites are for assisting in weed identification, the grower is encouraged to obtain confirmation of identification from a knowledgeable weed expert before implementing a weed control strategy. The following websites, among many others, offer photos and guides to the identification of weeds.

California, www.ipm.ucdavis.edu/PMG/weeds_common.html

Illinois, http://web.aces.uiuc.edu/weedid

Iowa, http://www.weeds.iastate.edu/weed-id/weedid.htm

Minnesota, http://www.extension.umn.edu/distribution/cropsystems/DC1352.html

Missouri, http://www.plantsci.missouri.edu/fishel/field_crops.htm

New Jersey, http://www.rce.rutgers.edu/weeds

Virginia, http://www.ppws.vt.edu/weedindex.htm

03
NOXIOUS WEEDS

Noxious weeds are plant species so injurious to agricultural crop interests that they are regulated or controlled by federal and/or state laws. Propagation, growing, or sales of these plants may be controlled. Some states divide noxious weeds into "prohibited" species, which may not be grown or sold, and "restricted" species, which may occur in the state and are considered nuisances or of economic concern for agriculture. States have different lists of plants considered noxious. The federal website below leads to state-based information about noxious weeds.

http://www.ars-grin.gov/cgi-bin/npgs/html/taxweed.pl (2005)

04
WEED CONTROL IN ORGANIC FARMING

Weeds can be a serious threat to vegetable production in organic systems. Weed control is one of the most costly activities in successful organic vegetable production. Some of the recommended main organic weed control strategies include:

Rotate crops.

Cover crops to compete with weeds in the non-crop season.

Be knowledgeable about potential weed contamination of manures, composts, and organic soil amendments.

Employ mechanical and manual control through cultivation, hoeing, mowing, hand weeding, etc.

Clean equipment to minimize transfer of weed propagules from one field to another.

Control weeds at the crop perimeter.

Completely till crop and weeds after final harvest.

Plan for fallow periods with mechanical destruction of weeds.

Use the stale seedbed technique when appropriate.

Use approved weed control materials selected from the Standards lists below.

Encourage crop competition due to optimum crop vigor, correct plant spacing, or shading of weeds.

Use approved soil mulching practices to smother weed seedlings around crops or in crop alleys.

Practice precise placement of fertilizers and irrigation water to minimize availability to weeds in walkways and row middles.

Websites offering information on weed control practices in organic vegetable production include:

- Organic Materials Review Institute (OMRI), http://www.omri.org
- National Organic Standards Board (NOSB), http://www.ams.usda.gov/nosb/index.htm
- National Organic Program (NOP), http://www.ams.usda.gov/nop/nop/nophome.html
- Weed Management Menu—Sustainable Farming Connection, http://www.ibiblio.org/farming-connection/weeds/home.htm

- National Sustainable Agriculture Information Service, http://www.attra.org
- G. Boyhan, D. Granberry, W. T. Kelley, and W. McLaurin, *Growing Vegetables Organically* (University of Georgia, 1999), http://pubs.caes.uga.edu/caespubs/pubcd/b1011-w.html.

05
COVER CROPS AND ROTATION IN WEED MANAGEMENT

Vegetable growers can take advantage of certain cover crops to help control weeds in vegetable production systems and crop rotation systems. Cover crops compete with weeds, reducing the growth and weed seed production capacity of weeds. Cover crops help build soil organic matter and can lead to more vigorous vegetable crops that compete more effectively with weeds. Rotation introduces weed populations to different crops with different weed control options and helps keep herbicide-resistant weed populations from becoming established.

Some websites for cover crops and rotation in vegetable production:

Michigan, www.kbs.msu.edu/extension/covercrops/home.htm

New York, http://www.nysaes.cornell.edu/recommends/4frameset.html

WEED CONTROL WITH HERBICIDES

Chemical weed control minimizes labor and is effective if used with care. The following precautions should be observed:

1. Do not use a herbicide unless the label states that it is registered for that particular crop. Be sure to use as directed by the manufacturer.
2. Use herbicides so that no excessive residues remain on the harvested product, which may otherwise be confiscated. Residue tolerances are established by the U.S. Environmental Protection Agency (EPA).
3. Note that some herbicides kill only certain weeds.
4. Make certain the soil is sufficiently moist for effective action of preemergence sprays. Do not expect good results in dry soil.
5. Keep in mind that postemergence herbicides are most effective when conditions favor rapid weed germination and growth.
6. Avoid using too much herbicide. Overdoses can injure the vegetable crop. Few crops, if any, are entirely resistant.
7. Use less herbicide on light sandy soils than on heavy clay soils. Muck soils require somewhat greater rates than do heavy mineral soils.
8. When using wettable powders, be certain the liquid in the tank is agitated constantly as spraying proceeds.
9. Use a boom and nozzle arrangement that fans out the material close to the ground in order to avoid drift.
10. Thoroughly clean spray tank after use.

CLEANING SPRAYERS AFTER APPLYING HERBICIDES

Sprayers must be kept clean to avoid injury to the crop on which they are to be used for applying insecticides or fungicides as well as to prevent possible deterioration of the sprayers after use of certain materials.

1. Rinse all parts of sprayer with water before and after any special cleaning operation is undertaken.

2. If in doubt about the effectiveness of water alone to clean the herbicide from the tank, pump, boom, hoses, and nozzles, use a cleaner. In some cases, it is desirable to use activated carbon to reduce contamination.
3. Fill the tank with water. Use one of the following materials for each 100 gal water: 5 lb paint cleaner (trisodium phosphate), 1 gal household ammonia, or 5 lb sal soda.
4. If hot water is used, let the solution stand in the tank for 18 hr. If cold water is used, leave it for 36 hr. Pump the solution through the sprayer.
5. Rinse the tank and parts several times with clear water.
6. If copper has been used in the sprayer before a weed control operation is performed, put 1 gal vinegar in 100 gal water and let the solution stay in the sprayer for 2 hr. Drain the solution and rinse thoroughly. Copper interferes with the effectiveness of some herbicides.

DETERMINING RATES OF APPLICATION OF WEED CONTROL MATERIALS

Commercially available herbicide formulations differ in their content of the active ingredient. The label indicates the amount of the active ingredient (lb/gal). Refer to this amount in the table to determine how much of the formulation you need in order to supply the recommended amount of the active ingredient per acre. For calibration of herbicide application equipment, see pages 328–339.

TABLE 7.1. HERBICIIDE DILUTION TABLE: QUANTITY OF LIQUID CONCENTRATES TO USE TO GIVE DESIRED DOSAGE OF ACTIVE CHEMICAL

Active Ingredient Content of Liquid Concentrate (lb/gal)	Active Ingredient Needed (lb/acre): 0.125	0.25	0.50	1	2	3	4
	Liquid Concentrate to Use (pint/acre)						
1	1.0	2.0	4.0	8.0	16.0	24.0	32.0
1½	0.67	1.3	2.6	5.3	10.6	16.0	21.3
2	0.50	1.0	2.0	4.0	8.0	12.0	16.0
3	0.34	0.67	1.3	2.7	5.3	8.0	10.7
4	0.25	0.50	1.0	2.0	4.0	6.0	8.0
5	0.20	0.40	0.80	1.6	3.2	4.8	6.4
6	0.17	0.34	0.67	1.3	2.6	4.0	5.3
7	0.14	0.30	0.60	1.1	2.3	3.4	4.6
8	0.125	0.25	0.50	1.0	2.0	3.0	4.0
9	0.11	0.22	0.45	0.9	1.8	2.7	3.6
10	0.10	0.20	0.40	0.8	1.6	2.4	3.2

Adapted from Spraying Systems Co., Catalog 36, Wheaton, Ill. (1978).

Additional Conversion Tables

http://pmep.cce.cornell.edu/facts-slides-self/facts/gen-peapp-conv-table.html
http://pubs.caes.uga.edu/caespubs/pubcd/b931.htm

SUGGESTED CHEMICAL WEED CONTROL PRACTICES

State recommendations for herbicides vary because the effect of herbicides is influenced by growing area, soil type, temperature, and soil moisture. Growers should consult local authorities for specific recommendations. The EPA has established residue tolerances for those herbicides that may leave injurious residues in or on a harvested vegetable and has approved certain materials, rates, and methods of application. Laws regarding vegetation and herbicides are constantly changing. Growers and commercial applicators should not use a chemical on a crop for which the compound is not registered. Herbicides should be used exactly as stated on the label regardless of information presented here. Growers are advised to give special attention to plant-back restrictions.

07
WEED CONTROL RECOMMENDATIONS

The Cooperative Extension Service in each state publishes recommendations for weed control practices. We present below some websites containing recommendations for weed control in vegetable crops. The list is not exhaustive, and these websites are presented for information purposes. Because weed control recommendations, especially recommended herbicides, may differ from state to state and year to year, growers are encouraged to consult the proper experts in their state for the latest information about weed control.

SELECTED WEBSITES FOR VEGETABLE WEED CONTROL RECOMMENDATIONS

- A. S. Culpepper, "Commercial Vegetables: Weed Control," in *2005 Georgia Pest Management Handbook, Commercial Edition,* http://pubs.caes.uga/caespubs/PMH/PMH-com-veggie-weed.pdf.
- D. W. Monks and W. E. Mitchem, "Chemical Weed Control in Vegetable Crops," in *2005 North Carolina Agricultural Chemicals Manual,* http://ipm.ncsu.edu/agchem/chptr8/817.pdf.
- R. D. William, "Weed Management in Vegetable Crops," in *Pacific Northwest Weed Management Handbook.* (2005), http://pnwpest.org/pnw/weeds.
- B. H. Zandstra, *2005 Weed Control Guide for Vegetable Crops* Michigan State University Extension Bulletin E-433 (Nov. 2004), http://web4.msue.msu.edu/veginfo/bulletins/E433_2005.pdf.
- *Commercial Vegetables Disease, Nematode, and Weed Control Recommendations for 2005* (Alabama), http://www.aces.edu/pubs/docs/A/ANR-0500-A/veg.pdf.
- *Weed Management,* (New York) Integrated Crop and Pest Management Guidelines for Commercial Vegetable Production, (2005), http://www.nysaes.cornell.edu/recommends/4frameset.html.
- S. Post, F. Hale, D. Robinson, R. Straw, and J. Wills, *The 2005 Tennessee Commercial Vegetable Disease, Insect, and Weed Control Guide,* http://www.utextension.utk.edu/publications/pbfiles/PB1282.pdf.
- Florida weed control information can be found at http://edis.ifas.ufl.edu/ and search for "weed control" and author = "W. M. Stall." This brings up weed control documents for individual vegetables.
- Ontario, Canada, http://www.omafra.gov.on.ca/english/environment/hort/references.htm (2005).

PART 8

HARVESTING, HANDLING, AND STORAGE

01
FOOD SAFETY

Additional information on food safety can be found on these websites:

http://www.jifsan.umd.edu/gaps.html
http://www.cfsan.fda.gov.html
http://www.cals.ncsu.edu/hort_sci/hsfoodsafety.html
http://www.ces.ncsu.edu/depts/foodsci/agentinfo/
http://foodsafe.msu.edu/
http://ucgaps.ucdavis.edu/
http://www.gaps.cornell.edu/
http://www.foodriskclearinghouse.umd.edu/
http://www.foodsafety.gov/
http://www.extension.iastate.edu/foodsafety/
http://www.cdc.gov/foodsafety/edu.htm

FOOD SAFETY ON THE FARM

Potential On-Farm Contamination Sources

- Soil
- Irrigation water
- Animal manure
- Inadequately composted manure
- Wild and domestic animals
- Inadequate field worker hygiene
- Harvesting equipment
- Transport containers (field to packing facility)
- Wash and rinse water
- Unsanitary handling during sorting and packaging, in packing facilities, in wholesale or retail operations, and at home
- Equipment used to soak, pack, or cut produce
- Ice
- Cooling units (hydrocoolers)
- Transport vehicles

- Improper storage conditions (temperatures)
- Improper packaging
- Cross-contamination in storage, display, and preparation

From *Food Safety Begins on the Farm: A Grower's Guide* (Ithaca, N.Y.: Cornell University), www.hort.cornell.edu/commercialvegetables/issues/foodsafe.html.

FOOD SAFETY IN VEGETABLE PRODUCTION

Plan Before Planting

- Select site for produce based on land history and location.
- Use careful manure handling.
- Keep good records.

Field Management Considerations

- Optimize irrigation water quality and methods.
- Avoid manure sidedressing.
- Practice good field sanitation.
- Exclude animals and wildlife.
- Emphasize worker training and hygiene.
- Keep records of the above activities.

From *Food Safety Begins on the Farm: A Grower's Guide* (Ithaca, N.Y.: Cornell University), www.hort.cornell.edu/commercialvegetables/issues/foodsafe.html.

FOOD SAFETY IN VEGETABLE HARVEST AND POSTHARVEST PRACTICES

Harvest Considerations

- Clean and sanitize storage facilities and produce contact surfaces prior to harvest.
- Clean harvesting aids each day.
- Emphasize worker hygiene and training.
- Emphasize hygiene to U-Pick customers.
- Keep animals out of the fields.

Postharvest Considerations

- Enforce good worker hygiene.
- Clean and sanitize packing area and lines daily.
- Maintain clean and fresh water.
- Cool produce quickly and maintain cold chain.
- Sanitize trucks before loading.
- Keep animals out of packinghouse and storage facilities.

From *Food Safety Begins on the Farm: A Grower's Guide* (Ithaca, N.Y.: Cornell University), www.hort.cornell.edu/commercialvegetables/issues/foodsafe.html.

Human Pathogens That May Be Associated with Fresh Vegetables

- Soil-associated pathogenic bacteria (*Clostridium botulinum, Listeria moncytogenes*)
- Fecal-associated pathogenic bacteria (*Salmonella* spp., *Shigella* spp., *E. coli* O157:H7, and others)
- Pathogenic parasites (*Cryptosporidium, Cyclospora*)
- Pathogenic viruses (Hepatitis, Norwalk virus, and others)

Basic Principles of Good Agricultural Practices (GAPs)

- Prevention of microbial contamination of fresh produce is favored over reliance on corrective actions once contamination has occurred.
- To minimize microbial food safety hazards in fresh produce, growers or packers should use GAPs in those areas over which they have a degree of control while not increasing other risks to the food supply or the environment.
- Anything that comes in contact with fresh produce has the potential to contaminate it. For most foodborne pathogens associated with produce, the major source of contamination is human or animal feces.
- Whenever water comes in contact with fresh produce, its source and quality dictate the potential for contamination.
- Practices using manure or municipal biosolid wastes should be closely managed to minimize the potential for microbial contamination of fresh produce.
- Worker hygiene and sanitation practices during production, harvesting, sorting, packing, and transport play a critical role in minimizing the potential for microbial contamination of fresh produce.

- Follow all applicable local, state, and federal laws and regulations, or corresponding or similar laws. regulations, or standards for operators outside the United States for agricultural practices.
- Accountability at all levels of the agricultural environment (farms, packing facility, distribution center, and transport operation) is important to a successful food safety program. There must be qualified personnel and effective monitoring to ensure that all elements of the program function correctly and to help track produce back through the distribution channels to the producer.

Adapted from James R. Gorny and Devon Zagory, "Food Safety," in *The Commercial Storage of Fruits, Vegetables, and Florist and Nursery Stocks* (USDA Agriculture Handbook 66, 2004), http://www.ba.ars.usda.gov/hb66/contents.html.

TABLE 8.1. SANITIZING CHEMICALS FOR PACKINGHOUSES

Compound	Advantages	Disadvantages
Chlorine (most widely used sanitizer in packinghouse water systems)	Relatively inexpensive. Broad spectrum—effective on many microbes. Practically no residue left on the commodity.	Corrosive to equipment. Sensitive to pH. Below 6.5 or above 7.5 reduces activity or increases noxious odors. Can irritate skin and damage mucous membranes.
Chlorine Dioxide	Activity is much less pH dependent than chlorine.	Must be generated on-site. Greater human exposure risk than chlorine. Off gassing of noxious gases is common. Concentrated gases can be explosive.
Peroxyacetic Acid	No known toxic residues or byproducts. Produces very little off gassing. Less affected by organic matter than chlorine. Low corrosiveness to equipment.	Activity is reduced in the presence of metal ions. Concentrated product is toxic to humans. Sensitive to pH. Greatly reduced activity at pH above 7–8.

TABLE 8.1. SANITIZING CHEMICALS FOR PACKINGHOUSES (*Continued*)

Compound	Advantages	Disadvantages
Ozone	Very strong oxidizer/ sanitizer. Can reduce pesticide residues in the water. Less sensitive to pH than chlorine (but breaks down much faster above ~pH 8.5). No known toxic residues or byproducts.	Must be generated on-site. Ozone gas is toxic to humans. Off gassing can be a problem. Treated water should be filtered to remove particulates and organic matter. Corrosive to equipment (including rubber and some plastics). Highly unstable in water—half-life ~15 min; may be less than 1 min in water with organic matter or soil.

Note: Although quaternary ammonia is an effective sanitizer with useful properties and can be used to sanitize equipment, it is not registered for contact with food.

Adapted from S. A. Sargent, M. A. Ritenour, and J. K. Brecht, "Handling, Cooling, and Sanitation Techniques for Maintaining Postharvest Quality," in *Vegetable Production Handbook* (University of Florida, 2005–2006).

TRACEABILITY OF PRODUCE IN THE UNITED STATES

Traceability has been a critical component of the produce industry for many years. Historically, the perishability of produce and the potential for deterioration during cross-country shipment demanded better recordkeeping to ensure correct payment to growers. Because produce must be packed in relatively small boxes to minimize damage, implementation of traceability has also been relatively low in cost. The industry is in a much better position to adapt to new concerns than industries where bulk sales are the norm and segregation and traceability involves new costs. Currently, two systems of information are involved in produce. First, there are physical labels on boxes and sometimes on pallets. For general business purposes, it is important to be able to identify the product in the boxes. Various state laws require box information, and marketing orders also often require additional box information. Pallet tags are completely voluntary. Second, a paper or electronic trail allows traceback between links in the marketing chain, though each link may use a different traceability system. U.S. and Canadian produce organizations are looking at ways to promote a universal traceability system. They recommend that shipper name, pallet tag number (if available), and lot number be part of the paperwork at each link. This would effectively combine information on boxes and the paper or electronic trail. Such a system would require developing a standardized system of barcodes or other machine-readable information as well as shipper and buyer investment in machines to apply and read codes. One of the challenges to developing a compelling technical solution that all market participants would use voluntarily is to ensure that all segments of the industry can afford the costs of the new system.

Perishable Agricultural Commodities Act

Key Legislation and Dates: Perishable Agricultural Commodities Act (PACA) was enacted in 1930.

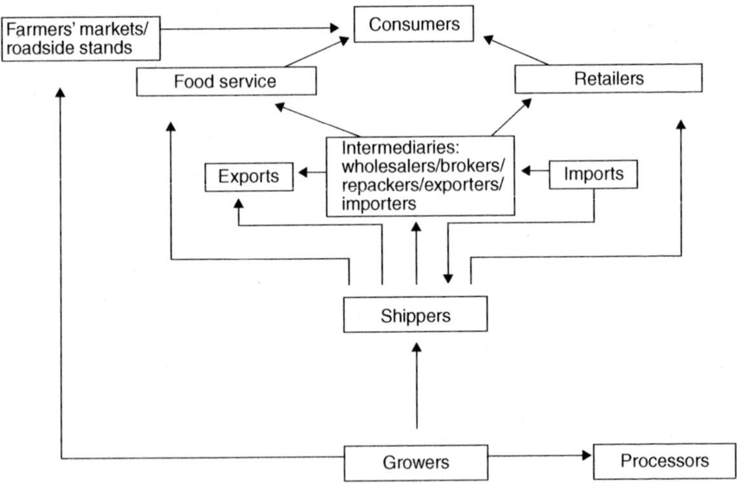

Figure 8.1. Tracing fresh produce through the food marketing system.

Objective: PACA was enacted to promote fair trading practices in the fruit and vegetable industry. The objective of the recordkeeping is to facilitate the marketing of fruit and vegetables, to verify claims, and to minimize misrepresentation of the condition of the item, particularly when long distances separate the traders.

Coverage: Fruit and vegetables.

Recordkeeping Required: PACA calls for complete and accurate recordkeeping and disclosure for shippers, brokers, and other first handlers of produce selling on behalf of growers. PACA has extensive recordkeeping requirements with respect to who buyers and sellers are, what quantities and kinds of produce is transacted, and when and how the transaction takes place. PACA regulations recognize that the varied fruit and vegetable industries have different recordkeeping needs, and the regulations allow for this variance. Records must be kept for 2 years from the closing date of the transaction.

Elise Golan et al., *Traceability in the U.S. Food Supply: Economic Theory and Industry Studies* (USDA ERS AER830, 2004), http://www.ers.usda.gov/publications/aer830.

02
GENERAL POSTHARVEST HANDLING PROCEDURES

Additional information on postharvest handling can be found on these websites:

Agriculture and Agri-Food Canada, www.agr.gc.ca

California Department of Food and Agriculture, www.cdfa.ca.gov

Environmental Protection Agency (EPA), Office of Pesticide Programs, www.epa.gov

Food and Agriculture Organization (FAO), Information Network on Postharvest Operations (InPhO), www.fao.org/inpho

Hawaii Agriculture Food Quality and Safety, http://www.hawaiiag.org/foodsaf.htm

Information Network on Postharvest Operations, www.fao.org/inpho

North Carolina State University Cooling and Handling Publications, http://www.bae.ncsu.edu/programs/extension/publicat/postharv/index.html

UC Davis Postharvest Technology Research and Information Center, http://postharvest.ucdavis.edu

USDA/Agricultural Marketing Service/Fruit and Vegetable Division, www.usda.gov/ams/fruitveg.htm

USDA/Economic Research Service, http://www.ers.usda.gov/

USDA Food Safety and Inspection Service, www.fsis.usda.gov

USDA National Agricultural Statistics Service, http://www.nass.usda.gov/

University of Florida Market Information System, http://marketing.ifas.ufl.edu/

University of Florida Postharvest Programs, http://postharvest.ifas.ufl.edu

TABLE 8.2. LEAFY, FLORAL, AND SUCCULENT VEGETABLES

Step	Function
1.	Harvesting mostly by hand; some harvesting aids are in use.
2.	Transport to packinghouse and unloading if not field packed.
3.	Cutting and trimming (by harvester or by different worker on mobile packing line or in packinghouse).
4.	Sorting and manual sizing (as above).
5.	Washing or rinsing.
6.	Wrapping (e.g., cauliflower, head lettuce) or bagging (e.g., celery).
7.	Packing in shipping containers (waxed fiberboard or plastic to withstand water or ice exposure for cooling).
8.	Palletization of shipping containers.
9.	Cooling methods. • Vacuum cooling: lettuce • Hydrovacuum cooling: cauliflower, celery • Hydrocooling: artichoke, celery, green onion, leaf lettuce, leek, spinach • Package ice: broccoli, parsley, spinach • Room cooling: artichoke, cabbage
10.	Transport, destination handling, retail handling.

TABLE 8.3. UNDERGROUND STORAGE ORGAN VEGETABLES

Step	Function
1.	Mechanical harvest (digging, lifting), except hand harvesting of sweet potato.
2.	Curing (in field) of potato, onion, garlic, and tropical crops.
3.	Field storage of potatoes and tropical storage organ vegetables in pits, trenches, or clamps.
4.	Collection into containers or into bulk trailers.
5.	Transport to packinghouse and unloading.
6.	Cleaning by dry brushing or with water.
7.	Sorting to eliminate defects.
8.	Sizing.
9.	Packing in bags or cartons; consumer packs placed within master containers.
10.	Palletization of shipping containers.
11.	Cooling methods. a. Hydrocooling of temperate storage roots and tubers. b. Room cooling of potato, onion, garlic, and tropical storage organs.
12.	Curing. a. Forced-air drying (onion and garlic). b. High temperature and relative humidity (potato and tropical storage organs).
13.	Storage. a. Ventilated storage of potato, onion, garlic, and sweet potato in cellars and warehouses. b. Temporary storage of temperate storage organ vegetables. c. Long-term storage of potato, onion, garlic, and tropical crops following curing.
14.	Fungicide treatment (sweet potato); sprout inhibitor (potato).
15.	Transport, destination handling, retail handling.

TABLE 8.4. IMMATURE FRUIT VEGETABLES

Step	Function
1.	Harvesting mostly by hand into buckets or trays; some harvesting aids are in use.
	a. Sweet corn, snap bean, and pea are also harvested mechanically.
	b. Field-packed vegetables are usually not washed, but they may be wiped with a moist cloth or spray-washed on a mobile packing line.
2.	For packinghouse operations, stacking buckets or trays on trailers or transferring to shallow pallet bins.
3.	Transporting harvested vegetables to packinghouse.
4.	Unloading by dry or wet dump.
5.	Washing or rinsing.
6.	Sorting to eliminate defects.
7.	Waxing cucumber and pepper.
8.	Sizing.
9.	Packing in shipping containers by weight or count.
10.	Palletizing shipping containers.
11.	Cooling methods:
	a. Hydrocooling: bean, pea, sweet corn.
	b. Forced-air cooling: chayote, cucumber, eggplant, okra, summer squash.
	c. Slush-ice cooling and vacuum cooling: sweet corn.
12.	Temporary storage.
13.	Transporting, destination handling, retail handling.

TABLE 8.5. MATURE FRUIT VEGETABLES

Step	Function
1.	Harvesting.
2.	Hauling to the packinghouse or processing plant.
3.	Cleaning.
4.	Sorting to eliminate defects.
5.	Waxing (tomato, pepper).
6.	Sizing and sorting into grades.
7.	Packing—shipping containers.
8.	Palletization and unitization.
9a.	Curing of winter squash and pumpkin.
9b.	Ripening of melon and tomato with ethylene.
10.	Cooling (hydrocooling, room cooling, forced-air cooling).
11.	Temporary storage.
12.	Loading into transport vehicles.
13.	Destination handling (distribution centers, wholesale markets, etc.).
14.	Delivery to retail.
15.	Retail handling.

TABLE 8.6. APPROXIMATE TIME FROM VEGETABLE PLANTING TO MARKET MATURITY UNDER OPTIMUM GROWING CONDITIONS

Vegetable	Time to Market Maturity[1] (days)		
	Early Variety	Late Variety	Common Variety
Bean, broad	—	—	120
Bean, bush	48	60	—
Bean, edible soy	84	115	—
Bean, pole	58	70	—
Bean, lima, bush	65	80	—
Bean, lima, pole	80	88	—
Beet	50	65	—
Broccoli[2]	60	85	—
Broccoli raab	—	—	70
Brussels sprouts[2]	90	120	—
Cabbage	70	120	—
Cardoon	—	—	120
Carrot	50	95	—
Cauliflower[2]	55	90	—
Celeriac	—	—	110
Celery[2]	90	125	—
Chard, Swiss	50	60	—
Chervil	—	—	60
Chicory	65	150	—
Chinese cabbage	50	80	—
Chive	—	—	90
Collards	70	85	—
Corn, sweet	60	95	—
Corn salad	—	—	60
Cress	—	—	45
Cucumber, pickling	48	58	—
Cucumber, slicing	55	70	—
Dandelion	—	—	85

TABLE 8.6. APPROXIMATE TIME FROM VEGETABLE PLANTING TO MARKET MATURITY UNDER OPTIMUM GROWING CONDITIONS (*Continued*)

Vegetable	Time to Market Maturity[1] (days)		
	Early Variety	Late Variety	Common Variety
Eggplant[2]	60	80	—
Endive	80	100	—
Florence fennel	—	—	100
Kale	—	—	55
Kohlrabi	50	60	—
Leek	—	—	120
Lettuce, butterhead	55	70	—
Lettuce, cos	65	75	—
Lettuce, head	70	85	—
Lettuce, leaf	40	50	—
Melon, cantaloupe	75	105	
Melon, casaba	—	—	110
Melon, honeydew	90	110	—
Melon, Persian	—	—	110
Okra	50	60	—
Onion, dry	90	150	—
Onion, green	45	70	—
Parsley	65	75	—
Parsley root	—	—	90
Parsnip	—	—	120
Pea	56	75	—
Pea, edible-podded	60	75	—
Pepper, hot[2]	60	95	—
Pepper, sweet[2]	65	80	—
Potato	90	120	—
Pumpkin	85	120	—
Radicchio	90	95	—
Radish	22	30	—
Radish, winter	50	65	—
Roselle	—	—	175
Rutabaga	—	—	90

TABLE 8.6. APPROXIMATE TIME FROM VEGETABLE PLANTING TO MARKET MATURITY UNDER OPTIMUM GROWING CONDITIONS (*Continued*)

	Time to Market Maturity[1] (days)		
Vegetable	Early Variety	Late Variety	Common Variety
Salsify	—	—	150
Scolymus	—	—	150
Scorzonera	—	—	150
Sorrel	—	—	60
Southern pea	65	85	—
Spinach	37	45	—
Squash, summer	40	50	—
Squash, winter	80	120	—
Sweet potato	120	150	—
Tomatillo	—	—	80
Tomato[2]	60	85	—
Tomato, processing	118	130	—
Turnip	35	50	—
Watercress	—	—	180
Watermelon	75	95	—

[1] Maturity may vary depending on season, latitude, production practices, variety, and other factors.
[2] Time from transplanting. See page 63–64.

TABLE 8.7. APPROXIMATE TIME FROM POLLINATION OF VEGETABLES TO MARKET MATURITY UNDER WARM GROWING CONDITIONS

Vegetable	Time to Market Maturity[1] (days)
Bean	7–10
Cantaloupe	42–46
Corn,[2] market	18–23
Corn,[2] processing	21–27
Cucumber, pickling (¾–1⅛ in. in diameter)	4–5
Cucumber, slicing	15–18
Eggplant (⅔ maximum size)	25–40
Okra	4–6
Pepper, green stage (about maximum size)	45–55
Pepper, red stage	60–70
Pumpkin, Connecticut Field	80–90
Pumpkin, Dickinson	90–110
Pumpkin, Small Sugar	65–75
Squash, summer, crookneck	6–7[3]
Squash, summer, straightneck	5–6[3]
Squash, summer, scallop	4–5[3]
Squash, summer, zucchini	3–4[3]
Squash, winter, banana	70–80
Squash, winter, Boston Marrow	60–70
Squash, winter, buttercup	60–70
Squash, winter, butternut	60–70
Squash, winter, Golden Delicious	60–70
Squash, winter, hubbard	80–90
Squash, winter, Table Queen or acorn	55–60
Strawberry	25–42
Tomato, mature green stage	35–45
Tomato, red ripe stage	45–60
Watermelon	42–45

[1] Maturity may vary depending on season, latitude, production practices, variety, and other factors.
[2] Days from 50% silking.
[3] For a weight of ¼–½ lb.

TABLE 8.8. ESTIMATING YIELDS OF CROPS

Predicting crop yields before the harvest aids in scheduling the harvests of various fields for total yields and allows harvesting to obtain highest yields of a particular grade or stage of maturity. To estimate yields, follow these steps:

1. Select and measure a typical 10-ft section of a row. If the field is variable or large, you may want to select several 10-ft sections.
2. Harvest the crop from the measured section or sections.
3. Weigh the entire sample for total yields, or grade the sample and weigh the graded sample for yield of a particular grade.
4. If you have harvested more than one 10-ft section, divide the yield by the number of sections harvested.
5. Multiply the sample weight by the conversion factor in the table for your row spacing. The value obtained will equal hundredweight (cwt) per acre.

Conversion Factors for Estimating Yields

Row Spacing (in.)	Multiply Sample Weight (lb) by Conversion Factor to Obtain cwt/acre
12	43.6
15	34.8
18	29.0
20	26.1
21	24.9
24	21.8
30	17.4
36	14.5
40	13.1
42	12.4
48	10.9

Example 1: A 10-ft sample of carrots planted in 12-in. rows yields 9 lb of No. 1 carrots.

$$9 \times 43.6 = 392.4 \text{ cwt/acre}$$

Example 2: The average yield of three 10-ft samples of No. 1 potatoes planted in 36-in. rows is 26 lb.

$$26 \times 14.5 = 377 \text{ cwt/acre}$$

TABLE 8.9. YIELDS OF VEGETABLE CROPS

Vegetable	Approximate Average Yield in the United States (cwt/acre)	Good Yield (cwt/acre)
Artichoke	125	160
Asparagus	30	45
Bean, fresh market	62	100
Bean, processing	75	135
Bean, lima, processing	25	40
Beet, fresh market	140	200
Beet, processing	320	400
Broccoli	145	200
Brussels sprouts	180	200
Cabbage, fresh market	320	450
Cabbage, processing	500	800
Carrot, fresh market	350	800
Carrot, processing	520	700
Cauliflower	170	200
Celeriac	—	200
Celery	650	750
Chard, Swiss	—	150
Chinese cabbage, napa	—	400
Chinese cabbage, pak choi	—	300
Collards	120	200
Corn, fresh market	110	200
Corn, processing	150	200
Cucumber, fresh market	185	300
Cucumber, processing	110	300
Eggplant	250	350
Endive, escarole	180	200
Garlic	165	200
Horseradish	—	80
Lettuce, head	360	400
Lettuce, leaf	200	325
Lettuce, romaine	300	350
Melon, cantaloupe	250	250
Melon, honeydew	205	250
Melon, Persian	130	160

TABLE 8.9. YIELDS OF VEGETABLE CROPS (*Continued*)

Vegetable	Approximate Average Yield in the United States (cwt/acre)	Good Yield (cwt/acre)
Okra	60	150
Onion	420	650
Pea, fresh market	40	60
Pea, processing (shelled)	30	45
Pepper, bell	290	375
Pepper, chile (fresh and dried)	100	230
Pepper, pimiento	—	100
Potato	315	400
Pumpkin	215	400
Radish	90	200
Rhubarb	—	200
Rutabaga	—	400
Snowpea	—	80
Southern pea	—	35
Spinach, fresh market	160	225
Spinach, processing	180	220
Squash, summer	160	300
Squash, winter	—	400
Sweet potato	145	300
Strawberry	400	600
Tomato, fresh market	290	400
Tomato, processing	700	900
Tomato, cherry	—	600
Turnip	—	400
Watermelon	260	500

TABLE 8.10. STATUS OF HAND VERSUS MECHANICAL HARVEST OF VEGETABLES

Acreage Hand Harvested (%)	Vegetable			
76–100	Artichoke	Asparagus	Broccoli[1]	Cabbage
	Cauliflower	Celery	Cucumber[1]	Lettuce
	Green onion	Collards	Cress	Dandelion
	Eggplant	Endive	Escarole	Fennel
	Kale	Kohlrabi	Mushroom[1]	Okra
	Pepper	Rapini	Rhubarb[1]	Romaine
	Sorrel	Squash	Watercress	Cassava
	Celeriac	Ginger	Parsley root	Parsnip
	Rutabaga	Salsify	Turnip	Taro[1]
	Jerusalem artichoke			
51–75	Sweet potato	Mustard greens	Parsley	Swiss chard
	Turnip greens			
26–50	Dry onion	Pumpkin[1]	Tomato[1]	
0–25	Beet[1]	Carrot	Potato[1]	Lima bean[1]
	Snap bean[1]	Sweet corn[1]	Spinach[1]	Horseradish[1]
	Pea[1]	Garlic	Brussels sprouts[1]	Malanga
	Boniato	Radish		

Adapted from A. A. Kader (ed.), *Postharvest Technology of Horticultural Crops*, 2nd ed., (University of California, Division of Agriculture and Natural Resources Publication 3311, 1992).

[1] More than 50% of the crop is processed.

TABLE 8.11. COMPARISON OF TYPICAL PRODUCT EFFECTS AND COST FOR COMMON COOLING METHODS

Product Effect	Forced-Air	Hydro	Vacuum	Water Spray	Ice	Room
Cooling Time (h)	1–10	0.1–1.0	0.3–2.0	0.3–2.0	0.1 to 0.3[1]	20–100
Product moisture loss (%)	0.1–2.0	0–0.5	2.0–4.0	No data	No data	0.1–2.0
Water contact with product	No	Yes	No	Yes	Yes, unless bagged	No
Potential for decay contamination	Low	High[2]	None	High[2]	Low	Low
Capital cost	Low	Low	Medium	Medium	High	High
Energy efficiency	Low	High	High	Medium	Low	Low
Water-resistant packaging needed	No	Yes	No	Yes	Yes	No
Portable	Sometimes	Rarely done	Common	Common	Common	No
Feasibility of in-line cooling	Rarely done	Yes	No	No	Rarely done	No

[1] Top icing can take much longer.

[2] Recirculated water must be constantly sanitized to minimize accumulation of decay pathogens.

Adapted from James F. Thompson, "Pre-cooling and Storage Facilities" in *The Commercial Storage of Fruits, Vegetables, and Nursery Stocks* (USDA Agriculture Handbook 66, 2004), http://www.ba.ars.usda.gov/hb66/contents.html.

TABLE 8.12. GENERAL COOLING METHODS FOR VEGETABLES

Method[1]	Vegetable	Comments
Room cooling	All vegetables	Too slow for most perishable commodities. Cooling rates vary extensively within loads, pallets, and containers.
Forced-air cooling (pressure cooling)	Strawberry, fruit-type vegetables, tubers, cauliflower	Much faster than room cooling; cooling rates uniform if properly used. Container venting and stacking requirements are critical to effective cooling.
Hydrocooling	Stems, leafy vegetables, some fruit-type vegetables	Very fast cooling; uniform cooling in bulk if properly used, but may vary extensively in packed shipping containers; daily cleaning and sanitation measures essential; product must tolerate wetting; water-tolerant shipping containers may be needed.
Package icing	Roots, stems, some flower-type vegetables, green onion, Brussels sprouts	Fast cooling; limited to commodities that can tolerate water-ice contact; water-tolerant shipping containers are essential.

Vacuum cooling	Leafy vegetables; some stem and flower-type vegetables	Commodities must have a favorable surface-to-mass ratio for effective cooling. Causes about 1% weight loss for each 10°F cooled. Adding water during cooling prevents this weight loss but equipment is more expensive, and water-tolerant shipping containers are needed.
Transit cooling: Mechanical refrigeration	All vegetables	Cooling in most available equipment is too slow and variable; generally not effective.
Top icing and channel icing	Some roots, stems, leafy vegetables, cantaloupe.	Slow and irregular, top-ice weight reduces net payload; water-tolerant shipping containers needed.

Adapted from A. A. Kader (ed.), *Postharvest Technology of Horticultural Crop,* 2nd ed. (University of California, Division of Agriculture and Natural Resources Publication 3311, 1992).

[1] For these methods to be effective, the product must be cooled continuously until reaching the consumer.

TABLE 8.13. SPECIFIC COOLING METHODS FOR VEGETABLES

	Size of Operation		
Vegetable	Large	Small	Remarks
Leafy Vegetables			
Cabbage	VC, FA	FA	
Iceberg lettuce	VC	FA	
Kale, collards	VC, R, WVC	FA	
Leaf lettuces, spinach, endive, escarole, Chinese cabbage, pak choi, romaine	VC, FA, WVC, HC	FA	
Root Vegetables			
With tops	HC, PI, FA	HC, FA	Carrots can be VC.
Topped	HC, PI	HC, PI, FA	
Irish potato	R w/evap coolers		With evap coolers, facilities should be adapted to curing.
Sweet potato	HC	R	
Stem and Flower Vegetables			
Artichoke	HC, PI	FA, PI	
Asparagus	HC	HC	
Broccoli, Brussels sprouts	HC, FA, PI	FA, PI	
Cauliflower	FA, VC	FA	

Celery, rhubarb	HC, WVC, VC	HC, FA	
Green onion, leek	PI, HC, WVC	PI	
Mushroom	FA, VC	FA	
Pod Vegetables			
Bean	HC, FA	FA	
Pea	FA, PI, VC	FA, PI	
Bulb Vegetables			
Dry onion	R	R, FA	Should be adapted to curing.
Garlic	R		
Fruit-type Vegetables			
Cucumber, eggplant	R, FA, FA-EC	FA, FA-EC	Fruit-type vegetables are chilling sensitive, but at varying temperature. See pages 444–448.
Melons			
Cantaloupe	HC, FA, PI	FA, FA-EC	
Crenshaw, honeydew, casaba	FA, R	FA, FA-EC	
Watermelon	FA, HC	FA, FA-EC	
Pepper	R, FA, FA-EC, VC	FA, FA-EC	
Summer squash, okra	R, FA, FA-EC	FA, FA-EC	
Sweet corn	HC, VC, PI	HC, FA, PI	
Tomatillo	R, FA, FA-EC	FA, FA-EC	
Tomato	R, FA, FA-EC		

TABLE 8.13. SPECIFIC COOLING METHODS FOR VEGETABLES (*Continued*)

Vegetable	Size of Operation		Remarks
	Large	Small	
Winter squash	R	R	
Fresh Herbs			
Not packaged	HC, FA	FA, R	Can be easily damaged by water beating in HC.
Packaged	FA	FA, R	
Cactus			
Leaves (nopalitos)	R	FA	
Fruit (tunas or prickly pears)	R	FA	

Adapted from A. A. Kader (ed.), *Postharvest Technology of Horticultural Crops,* 3rd ed. (University of California, Division of Agriculture and Natural Resources Publication 3311, 2002).

R = Room Cooling; FA = Forced-air Cooling; HC = Hydrocooling; VC = Vacuum Cooling; WVC = Water Spray Vacuum Cooling; FA-EC = Forced-air Evaporative Cooling; PI = Package Icing.

TABLE 8.14. RELATIVE PERISHABILITY AND POTENTIAL STORAGE LIFE OF FRESH VEGETABLES IN AIR AT NEAR OPTIMUM STORAGE TEMPERATURE AND RELATIVE HUMIDITY

Potential Storage Life (weeks)			
<2	2–4	4–8	8–16
Asparagus	Artichoke	Beet	Garlic
Bean sprouts	Green bean	Carrot	Onion
Broccoli	Brussels sprouts	Potato (immature)	Potato (mature)
Cantaloupe	Cabbage	Radish	Pumpkin
Cauliflower	Celery		Winter squash
Green onion	Eggplant		Sweet potato
Leaf lettuce	Head lettuce		Taro
Mushroom	Mixed melons		Yam
Pea	Okra		
Spinach	Pepper		
Sweet corn	Summer squash		
Tomato (ripe)	Tomato (partially ripe)		
Fresh-cut vegetables			

Adapted from A. A. Kader (ed.), *Postharvest Technology of Horticultural Crops,* 3rd ed. (University of California, Division of Agriculture and Natural Resources Publication 3311, 2002).

TABLE 8.15. RECOMMENDED TEMPERATURE AND RELATIVE HUMIDITY CONDITIONS AND APPROXIMATE STORAGE LIFE OF FRESH VEGETABLES

For more detailed information about specific commodities, go to one of the following websites:

http://www.ba.ars.usda.gov/hb66/index.html
http://postharvest.ucdavis.edu/produce/producefacts/index.shtml

	Storage Conditions		
Vegetable	Temperature (°F)	Relative Humidity (%)	Approximate Storage Life
Amaranth	32–36	95–100	10–14 days
Anise	32–36	90–95	2–3 weeks
Artichoke, globe	32	95–100	2–3 weeks
Artichoke, Jerusalem	31–32	90–95	4 months
Asparagus	36	95–100	2–3 weeks
Bean, fava	32	90–95	1–2 weeks
Bean, lima	37–41	95	5–7 days
Bean, snap	40–45	95	7–10 days
Bean, yardlong	40–45	95	7–10 days
Beet, bunched	32	98–100	10–14 days
Beet, topped	32	98–100	4 months
Bitter melon	50–54	85–90	2–3 weeks
Bok choy	32	95–100	3 weeks
Boniato	55–60	85–90	4–5 months
Broccoli	32	95–100	10–14 days
Brussels sprouts	32	95–100	3–5 weeks
Cabbage, early	32	98–100	3–6 weeks
Cabbage, late	32	95–100	5–6 months
Cabbage, Chinese	32	95–100	2–3 months
Cactus, leaves	41–50	90–95	2–3 weeks
Cactus, pear	41	90–95	3 weeks
Calabaza	50–55	50–70	2–3 months
Carrot, bunched	32	98–100	10–14 days
Carrot, topped	32	98–100	6–8 months
Cassava	32–41	85–90	1–2 months

TABLE 8.15. RECOMMENDED TEMPERATURE AND RELATIVE HUMIDITY CONDITIONS AND APPROXIMATE STORAGE LIFE OF FRESH VEGETABLES (*Continued*)

Vegetable	Storage Conditions		Approximate Storage Life
	Temperature (°F)	Relative Humidity (%)	
Cauliflower	32	95–98	3–4 weeks
Celeriac	32	98–100	6–8 months
Celery	32	98–100	1–2 months
Chard	32	95–100	10–14 days
Chayote	45	85–90	4–6 weeks
Chicory, witloof	36–38	95–98	2–4 weeks
Chinese broccoli	32	95–100	10–14 days
Collards	32	95–100	10–14 days
Cucumber, slicing	50–54	85–90	10–14 days
Cucumber, pickling	40	95–100	7 days
Daikon	32–34	95–100	4 months
Eggplant	50	90–95	1–2 weeks
Endive, escarole	32	95–100	2–4 weeks
Garlic	32	65–70	6–7 months
Ginger	55	65	6 months
Greens, cool-season	32	95–100	10–14 days
Greens, warm-season	45–50	95–100	5–7 days
Horseradish	30–32	98–100	10–12 months
Jicama	55–65	85–90	1–2 months
Kale	32	95–100	2–3 weeks
Kohlrabi	32	98–100	2–3 months
Leek	32	95–100	2 months
Lettuce	32	98–100	2–3 weeks
Malanga	45	70–80	3 months
Melon			
Cantaloupe	36–41	95	2–3 weeks
Casaba	45–50	85–90	3–4 weeks
Crenshaw	45–50	85–90	2–3 weeks
Honeydew	41–50	85–90	3–4 weeks

TABLE 8.15. **RECOMMENDED TEMPERATURE AND RELATIVE HUMIDITY CONDITIONS AND APPROXIMATE STORAGE LIFE OF FRESH VEGETABLES (*Continued*)**

Vegetable	Storage Conditions		Approximate Storage Life
	Temperature (°F)	Relative Humidity (%)	
Persian	45–50	85–90	2–3 weeks
Watermelon	50–59	90	2–3 weeks
Mushroom	32	90	7–14 days
Mustard greens	32	90–95	7–14 days
Okra	45–50	90–95	7–10 days
Onion, dry	32	65–70	1–8 months
Onion, green	32	95–100	3 weeks
Parsley	32	95–100	8–10 weeks
Parsnip	32	98–100	4–6 months
Pea, English, snow, snap	32–34	90–98	1–2 weeks
Pepino	41–50	95	4 weeks
Pepper, chile	41–50	85–95	2–3 weeks
Pepper, sweet	45–50	95–98	2–3 weeks
Potato, early[1]	50–59	90–95	10–14 days
Potato, late[2]	40–54	95–98	5–10 months
Pumpkin	54–59	50–70	2–3 months
Radicchio	32–34	95–100	3–4 weeks
Radish, spring	32	95–100	1–2 months
Radish, winter	32	95–100	2–4 months
Rhubarb	32	95–100	2–4 weeks
Rutabaga	32	98–100	4–6 months
Salisfy	32	95–98	2–4 months
Scorzonera	32–34	95–98	6 months
Shallot	32–36	65–70	—
Southern pea	40–41	95	6–8 days
Spinach	32	95–100	10–14 days
Sprouts			
Alfalfa	32	95–100	7 days
Bean	32	95–100	7–9 days

TABLE 8.15. RECOMMENDED TEMPERATURE AND RELATIVE HUMIDITY CONDITIONS AND APPROXIMATE STORAGE LIFE OF FRESH VEGETABLES (*Continued*)

Vegetable	Storage Conditions: Temperature (°F)	Storage Conditions: Relative Humidity (%)	Approximate Storage Life
Radish	32	95–100	5–7 days
Squash, summer	45–50	95	1–2 weeks
Squash, winter[3]	54–59	50–70	2–3 months
Strawberry	32	90–95	5–7 days
Sweet corn, shrunken-2	32	90–95	10–14 days
Sweet corn, sugary	32	95–98	5–8 days
Sweet potato[4]	55–59	85–95	4–7 months
Tamarillo	37–40	85–95	10 weeks
Taro	45–50	85–90	4 months
Tomatillo	45–55	85–90	3 weeks
Tomato, mature green	50–55	90–95	2–5 weeks
Tomato, firm ripe	46–50	85–90	1–3 weeks
Turnip greens	32	95–100	10–14 days
Turnip root	32	95	4–5 months
Water chestnut	32–36	85–90	2–4 months
Watercress	32	95–100	2–3 weeks
Yam	59	70–80	2–7 months

Adapted from A. A. Kader (ed.), *Postharvest Technology of Horticultural Crops*, 3rd ed. (University of California, Division of Agriculture and Natural Resources Publication 3311, 2002).

[1] Winter, spring, or summer-harvested potatoes are usually not stored. However, they can be held 4–5 months at 40°F if cured 4 or more days at 60–70°F before storage. Potatoes for chips should be held at 70°F or conditioned for best chip quality.

[2] Fall-harvested potatoes should be cured at 50–60°F and high relative humidity for 10–14 days. Storage temperatures for table stock or seed should be lowered gradually to 38–40°F. Potatoes intended for processing should be stored at 50–55°F; those stored at lower temperatures or with a high reducing sugar content should be conditioned at 70°F for 1–4 weeks or until cooking tests are satisfactory.

[3] Winter squash varieties differ in storage life.

[4] Sweet potatoes should be cured immediately after harvest by holding at 85°F and 90–95% relative humidity for 4–7 days.

TABLE 8.16. POSTHARVEST HANDLING OF FRESH CULINARY HERBS

Herb	Storage Conditions		Ethylene Sensitivity	Approximate Storage Life
	Temperature (°F)	Relative Humidity (%)		
Basil	50	90	High	7 days
Chives	32	95–100	Medium	—
Cilantro	32–34	95–100	High	2 weeks
Dill	32	95–100	High	1–2 weeks
Epazote	32–41	90–95	Medium	1–2 weeks
Mint	32	95–100	High	2–3 weeks
Oregano	32–41	90–95	Medium	1–2 weeks
Parsley	32	95–100	High	1–2 months
Perilla	50	95	Medium	7 days
Sage	32–50	90–95	Medium	2–3 weeks
Thyme	32	90–95	—	2–3 weeks

Adapted from A. A. Kader (ed.), *Postharvest Technology of Horticultural Crops,* 3rd ed. (University of California, Division of Agriculture and Natural Resources Publication 3311, 2002).

TABLE 8.17. RESPIRATION RATES OF FRESH CULINARY HERBS

Herb	Respiration Rates (mg/kg/hr of CO_2)		
	32°F	50°F	68°F
Basil	36	71	167
Chervil	12	80	170
Chinese chive	54	99	432
Chives	22	110	540
Coriander	22	nd	nd
Dill	22	103	nd
Fennel	19[1]	nd	32
Ginger	nd	nd	6[2]
Ginseng	6	15	nd
Marjoram	28	68	nd
Mint	20	76	252
Oregano	22	101	176
Sage	36	103	157
Tarragon	40	94	234
Thyme	12	25	52

Adapted from *The Commercial Storage of Fruits, Vegetables, and Florist and Nursery Stocks* (USDA, ARS Agriculture Handbook 66, 2004), http://www.ba.ars.usda.gov/hb66/contents.html.

[1] At 36°F

[2] At 55°F

TABLE 8.18. AVERAGE RESPIRATION RATES OF VEGETABLES AT VARIOUS TEMPERATURES

Vegetable	Respiration Rate (mg/kg/hr of CO_2)					
	32°F	41°F	50°F	59°F	68°F	77°F
Artichoke, globe	30	43	71	110	193	nd[1]
Artichoke, Jerusalem	10	12	19	50	nd	nd
Asparagus[2]	60	105	215	235	270	nd
Bean, snap	20	34	58	92	130	nd
Bean, long	40	46	92	202	220	nd
Beet	5	11	18	31	60	nd
Broccoli	21	34	81	170	300	nd
Brussels sprouts	40	70	147	200	276	nd
Cabbage	5	11	18	28	42	62
Cabbage, Chinese	10	12	18	26	39	nd
Carrot, topped	15	20	31	40	25	nd
Cauliflower	17	21	34	46	79	92
Celeriac	7	13	23	35	45	nd
Celery	15	20	31	40	71	nd
Chicory	3	6	13	21	37	nd
Cucumber	nd	nd	26	29	31	37
Eggplant, American	nd	nd	nd	69[3]	nd	nd
Eggplant, Japanese	nd	nd	nd	131[3]	nd	nd

Eggplant, white egg	nd	nd	nd	113[3]	nd	nd
Endive	45	52	73	100	133	200
Garlic	8	16	24	22	20	nd
Jicama	6	11	14	nd	6	nd
Kohlrabi	10	16	31	46	nd	nd
Leek	15	25	60	96	110	115
Lettuce, head	12	17	31	39	56	82
Lettuce, leaf	23	30	39	63	101	147
Luffa	14	27	36	63	79	nd
Melon						
Cantaloupe	6	10	15	37	55	67
Honeydew	nd	8	14	24	30	33
Watermelon	nd	4	8	nd	21	nd
Mushroom	35	70	97	nd	264	nd
Nopalito	nd	18	40	56	74	nd
Okra	nd	40	91	146	261	345
Onion, dry	3	5	7	7	8	nd
Pak choi	6	11	20	39	56	nd
Parsley	30	60	114	150	199	274
Parsnip	12	13	22	37	nd	nd
Pea, English	38	64	86	175	271	313
Pea, edible-podded	39	64	89	176	273	nd
Pepper	nd	7	12	27	34	nd
Potato, cured	nd	12	16	17	22	nd
Prickly pear	nd	nd	nd	nd	32	nd
Radicchio	8	13[4]	23[5]	nd	nd	45
Radish, topped	16	20	34	74	130	172
Radish, with tops	6	10	16	32	51	75
Rhubarb	11	15	25	40	49	nd
Rutabaga	5	10	14	26	37	nd

TABLE 8.18. AVERAGE RESPIRATION RATES OF VEGETABLES AT VARIOUS TEMPERATURES (*Continued*)

Vegetable	Respiration Rate (mg/kg/hr of CO_2)					
	32°F	41°F	50°F	59°F	68°F	77°F
Salsify	25	43	49	nd	193	nd
Spinach	21	45	110	179	230	nd
Squash, summer	25	32	67	153	164	nd
Sweet corn	41	63	105	159	261	359
Swiss chard	19[6]	nd	nd	nd	29	nd
Southern pea, pods	24[6]	25	nd	nd	148	nd
Southern pea, peas	29[6]	nd	nd	nd	126	nd
Tomatillo	nd	13	16	nd	32	nd
Tomato	nd	nd	15	22	35	43
Turnip, topped	8	10	16	23	25	nd
Watercress	22	50	110	175	322	377

Adapted from *The Commercial Storage of Fruits, Vegetables, and Florist and Nursery Stocks* (USDA Agriculture Handbook 66, 2004), http://www.ba.ars.usda.gov/hb66/contents.html.

[1] Not determined
[2] 1 day after harvest
[3] At 55°F
[4] At 43°F
[5] At 45°F
[6] At 36°F

TABLE 8.19. RECOMMENDED CONTROLLED ATMOSPHERE OR MODIFIED ATMOSPHERE CONDITIONS DURING TRANSPORT AND/OR STORAGE OF SELECTED VEGETABLES

Vegetable	Temperature [1] (°F)		Controlled Atmosphere[2] (%)		Application
	Optimum	Range	O_2	CO_2	
Artichoke	32	32–41	2–3	2–3	Moderate
Asparagus	36	34–41	air	10–14	High
Bean, green	46	41–50	2–3	4–7	Slight
Bean, processing	46	41–50	8–10	20–30	Moderate
Broccoli	32	32–41	1–2	5–10	High
Brussels sprouts	32	32–41	1–2	5–7	Slight
Cabbage	32	32–41	2–3	3–6	High
Cantaloupe	37	36–41	3–5	10–20	Moderate
Cauliflower	32	32–41	2–3	3–4	Slight
Celeriac	32	32–41	2–4	2–3	Slight
Celery	32	32–41	1–4	3–5	Slight
Chinese cabbage	32	32–41	1–2	0–5	Slight
Cucumber, slicing	54	46–54	1–4	0	Slight
Cucumber, processing	39	34–39	3–5	3–5	Slight
Herbs [3]	34	32–41	5–10	4–6	Moderate
Leek	32	32–41	1–2	2–5	Slight
Lettuce, crisphead	32	32–41	1–3	0	Moderate
Lettuce, cut or shredded	32	32–41	1–5	5–20	High
Lettuce, leaf	32	32–41	1–3	0	Moderate
Mushroom	32	32–41	3–21	5–15	Moderate
Okra	50	45–54	air	4–10	Slight
Onion, dry	32	32–41	1–2	0–10	Slight
Onion, green	32	32–41	2–3	0–5	Slight
Parsley	32	32–41	8–10	8–10	Slight
Pea, sugar	32	32–50	2–3	2–3	Slight
Pepper, bell	46	41–54	2–5	2–5	Slight
Pepper, chile	46	41–54	3–5	0–5	Slight
Pepper, processing	41	41–54	3–5	10–20	Moderate
Radish, topped	32	32–41	1–2	2–3	Slight

TABLE 8.19. **RECOMMENDED CONTROLLED ATMOSPHERE OR MODIFIED ATMOSPHERE CONDITIONS DURING TRANSPORT AND/OR STORAGE OF SELECTED VEGETABLES (*Continued*)**

	Temperature [1] (°F)		Controlled Atmosphere[2] (%)		
Vegetable	Optimum	Range	O_2	CO_2	Application
Spinach	32	32–41	7–10	5–10	Slight
Sweet corn	32	32–41	2–4	5–10	Slight
Tomato, mature, green	54	54–59	3–5	3–5	Moderate
Tomato ripe	50	50–59	3–5	3–5	Moderate
Witloof chicory	32	32–41	3–4	4–5	Slight

Adapted from A. A. Kader (ed.), *Postharvest Technology of Horticultural Crops,* 3rd ed. (University of California, Division of Agriculture and Natural Resources Publication 3311, 2002).

[1] Usual and/or recommended range. A relative humidity of 90–98% is recommended.

[2] Specific CA recommendations depend on varieties, temperature, and duration of storage.

[3] Herbs: chervil, chives, coriander, dill, sorrel, and watercress.

TABLE 8.20. OPTIMUM CONDITIONS FOR CURING ROOT, TUBER, AND BULB VEGETABLES PRIOR TO STORAGE

Vegetable	Temperature (°F)	RH (%)	Duration (days)
Cassava	86–95	85–90	4–7
Malanga	86–95	90–95	7
Potato			
Early crop	59–68	90–95	4–5
Late crop	50–59	90–95	10–15
Sweet potato	84–90	80–90	4–7
Taro	93–97	95	5
Water chestnut	86–90	95–100	3
Yam	86–95	85–95	4–7
Garlic and onion	ambient (75–90 best)		5–10 (field drying)
	93–113	60–75	0.5–3 (forced heated air)

TABLE 8.21. EFFECT OF TEMPERATURE ON THE RATE OF DETERIORATION OF VEGETABLES NOT SENSITIVE TO CHILLING INJURY

Temperature (T) (°F)	Assumed Q_{10}[1]	Relative Rate of Deterioration	Relative Shelf Life	Loss per Day (%)
32		1.0	100	1
50	3.0	3.0	33	3
68	2.5	7.5	13	8
86	2.0	15.0	7	14
104	1.5	22.5	4	25

Adapted from A. A. Kader (ed.), *Postharvest Technology of Horticultural Crops*, 3rd ed. (University of California, Division of Agriculture and Natural Resources Publication 3311, 2002).

[1] $Q_{10} = \frac{\text{rate of deterioration at T + 10°C}}{\text{rate of deterioration at T}}$

TABLE 8.22. MOISTURE LOSS FROM VEGETABLES

High	Medium	Low
Broccoli	Artichoke	Eggplant
Cantaloupe	Asparagus	Garlic
Chard	Bean, snap	Ginger
Green onion	Beet[1]	Melons
Kohlrabi	Brussels sprouts	Onion
Leafy greens	Cabbage	Potato
Mushroom	Carrot[1]	Pumpkin
Oriental vegetables	Cassava[2]	Winter squash
Parsley	Cauliflower	
Strawberry	Celeriac[1]	
	Celery	
	Sweet corn[3]	
	Cucumber[2]	
	Endive	
	Escarole	
	Leek	
	Lettuce	
	Okra	
	Parsnip[1]	
	Pea	
	Pepper	
	Radish[1]	
	Rutabaga[1,2]	
	Sweet potato	
	Summer squash	
	Tomato[2]	
	Yam	

Adapted from B. M. McGregor, *Tropical Products Transport Handbook,* USDA Agricultural Handbook 668 (1987).

[1] Root crops with tops have a high rate of moisture loss.

[2] Waxing reduces the rate of moisture loss.

[3] Husk removal reduces water loss.

TABLE 8.23. STORAGE SPROUT INHIBITORS

Sprout inhibitors are most effective when used in conjunction with good storage; their use cannot substitute for poor storage or poor storage management. However, storage temperatures may be somewhat higher when sprout inhibitors are used than when they are not. Follow label directions.

Vegetable	Material	Application
Potato (do not use on seed potatoes)	Maleic hydrazide	When most tubers are 1½–2 in. in diameter. Vines must remain green for several weeks after application.
	Chlorprophan (CIPC)	In storage, 2–3 weeks after harvest as an aerosol treatment. Do not store seed potatoes in a treated storage.
		During washing, as an emulsifiable concentrate added to wash water to prevent sprouting during marketing.
Onion	Maleic hydrazide	Apply when 50% of the tops are down, the bulbs are mature, the necks soft, and 5–8 leaves are still green.

TABLE 8.24. SUSCEPTIBILITY OF VEGETABLES TO CHILLING INJURY

Vegetable	Approximate Lowest Safe Temperature (°F)	Appearance When Stored Between 32°F and Safe Temperature[1]
Asparagus	32–36	Dull, gray-green, and limp tips
Bean, lima	34–40	Rusty brown specks, spots, or areas
Bean, snap	45	Pitting and russeting
Chayote	41–50	Dull brown discoloration, pitting, flesh darkening
Cucumber	45	Pitting, water-soaked spots, decay
Eggplant	45	Surface scald, alternaria rot, blackening of seeds
Ginger	45	Softening, tissue breakdown, decay
Jicama	55–65	Surface decay, discoloration
Melon		
Cantaloupe	36–41	Pitting, surface decay
Casaba	45–50	Pitting, surface decay, failure to ripen
Crenshaw	45–50	Pitting, surface decay, failure to ripen
Honeydew	45–50	Reddish tan discoloration, pitting, surface decay, failure to ripen
Persian	45–50	Pitting, surface decay, failure to ripen
Watermelon	40	Pitting, objectionable flavor
Okra	45	Discoloration, water-soaked areas, pitting, decay

Pepper, sweet	45	Sheet pitting, alternaria rot on fruit and calyx, darkening of seed
Potato	38	Mahogany browning, sweetening
Pumpkin and hard-shell squash	50	Decay, especially alternaria rot
Sweet potato	55	Decay, pitting, hard core when cooked
Tamarillo	37–40	Surface pitting, discoloration
Taro	50	Internal browning, decay
Tomato, ripe	45–50	Watersoaking and softening, decay
Tomato, mature, green	55	Poor color when ripe, alternaria rot

Adapted from Chien Yi Wang, "Chilling and Freezing Injury," in *The Commercial Storage of Fruits, Vegetables, and Florist and Nursery Stock* (USDA Agriculture Handbook 66, 2004), http://www.ba.ars.usda.gov/hb66/index.html.

[1] Severity of injury is related to temperature and time.

TABLE 8.25. VEGETABLES CLASSIFIED ACCORDING TO CHILLING INJURY SUSCEPTIBILITY

Not Susceptible to Chilling Injury	Susceptible to Chilling Injury
Artichoke	Bean, snap
Asparagus	Cantaloupe
Bean, lima	Cassava
Beet	Cucumber
Broccoli	Eggplant
Brussels sprouts	Ginger
Cabbage	Okra
Carrot	Pepper
Cauliflower	Pepino
Celery	Prickly pear
Endive	Pumpkin
Garlic	Squash
Lettuce	Sweet potato
Mushroom	Tamarillo
Onion	Taro
Parsley	Tomato
Parsnip	Watermelon
Pea	Yam
Radish	
Spinach	
Strawberry	
Sweet corn	
Turnip	

Adapted from A. A. Kader (ed.), *Postharvest Technology of Horticultural Crops,* 3rd ed. (University of California, Division of Agriculture and Natural Resources Publication 3311, 2002).

TABLE 8.26. RELATIVE SUSCEPTIBILITY OF VEGETABLES TO CHILLING INJURY

Most Susceptible	Moderately Susceptible	Least Susceptible
Asparagus	Broccoli	Beet
Bean, snap	Carrot	Brussels sprouts
Cucumber	Cauliflower	Cabbage, mature and savoy
Eggplant	Celery	Kale
Lettuce	Onion, dry	Kohlrabi
Okra	Parsley	Parsnip
Pepper, sweet	Pea	Rutabaga
Potato	Radish	Salsify
Squash, summer	Spinach	Turnip
Sweet potato	Squash, winter	
Tomato		

Adapted from Chien Yi Wang, "Chilling and Freezing Injury," in *The Commercial Storage of Fruits, Vegetables, and Florist and Nursery Stocks* (USDA, Agriculture Handbook 66, 2004), http://www.ba.ars.usda.gov/hb66/index.html.

TABLE 8.27. CHILLING THRESHOLD TEMPERATURES AND VISUAL SYMPTOMS OF CHILLING INJURY FOR SOME SUBTROPICAL AND TROPICAL STORAGE ORGAN VEGETABLES

Vegetable	Chilling Threshold (°F)	Symptoms
Cassava	41–46	Internal breakdown, increased water loss, failure to sprout, increased decay, loss of eating quality
Ginger	54	Accelerated softening and shriveling, oozes moisture from the surface, decay
Jicama	55–59	External decay, rubbery and translucent flesh with grown discoloration, increased water loss
Malanga	45	Tissue breakdown and internal discoloration, increased water loss, increased decay, undesirable flavor changes
Potato	39	Mahogany browning, reddish-brown areas in the flesh, adverse effects on cooking quality
Sweet potato	54	Internal brown-black discoloration, adverse effects on cooled quality, hard core, accelerated decay
Taro	45–50	Tissue breakdown and internal discoloration, increased water loss, increased decay, undesirable flavor changes
Yam	55	Tissue softening, internal discoloration (grayish flecked with reddish brown), shriveling, decay

TABLE 8.28. SYMPTOMS OF FREEZING INJURY ON SOME VEGETABLES

Vegetable	Symptoms
Artichoke	Epidermis becomes detached and forms whitish to light tan blisters. When blisters are broken, underlying tissue turns brown.
Asparagus	Tip becomes limp and dark; the rest of the spear is water-soaked. Thawed spears become mushy.
Beet	External and internal water-soaking and, sometimes, blackening of conducting tissue.
Broccoli	The youngest florets in the center of the curd are most sensitive to freezing injury. They turn brown and give strong off-odors on thawing.
Cabbage	Leaves become water-soaked, translucent, and limp on thawing; separated epidermis.
Carrot	A blistered appearance; jagged lengthwise cracks. Interior becomes water-soaked and darkened on thawing.
Cauliflower	Curds turn brown and have a strong off-odor when cooked.
Celery	Leaves and petioles appear wilted and water-soaked on thawing. Petioles freeze more readily than leaves.
Garlic	Thawed cloves appear water-soaked, grayish yellow.
Lettuce	Blistering, dead cells of the separated epidermis on outer leaves become tan; increased susceptibility to physical damage and decay.
Onion	Thawed bulbs are soft, grayish yellow, and water-soaked in cross section; often limited to individual scales.
Pepper, bell	Dead, water-soaked tissue in part of or all pericarp surface; pitting, shriveling, decay follow thawing.
Potato	Freezing injury may not be externally evident but shows as gray or bluish gray patches beneath the skin. Thawed tubers become soft and watery.
Radish	Thawed tissues appear translucent; roots soften and shrivel.
Sweet potato	A yellowish brown discoloration of the vascular ring, and a yellowish green water-soaked appearance of other tissues. Roots soften and become susceptible to decay.
Tomato	Water-soaked and soft on thawing. In partially frozen fruits, the margin between healthy and dead tissue is distinct, especially in green fruits.
Turnip	Small water-soaked spots or pitting on the surface. Injured tissues appear tan or gray and give off an objectionable odor.

Adapted from A. A. Kader, J. M. Lyons, and L. L. Morris, "Postharvest Responses of Vegetables to Preharvest Field Temperature," *HortScience* 9 (1974):523–527.

TABLE 8.29. SOME POSTHARVEST PHYSIOLOGICAL DISORDERS OF VEGETABLES, ATTRIBUTABLE DIRECTLY OR INDIRECTLY TO PREHARVEST FIELD TEMPERATURES

Vegetable	Disorder	Symptoms and Development
Asparagus	Feathering	Bracts of the spears are partly spread as a result of high temperature.
Brussels sprouts	Black leaf speck	Becomes visible after storage for 1–2 weeks at low temperature. Has been attributed in part to cauliflower mosaic virus infection in the field, which is influenced by temperature and other environmental factors.
	Tip burn	Leaf margins turn light tan to dark brown.
Cantaloupe	Vein tract browning	Discoloration of unnetted longitudinal stripes; related partly to high temperature and virus diseases.
Garlic	Waxy breakdown	Enhanced by high temperature during growth; slightly sunken, light yellow areas in fleshy cloves, then the entire clove becomes amber, slightly translucent, and waxy but still firm.
Lettuce	Tip burn	Light tan to dark brown margins of leaves. Has been attributed to several causes, including field temperature; it can lead to soft rot development during postharvest handling.
	Rib discoloration	More common in lettuce grown when day temperatures exceed 81°F or when night temperatures are between 55–64°F than in lettuce grown during cooler periods.
	Russet spotting	Small tan, brown, or olive spots randomly distributed over the affected leaf; a postharvest disorder of lettuce induced by ethylene. Lettuce is more susceptible to russet spotting when harvested after high field temperatures (above 86°F) for 2 days or more during the 10 days before harvest.

TABLE 8.29. SOME POSTHARVEST PHYSIOLOGICAL DISORDERS OF VEGETABLES, ATTRIBUTABLE DIRECTLY OR INDIRECTLY TO PREHARVEST FIELD TEMPERATURES (*Continued*)

Vegetable	Disorder	Symptoms and Development
	Rusty brown discoloration	Rusty brown discoloration has been related to internal rib necrosis associated with lettuce mosaic virus infection, which is influenced by field temperature and other environmental factors.
Onion	Translucent scale	Grayish water-soaked appearance of the outer two or three fleshy scales of the bulb; translucency makes venation distinct. In severe cases, the entire bulb softens, and off-odors may develop.
Potato	Blackheart	May occur in the field during excessively hot weather in waterlogged soils. Internal symptom is dark gray to purplish or black discoloration, usually in the center of the tuber.
Radish	Pithiness	Textured white spots or streaks in cross section, large air spaces near the center, tough and dry roots. Results from high temperature.

Adapted from A. A. Kader, J. M. Lyons, and L. L. Morris, "Postharvest Responses of Vegetables to Preharvest Field Temperature," *HortScience* 9 (1974):523–527.

TABLE 8.30. SYMPTOMS OF SOLAR INJURY ON SOME VEGETABLES

Vegetable	Symptoms
Bean, snap	Very small brown or reddish spots on one side of the pod coalesce and become water-soaked and slightly shrunken.
Cabbage	Blistering of some outer leaves leads to a bleached, papery appearance. Desiccated leaves are susceptible to decay.
Cauliflower	Discoloration of curds from yellow to brown to black (solar browning).
Cantaloupe	Sunburn: dry, sunken, and white to light tan areas. In milder sunburn, ground color is green or spotty brown.
Lettuce	Papery areas on leaves, especially the cap leaf, develop during clear weather when air temperatures are higher than 77°F; affected areas become focus for decay.
Honeydew melon	White to gray area at or near the top, may be slightly wrinkled, undesirable flavor, or brown blotch, which is tan to brown discolored areas caused by death of epidermal cells due to excessive ultraviolet radiation.
Onion and garlic	Sunburn: dry scales are wrinkled, sometimes extending to one or two fleshy scales. Injured area may be bleached depending on the color of the bulb.
Pepper, bell	Dry and papery areas, yellowing, and, sometimes, wilting.
Potato	Sunscald: water and blistered areas on the tuber surface. Injured areas become sunken and leathery, and subsurface tissue rapidly turns dark brown to black when exposed to air.
Tomato	Sunburn (solar yellowing): affected areas on the fruit become whitish, translucent, thin-walled, a netted appearance may develop. Mild solar injury may not be noticeable at harvest but becomes more apparent after harvest as uneven ripening.

Adapted from A. A. Kader, J. M. Lyons, and L. L. Morris, "Postharvest Responses of Vegetables to Preharvest Field Temperature," *HortScience* 9 (1974):523–527.

TABLE 8.31. CLASSIFICATION OF HORTICULTURAL COMMODITIES ACCORDING TO ETHYLENE PRODUCTION RATES

Very Low	Low	Moderate	High	Very High
Artichoke	Blackberry	Banana	Apple	Cherimoya
Asparagus	Blueberry	Fig	Apricot	Mammee apple
Cauliflower	Casaba melon	Guava	Avocado	Passion fruit
Cherry	Cranberry	Honeydew melon	Cantaloupe	Sapote
Citrus	Cucumber	Mango	Feijoa	
Grape	Eggplant	Plantain	Kiwi fruit (ripe)	
Jujube	Okra	Tomato	Nectarine	
Leafy vegetables	Olive		Papaya	
Most cut flowers	Pepper		Peach	
Pomegranate	Persimmon		Pear	
Potato	Pineapple		Plum	
Root vegetables	Pumpkin			
Strawberry	Raspberry			
	Tamarillo			
	Watermelon			

Adapted from A. A. Kader (ed.), *Postharvest Technology of Horticultural Crops*, 3rd ed. (University of California, Division of Agriculture and Natural Resources Publication 3311, 2002).

COMPATIBILITY OF FRESH PRODUCE IN MIXED LOADS UNDER VARIOUS RECOMMENDED TRANSIT CONDITIONS

Shippers and receivers of fresh fruits and vegetables frequently prefer to handle shipments that consist of more than one commodity. In mixed loads, it is important to combine only those commodities that are compatible in their requirements for temperature, modified atmosphere, relative humidity, protection from odors, and protection from physiologically active gases such as ethylene.

TABLE 8.32. RECOMMENDED TRANSIT CONDITIONS FOR COMPATIBLE GROUPS

Temp.: 55–60°F; Relative humidity: 85–95%; Ice: No contact with commodity	Temp.: 36–41°F; Relative humidity: 90–95%; Ice: Contact cantaloupe only	Temp.: 40–45°F; Relative humidity: about 95%; Ice: No contact with commodity	Temp.: 40–55°F; Relative humidity: 85–90%; Ice: No contact with commodity
Avocado	Cranberry	Snap bean	Cucumber
Banana	Lemon	Lychee	Eggplant
Grapefruit (AZ and CA; FL before Jan. 1)	Cantaloupe	Okra	Ginger (not with eggplant)
Guava	Orange	Pepper, green (not with bean)	Grapefruit (Fla. after Jan. 1; and Tex.)
Mango	Tangerine	Pepper, red	Lime
Casaba melon		Summer squash	Potato
Crenshaw melon		Tomato, pink	Pumpkin
Honeydew melon		Watermelon	Watermelon
Persian melon			Winter squash
Olive			
Papaya			
Pineapple (not with avocado)			
Tomato, green			
Tomato, pink			
Watermelon			

TABLE 8.32. RECOMMENDED TRANSIT CONDITIONS FOR COMPATIBLE GROUPS (*Continued*)

Temp.: 32–34°F; Relative humidity: 95–100%; Ice: No contact with asparagus, fig, grape, mushroom	Temp.: 32–34°F; Relative humidity: 95–100%; Ice: Contact with all commodities	Temp.: 55–65°F; Relative humidity: 85–90%; Ice: No contact with any commodity	Temp.: 32–34°F; Relative humidity: 65–75%; Ice: No contact with any commodity
Artichoke	Broccoli	Ginger	Garlic
Asparagus	Brussels sprouts	Sweet potato	Onion, dry
Beet	Cabbage		
Carrot	Cauliflower		
Endive, escarole	Celeriac		
Fig	Celery		
Grape	Horseradish		
Greens	Kohlrabi		
Leek (not with fig or grape)	Onion, green (not with rhubarb, fig, or grape; probably not with mushroom or sweet corn)		
Lettuce	Radish		
Mushroom	Rutabaga		
Parsley	Turnip		
Parsnip			
Pea			
Rhubarb			
Salsify			
Spinach			
Sweet corn			
Watercress			

Adapted from W. J. Lipton, *Compatibility of Fruits and Vegetables During Transport in Mixed Loads,* USDA, ARS, Marketing Research Report 1070 (1977).

TABLE 8.33. COMPATIBLE FRESH FRUITS AND VEGETABLES DURING 10-DAY STORAGE

GROUP 1A (32°–36°) AND 90–98% RH VEGETABLES				
Alfalfa sprouts	Cabbage[1]	Endive,[1] chicory	Lettuce[1]	Rhubarb
Amaranth[1]	Carrot[1]	Escarole[1]	Mint[1]	Salsify
Anise[1]	Cauliflower[1]	Fennel[1]	Mushroom[1]	Scorzonera
Artichoke[1]	Celeriac	Garlic	Mustard greens[1]	Shallot[1]
Asparagus[1]	Celery[1]	Green onion[1]	Pak choi[1]	Snow pea[1]
Beans: fava, lima	Chard[1]	Herbs[1] (not basil)	Parsley[1]	Spinach[1]
Bean sprouts	Chinese cabbage[1]	Horseradish	Parsnip[1]	Swiss chard[1]
Beet	Chinese turnip	Jerusalem artichoke	Pea[1]	Turnip
Belgian endive[1]	Collards[1]	Kailon	Radicchio[1]	Turnip greens[1]
Broccoflower[1]	Corn: sweet, baby	Kale[1]	Radish	Water chestnut
Broccoli[1]	Cut vegetables	Kohlrabi	Rutabaga	Watercress[1]
Brussels sprouts[1]	Daikon[1]	Leek[1]		

GROUP 1B (32°–36°) AND 85–95% RH FRUITS AND MELONS				
Apple	Caimito	Date	Loganberry	Plumcot
Apricot	Cantaloupe	Dewberry	Longan	Pomegranate
Avocado, ripe	Cashew apple	Elderberry	Loquat	Prune
Barbados cherry	Cherry	Fig	Pear: Asian, European	Quince
Blackberry	Coconut	Gooseberry	Persimmon[1]	Raspberry
Blueberry	Currant	Grape	Plum	Strawberry
Boysenberry	Cut fruits	Kiwifruit[1]		

TABLE 8.33. COMPATIBLE FRESH FRUITS AND VEGETABLES DURING 10-DAY STORAGE (*Continued*)

GROUP 2 (55°–65°) AND 85–95% RH VEGETABLES		FRUITS AND MELONS		
Beans: snap, green, wax	Long bean	Babaco	Juan canary melon	Pepino
Cactus leaves (nopales)[1]	Malanga[1]	Calamondin	Lemon[1]	Pummelo
Calabaza	Pepper: bell, chili	Carambola	Lime[1]	Tamarillo
Chayote[1]	Southern pea	Cranberry	Limequat	Tamarind
Cucumber[1]	Squash: summer[1]	Custard apple	Mandarin	Tangelo
Eggplant[1]	Tomatillo	Durian	Olive	Tangerine
Kiwano (horned melon)	Winged bean	Feijoa	Orange	Ugli fruit
		Granadilla	Passion fruit	Watermelon
		Grapefruit[1]		

GROUP 3 (55°–65°) AND 85–95% RH VEGETABLES		FRUITS AND MELONS		
Bitter melon	Potato	Atemoya	Honeydew melon	Persian melon
Boniato[1]	Pumpkin	Banana	Jaboticaba	Plantain
Cassava	Squash: winter[1]	Breadfruit	Jackfruit	Rambutan
Dry onion	Sweet potato[1]	Canistel	Mamey sapote	Sapodilla
Ginger	Taro (dasheen)	Casaba melon	Mango	Sapote
Jicama	Yam	Cherimoya	Mangosteen	Soursop
		Crenshaw melon	Papaya	

Adapted from A. A. Kader (ed.), *Postharvest Technology of Horticultural Crops,* 3rd ed. (University of California. Division of Agriculture and Natural Resources Publications 3311, 2002).

Note: Ethylene level should be kept below 1 ppm in storage areas.

[1] Products sensitive to ethylene damage.

07
POSTHARVEST DISEASES

INTEGRATED CONTROL OF POSTHARVEST DISEASES

Effective and consistent control of storage diseases depends on integration of the following practices:

- Select disease resistant cultivars where possible.
- Maintain correct crop nutrition by use of leaf and soil analysis.
- Irrigate based on crop requirements and avoid overhead irrigation.
- Apply preharvest treatments to control insects and diseases.
- Harvest the crop at the correct maturity for storage.
- Apply postharvest treatments to disinfest and control diseases and disorders on produce.
- Maintain good sanitation in packing areas and keep dump water free of contamination.
- Store produce under conditions least conducive to growth of pathogens.

TABLE 8.34. IMPORTANT POSTHARVEST DISEASES OF VEGETABLES

Vegetable	Disease	Causal Agent	Fungal Class/Type
Artichoke	Gray mold	*Botrytis cinerea*	Hyphomycete
	Watery soft rot	*Sclerotinia sclerotiorum*	Discomycete
Asparagus	Bacterial soft rot	*Erwinia or Pseudomonas* spp.	Bacteria
	Fusarium rot	*Fusarium* spp.	Hyphomycete
	Phytophthora rot	*Phytophthora* spp.	Oomycete
	Purple spot	*Stemphylium* spp.	Hyphomycete
Bulbs	Bacterial soft rot	*Erwinia caratovora*	Bacterium
	Black rot	*Aspergillus niger*	Hyphomycete
	Blue mold rot	*Penicillium* spp.	Hyphomycete
	Fusarium basal rot	*Fusarium oxysporum*	Hyphomycete
	Neck rot	*Botrytis* spp.	Hyphomycete
	Purple blotch	*Alternaria porri*	Hyphomycete
	Sclerotium rot	*Sclerotium rolfsii*	Agonomycete
	Smudge	*Colletotrichum circinans*	Coelomycete
Carrot	Bacterial soft rot	*Erwinia* spp. *Pseudomonas* spp.	Bacteria
	Black rot	*A. radicina*	Hyphomycete
	Cavity spot	Disease complex	Soil fungi
	Chalaropsis rot	*Chalara* spp.	Hyphomycetes
	Crater rot	*R. carotae*	Agonomycete
	Gray mold rot	*B. cinerea*	Hyphomycete
	Sclerotium rot	*S. rolfsii*	Agonomycete
	Watery soft rot	*Sclerotinia* spp.	Discomycete

Celery	Bacterial soft rot	*Erwinia* or *Pseudomonas* spp.	Bacteria
	Brown spot	*Cephalosporium apii*	Hyphomycete
	Cercospora spot	*Cercospora apii*	Hyphomycete
	Gray mold	*Botrytis cinerea*	Hyphomycete
	Licorice rot	*Mycocentrospora acerina*	Hyphomycete
	Phoma rot	*Phoma apiicola*	Coelomycete
	Pink rot	*Sclerotinia* spp.	Discomycete
	Septoria spot	*Septoria apiicola*	Coelomycete
Crucifers	Alternaria leaf spot	*Alternaria* spp.	Hyphomycete
	Bacterial soft rot	*E. caratovora*	Bacterium
	Black rot	*Xanthomonas campestris*	Bacterium
	Downy mildew	*Peronospora parasitica*	Oomycete
	Rhizoctonia rot	*Rhizoctonia solani*	Agonomycete
	Ring spot	*Mycosphaerella brassicicola*	Loculoascomyete
	Virus diseases	Cauliflower mosaic virus Turnip mosaic virus	Virus
	Watery soft rot	*Sclerotinia* spp.	Discomycete
	White blister	*Albugo candida*	Oomycete

TABLE 8.34. IMPORTANT POSTHARVEST DISEASES OF VEGETABLES (*Continued*)

Vegetable	Disease	Causal Agent	Fungal Class/Type
Cucurbits	Anthracnose	*Colletotrichum* spp.	Coelomycete
	Bacterial soft rot	*Erwinia* spp.	Bacterium
	Black rot	*Didymella bryoniae*	Loculoascomycete
	Botryodiplodia rot	*Botryodiplodia theobromae*	Coelomycete
	Charcoal rot	*Macrophomina phaseolina*	Coelomycete
	Fusarium rot	*Fusarium* spp.	Hyphomycete
	Leak	*Pythium* spp.	Oomycete
	Rhizopus rot	*Rhizopus* spp.	Zygomycete
	Sclerotium rot	*Sclerotium rolfsii*	Agonomycete
	Soil rot	*R. solani*	Agonomycete
Legumes	Alternaria blight	*A. alternata*	Hyphomycete
	Anthracnose	*Colletotrichum* spp.	Coelomycete
	Ascochyta pod spot	*Ascochyta* spp.	Coelomycetes
	Bacterial blight	*Pseudomonas* spp. *Xanthomonas* spp.	Bacteria
	Chocolate spot	*B. cinerea*	Hyphomycete
	Cottony leak	*Pythium* spp.	Oomycete
		Mycosphaerella blight *M. pinodes*	Loculoascomycete
	Rust	*Uromyces* spp.	Hemibasidiomycete
	Sclerotium rot	*S. rolfsii*	Agonomycete
	Soil rot	*R. solani*	Agonomycete
	White mold	*Sclerotinia* spp.	Discomycete

Lettuce	Bacterial rot	*Erwinia, Pseudomonas, Xanthomonas* spp.	Bacteria
	Gray mold rot	*B. cinerea*	Hyphomycete
	Rhizoctonia rot	*R. solani*	Agonomycete
	Ringspot	*Microdochium panattonianum*	Hyphomycete
	Septoria spot	*S. lactucae*	Coelomycete
	Stemphylium spot	*Stemphylium herbarum*	Hyphomycete
	Watery soft rot	*Sclerotinia* spp.	Discomycete
Potato	Bacterial soft rot	*Erwinia* spp.	Bacteria
	Blight	*Phytophthora infestans*	Oomycete
	Charcoal rot	*S. bataticola*	Agonomycete
	Common scab	*Streptomyces scabies*	Actinomycete
	Fusarium rot	*Fusarium* spp.	Hyphomycete
	Gangrene	*Phoma exigua*	Coelomycete
	Ring rot	*Clavibacter michiganensis*	Bacterium
	Sclerotium rot	*S. rolfsii*	Agonomycete
	Silver scurf	*Helminthosporium solani*	Hyphomycete
	Watery wound rot	*Pythium* spp.	Oomycete

TABLE 8.34. IMPORTANT POSTHARVEST DISEASES OF VEGETABLES (*Continued*)

Vegetable	Disease	Causal Agent	Fungal Class/Type
Solanaceous Fruits	Alternaria rot	*A. alternata*	Hyphomycete
	Anthracnose	*Colletotrichum* spp.	Coelomycete
	Bacterial canker	*C. michiganensis*	Bacterium
	Bacterial speck	*Pseudomonas syringae*	Bacterium
	Bacterial spot	*X. campestris*	Bacterium
	Fusarium rot	*Fusarium* spp.	Hyphomycetes
	Gray mold rot	*B. cinerea*	Hyphomycete
	Light blight	*P. infestans*	Oomycete
	Phoma rot	*Phoma lycopersici*	Hyphomycete
	Phomopsis rot	*Phomopsis* spp.	Coelomycetes
	Phytophthora rot	*Phytophthora* spp.	Oomycete
	Pleospora rot	*Stemphylium herbarum*	Hyphomycete
	Rhizopus rot	*Rhizopus* spp.	Zygomycetes
	Sclerotium rot	*S. rolfsii*	Agonomycete
	Soil rot	*R. solani*	Agonomycete
	Sour rot	*Geotrichum candidum*	Hyphomycete
	Watery soft rot	*Sclerotinia* spp.	Discomycetes
Sweet potato	Black rot	*Ceratocystis fimbriata*	Pyrenomycete
	Fusarium rot	*Fusarium* spp.	Hyphomycete
	Rhizopus rot	*Rhizopus* spp.	Zygomycetes
	Soil rot	*Streptomyces ipomoeae*	Actinomycete
	Scurf	*Monilochaetes infuscans*	Hyphomycete

Adapted from Peter L. Sholberg and William S. Conway, "Postharvest Pathology," in *The Commercial Storage of Fruits, Vegetables, and Florist and Nursery Stocks* (2004), http://www.ba.ars.usda.gov/hb66/contents.html.

08
VEGETABLE QUALITY

Quality is defined as "any of the features that make something what it is" or "the degree of excellence or superiority." The word *quality* is used in various ways in reference to fresh fruits and vegetables, such as *market* quality, *edible* quality, *dessert* quality, *shipping* quality, *table* quality, *nutritional* quality, *internal* quality, and *appearance* quality.

Quality of fresh vegetables is a combination of characteristics, attributes, and properties that give the vegetables value to humans for food and enjoyment. Producers are concerned that their commodities have good appearance and few visual defects, but for them a useful variety also must score high on yield, disease resistance, ease of harvest, and shipping quality. To receivers and market distributors, appearance quality is most important; they are also keenly interested in firmness and long storage life. Consumers consider good-quality vegetables those that look good, are firm, and offer good flavor and nutritive value. Although consumers buy on the basis of appearance and feel, their satisfaction and repeat purchases depend on good edible quality.

TABLE 8.35. QUALITY ASSURANCE RECORDS FOR VEGETABLES

Field packing

- Maturity/ripeness stage and uniformity
- Harvest method (hand or mechanical)
- Temperature of product (harvest during cool times of the day and keep product shaded)
- Uniformity of packs (size, trimming, maturity)
- Well-constructed boxes; palletization and unitization
- Condition of field boxes or bins (no rough or dirty surfaces)
- Cleaning and sanitization of bins and harvest equipment

Packinghouse

- Time from harvest to arrival
- Shaded receiving area
- Uniformity of harvest (size, trimming, maturity)
- Washing/hydrocooling operation (sanitization)
- Water changed daily and constant sanitizer levels maintained in dump tanks

TABLE 8.35. QUALITY ASSURANCE RECORDS FOR VEGETABLES (*Continued*)

- Decay control chemical usage
- Sorting for size, color, quality, etc.
- Product that falls on the floor discarded
- Culls checks for causes of rejection and for sorting accuracy
- Facilities and equipment sanitized regularly
- Well-constructed boxes; palletization and unitization

Cooler

- Time from harvest to cooler
- Time from arrival to start of cooling
- Package design (ventilation)
- Speed of cooling and final temperature
- Temperature of product after cooling
- Temperature of holding room
- Time from cooling to loading

Loading Trailer

- First-in, first-out truck loading
- Temperature of product
- Boxes palletized and unitized
- Truck condition (clean, undamaged, precooled)
- Loading pattern; palletization and unitization
- Duration of transport
- Temperature during transport (thermostat setting and use of recorders)

Arrival at Distribution Center

- Transit time
- Temperature of product
- Product condition and uniformity
  - Uniformity of packs (size, trimming, maturity)
  - Ripeness stage, firmness
  - Decay incidence/type
- Refrigeration maintained during cooling

TABLE 8.36. QUALITY COMPONENTS OF FRESH VEGETABLES

Main Factors	Components
Appearance (visual)	*Size:* dimensions, weight, volume *Shape and form:* diameter/depth ratio, compactness, uniformity *Color:* uniformity, intensity *Gloss:* nature of surface wax *Defects,* external and internal: morphological, physical and mechanical, physiological, pathological, entomological
Texture (feel)	Firmness, hardness, softness Crispness Succulence, juiciness Mealiness, grittiness Toughness, fibrousness
Flavor (taste and smell)	Sweetness Sourness (acidity) Astringency Bitterness Aroma (volatile compounds) Off-flavors and off-odors
Nutritional value	Carbohydrates (including dietary fiber) Proteins Lipids Vitamins Essential elements
Safety	Naturally occurring toxicants Contaminants (chemical residues, heavy metals) Mycotoxins Microbial contamination

Adapted from A. A. Kader (ed.), *Postharvest Technology of Horticultural Crops,* 3rd ed. (University of California, Division of Agriculture and Natural Resources Publication 3311, 2002).

09

U.S. STANDARDS FOR GRADES OF VEGETABLES

Grade standards issued by the U.S. Department of Agriculture (USDA) are currently in effect for most vegetables for fresh market and for processing. Some standards have been unchanged since they became effective, whereas others have been revised quite recently.

For the U.S. Standards for Grades of Fresh and Processing Vegetables, go to:

http://www/ams.usda.gov/standards/stanfrfv.htm

TABLE 8.37. QUALITY FACTORS FOR FRESH VEGETABLES IN THE U.S. STANDARDS FOR GRADES

Vegetable	Date Issued	Quality Factors
Anise, sweet	1973	Firmness, tenderness, trimming, blanching, and freedom from decay and damage caused by growth cracks, pithy branches, wilting, freezing, seedstems, insects, and mechanical means
Artichoke	1969	Stem length, shape, overmaturity, uniformity of size, compactness, and freedom from decay and defects
Asparagus	1966	Freshness (turgidity), trimming, straightness, freedom from damage and decay, diameter of stalks, percent green color
Bean, lima	1938	Uniformity, maturity, freshness, shape, and freedom from damage (defect) and decay
Bean, snap	1936	Uniformity, size, maturity, firmness, and freedom from defect and decay
Beet, bunched or topped	1955	Root shape, trimming of rootlets, firmness (turgidity), smoothness, cleanness, minimum size (diameter), and freedom from defect
Beet, greens	1959	Freshness, cleanness, tenderness, and freedom from decay, other kinds of leaves, discoloration, insects, mechanical injury, and freezing injury
Broccoli	1943	Color, maturity, stalk diameter and length, compactness, base cut, and freedom from defects and decay
Brussels sprouts	1954	Color, maturity (firmness), no seedstems, size (diameter and length), and freedom from defect and decay
Cabbage	1945	Uniformity, solidity (maturity or firmness), no seedstems, trimming, color, and freedom from defect and decay
Cantaloupe	1968	Soluble solids (>9 percent), uniformity of size, shape, ground color and netting; maturity and turgidity; and freedom from wet slip, sunscald, and other defects

TABLE 8.37. QUALITY FACTORS FOR FRESH VEGETABLES IN THE U.S. STANDARDS FOR GRADES (*Continued*)

Vegetable	Date Issued	Quality Factors
Carrot, bunched	1954	Shape, color, cleanness, smoothness, freedom from defects, freshness, length of tops, and root diameter
Carrot, topped	1965	Uniformity, turgidity, color, shape, size, cleanness, smoothness, and freedom from defect (growth cracks, pithiness, woodiness, internal discoloration)
Carrots with short trimmed tops	1954	Roots; firmness, color, smoothness, and freedom from defect (sunburn, pithiness, woodiness, internal discoloration, and insect and mechanical injuries) and decay; leaves: (cut to <4 inches) freedom from yellowing or other discoloration, disease, insects, and seedstems
Cauliflower	1968	Curd cleanness, compactness, white color, size (diameter), freshness and trimming of jacket leaves, and freedom from defect and decay
Celery	1959	Stalk form, compactness, color, trimming, length of stalk and midribs, width and thickness of midribs, no seedstems, and freedom from defect and decay
Collard greens or broccoli greens	1953	Freshness, tenderness, cleanness, and freedom from seedstems, discoloration, freezing injury, insects, and diseases
Corn, sweet	1992	Uniformity of color and size, freshness, milky kernels, cob length, freedom from insect injury, discoloration, and other defects, coverage with fresh husks
Cucumber	1958	Color, shape, turgidity, maturity, size (diameter and length), and freedom from defect and decay
Cucumber, greenhouse	1985	Freshness, shape, firmness, color, size (11 in. or longer), and freedom from decay, cuts, bruises, scars, insect injury, and other defects
Dandelion greens	1955	Freshness, cleanness, tenderness, and freedom from damage caused by seedstems, discoloration, freezing, diseases, insects, and mechanical injury

Eggplant	1953	Color, turgidity, shape, size, and freedom from defect and decay
Endive, escarole, or chicory	1964	Freshness, trimming, color (blanching), no seedstems, and freedom from defect and decay
Garlic	1944	Maturity, curing, compactness, well-filled cloves, bulb size, and freedom from defect
Honeydew and honey ball melons	1967	Maturity, firmness, shape, and freedom from decay and defect (sunburn, bruising, hail spots, and mechanical injuries)
Horseradish roots	1936	Uniformity of shape and size, firmness, smoothness, and freedom from hollow heart, other defects, and decay
Kale	1934	Uniformity of growth and color, trimming, freshness, and freedom from defect and decay
Lettuce, crisp head	1975	Turgidity, color, maturity (firmness), trimming (number of wrapper leaves), and freedom from tip burn, other physiological disorders, mechanical damage, seedstems, other defects, and decay
Lettuce, greenhouse leaf	1964	Well-developed, well-trimmed, and freedom from coarse stems, bleached or discolored leaves, wilting, freezing, insects, and decay
Lettuce, romaine	1960	Freshness, trimming, and freedom from decay and damage caused by seedstems, broken, bruised, or discolored leaves, tipburn, and wilting
Mushroom	1966	Maturity, shape, trimming, size, and freedom from open veils, disease, spots, insect injury, and decay

TABLE 8.37. QUALITY FACTORS FOR FRESH VEGETABLES IN THE U.S. STANDARDS FOR GRADES (*Continued*)

Vegetable	Date Issued	Quality Factors
Mustard greens and turnip greens	1953	Freshness, tenderness, cleanness, and freedom from damage caused by seedstems, discoloration, freezing, disease, insects, or mechanical means; roots (if attached): firmness and freedom from damage
Okra	1928	Freshness, uniformity of shape and color, and freedom from defect and decay
Onion, dry		
Creole	1943	Maturity, firmness, shape, size (diameter), and freedom from decay, wet sunscald, doubles, bottlenecks, sprouting, and other defects
Bermuda-Granex-Grano	1995	Maturity, firmness, shape, size (diameter) and freedom from decay, wet sunscald, doubles, bottlenecks, seedstems, sprouting and other defects
Other varieties	1995	Maturity, firmness, shape, size (diameter), and freedom from decay, wet sunscald, doubles, bottlenecks, sprouts and other defects.
Onion, green	1947	Turgidity, color, form, cleanness, bulb trimming, no seedstems, and freedom from defect and decay
Onion sets	1940	Maturity, firmness, size, and freedom from decay and damage caused by tops, sprouting, freezing, mold, moisture, dirt, disease, insects, or mechanical means
Parsley	1930	Freshness, green color, and freedom from defects, seedstems, and decay
Parsnip	1945	Turgidity, trimming, cleanness, smoothness, shape, freedom from defects and decay, and size (diameter)

Pea, fresh	1942	Maturity, size, shape, freshness, and freedom from defects and decay
Pepper, sweet	1989	Maturity, color, shape, size, firmness, and freedom from defects (sunburn, sunscald, freezing injury, hail, scars, insects, mechanical damage) and decay
Potato	1991	Uniformity, maturity, firmness, cleanness, shape, size, and freedom from sprouts, scab, growth cracks, hollowheart, blackheart, greening, and other defects
Radish	1968	Tenderness, cleanness, smoothness, shape, size, and freedom from pithiness and other defects; tops of bunched radishes fresh and free from damage
Rhubarb	1966	Color, freshness, straightness, trimming, cleanness, stalk diameter and length, and freedom from defect
Shallot, bunched	1946	Firmness, form, tenderness, trimming, cleanness, and freedom from decay and damage caused by seedstems, disease, insects, mechanical and other means; tops: freshness, green color, and no mechanical damage
Southern pea	1956	Maturity, pod shape, and freedom from discoloration and other defects
Spinach bunches	1987	Freshness, cleanness, trimming, and freedom from decay and damage caused by coarse stalks or seedstems, discoloration, insects, or mechanical means
Spinach leaves	1946	Color, turgidity, cleanness, trimming, and freedom from seedstems, coarse stalks, and other defects
Squash, summer	1984	Immaturity, tenderness, shape, firmness, and freedom from decay, cuts, bruises, scars, and other defects

TABLE 8.37. QUALITY FACTORS FOR FRESH VEGETABLES IN THE U.S. STANDARDS FOR GRADES (*Continued*)

Vegetable	Date Issued	Quality Factors
Squash, winter, and pumpkin	1983	Maturity, firmness, freedom from discoloration, cracking, dry rot, insect damage, and other defects; uniformity of size
Strawberry	1965	Maturity ($>\frac{1}{2}$ or $>\frac{3}{4}$ of surface showing red or pink color, depending on grade), firmness, attached calyx, size, and freedom from defect and decay
Sweet potato	1963	Firmness, smoothness, cleanness, shape, size, and freedom from mechanical damage, growth cracks, internal breakdown, insect damage, other defects, and decay
Tomato	1991	Maturity and ripeness (color chart), firmness, shape, size, and freedom from defect (puffiness, freezing injury, sunscald, scars, catfaces, growth cracks, insect injury, and other defects) and decay
Tomato, greenhouse	1966	Maturity, firmness, shape, size, and freedom from decay, sunscald, freezing injury, bruises, cuts, shriveling, puffiness, catfaces, growth cracks, scars, disease, and insects
Turnip and rutabaga	1955	Uniformity of root color, size, and shape, trimming, freshness, and freedom from defects (cuts, growth cracks, pithiness, woodiness, water core, dry rot)
Watermelon	2006	Maturity and ripeness (optional internal quality criteria: soluble solids content 10% or more very good, $>8\%$ good), shape, uniformity of size (weight), and freedom from anthracnose, decay, sunscald, and whiteheart

Adapted from A. A. Kader (ed.), *Postharvest Technology of Horticultural Crops,* 3rd ed. (University of California, Division of Agriculture and Natural Resources Publication 3311, 2002).

TABLE 8.38. QUALITY FACTORS FOR PROCESSING VEGETABLES IN THE U.S. STANDARDS FOR GRADES

Vegetable	Date Issued	Quality Factors
Asparagus, green	1972	Freshness, shape, green color, size (spear length), and freedom from defect (freezing damage, dirt, disease, insect injury, and mechanical injuries) and decay
Bean, shelled lima	1953	Tenderness green color, and freedom from decay and from injury caused by discoloration, shriveling, sunscald, freezing, heating, disease, insects, or other means
Bean, snap	1985	Freshness, tenderness, shape, size, and freedom from decay and from damage caused by scars, rust, disease, insects, bruises, punctures, broken ends, or other means
Beet	1945	Firmness, tenderness, shape, size, and freedom from soft rot, cull material, growth cracks, internal discoloration, white zoning, rodent damage, disease, insects, and mechanical injury
Broccoli	1959	Freshness, tenderness, green color, compactness, trimming, and freedom from decay and damage caused by discoloration, freezing, pithiness, scars, dirt, or mechanical means
Cabbage	1944	Firmness, trimming, and freedom from soft rot, seedstems, and from damage caused by bursting, discoloration, freezing, disease, birds, insects, or mechanical or other means
Carrot	1984	Firmness, color, shape, size (root length), smoothness, not woody, and freedom from soft rot, cull material, and from damage caused by growth cracks, sunburn, green core, pithy core, water core, internal discoloration, disease, or mechanical means

TABLE 8.38. QUALITY FACTORS FOR PROCESSING VEGETABLES IN THE U.S. STANDARDS FOR GRADES (*Continued*)

Vegetable	Date Issued	Quality Factors
Cauliflower	1959	Freshness, compactness, color, and freedom from jacket leaves, stalks, and other cull material, decay, and damage caused by discoloration, bruising, fuzziness, enlarged bracts, dirt, freezing, hail, or mechanical means
Corn, sweet	1962	Maturity, freshness, and freedom from damage by freezing, insects, birds, disease, cross-pollination, or fermentation
Cucumber, pickling	1936	Color, shape, freshness, firmness, maturity, and freedom from decay and from damage caused by dirt, freezing, sunburn, disease, insects, or mechanical or other means
Mushroom	1964	Freshness, firmness, shape, and freedom from decay, disease spots, and insects, and from damage caused by insects, bruising, discoloration, or feathering
Okra	1965	Freshness, tenderness, color, shape, and freedom from decay and insects, and from damage caused by scars, bruises, cuts, punctures, discoloration, dirt or other means
Onion	1944	Maturity, firmness, and freedom from decay, sprouts, bottlenecks, scallions, seedstems, sunscald, roots, insects, and mechanical injury
Pea, fresh shelled for canning/freezing	1946	Tenderness, succulence, color, and freedom from decay, scald, rust, shriveling, heating, disease, and insects
Pepper, sweet	1948	Firmness, color, shape, and freedom from decay, insects, and damage by any means that results in 5–20% trimming (by weight) depending on grade
Potato	1983	Shape, smoothness, freedom from decay and defect (freezing injury, blackheart, sprouts), size, specific gravity, glucose content, and fry color

Potato for chipping	1978	Firmness, cleanness, shape, freedom from defect (freezing, blackheart, decay, insect injury, and mechanical injury), size; optional tests for specific gravity and fry color included
Southern pea	1965	Pods: maturity, freshness, and freedom from decay; seeds: freedom from scars, insects, decay, discoloration, splits, cracked skin, and other defects
Spinach	1956	Freshness, freedom from decay, grass weeds, and other foreign material, and freedom from damage caused by seedstems, discoloration, coarse stalks, insects, dirt, or mechanical means
Sweet potato for canning/freezing	1959	Firmness, shape, color, size, and freedom from decay and defect (freezing injury, scald, cork, internal discoloration, bruises, cuts, growths cracks, pithiness, stringiness, and insect injury)
Sweet potato for dicing/pulping	1951	Firmness, shape size, and freedom from decay and defect (scald, freezing injury, cork, internal discoloration, pithiness, growth cracks, insect damage, and stringiness)
Tomato	1983	Firmness, ripeness (color as determined by a photoelectric instrument), and freedom from insect damage, freezing, mechanical damage, decay, growth cracks, sunscald, gray wall, and blossom-end rot
Tomato, green	1950	Firmness, color (green), and freedom from decay and defect (growth cracks, scars, catfaces, sunscald, disease, insects, or mechanical damage)
Tomato, Italian type for canning	1957	Firmness, color uniformity, and freedom from decay and defect (growth cracks, sunscald, freezing, disease, insects, or mechanical injury)

Adapted from A. A. Kader (ed.), *Postharvest Technology of Horticultural Crops,* 3rd ed. (University of California, Division of Agriculture and Natural Resources Publication 3311, 2002).

INTERNATIONAL STANDARDS

International standards for vegetables published by the Organization for Economic Cooperation and Development (OECD) are available from the OECD Bookshop at http://www.oecdbookshop.org in the series International Standardization of Fruit and Vegetables.

Ontario, Canada, Grading and Packing Manuals are available at:

http://www.gov.on.ca/omafra/english/food/inspection/fruitveg/intro.htm

10
MINIMALLY PROCESSED VEGETABLES

Helpful website: http://www.fresh-cuts.org

TABLE 8.39. BASIC REQUIREMENTS FOR PREPARATION OF MINIMALLY PROCESSED VEGETABLES

High-quality raw material
Variety selection
Production practices
Harvest and storage conditions
Strict hygiene and good manufacturing practices
Use of Hazard Analysis Critical Control Points principles
Sanitation of processing line, product, and workers
Low temperatures during processing
Careful cleaning and/or washing before and after peeling
Good-quality water (sensory, microbiological, pH)
Use of mild processing aids in wash water for disinfection or prevention of browning and texture loss
Chlorine, ozone, and other disinfectants
Antioxidant chemicals such as ascorbic acid, citric acid, etc.
Calcium salts to reduce textural changes
Minimal damage during peeling, cutting, slicing, and shredding operations
Sharp knives and blades on cutters
Elimination of defective and damaged pieces
Gentle draining, spin- or air-drying to remove excess moisture
Correct packing materials and packaging methods
Selection of plastic films to ensure adequate O_2 levels to avoid fermentation
Correct temperature during distribution and handling
All minimally processed products kept at 32–41°F

Adapted from A. A. Kader (ed.), *Postharvest Technology of Horticultural Crops,* 3rd ed. (University of California, Division of Agriculture and Natural Resources Publication 3311. 2002).

TABLE 8.40. ADVANTAGES, DISADVANTAGES, AND REQUIREMENTS OF FRESH-CUT VEGETABLE PRODUCTS PREPARED AT DIFFERENT LOCATIONS

Location of Processing	Characteristics and Requirements
Source of production	Raw product processed fresh when it is of the highest quality. Processed product requires a minimum of 14 days postprocessing shelf life. Good temperature management critical. Economy of scale. Avoid long-distance transport of unusable product. Vacuum- and gas-flushing common; differentially permeable films.
Regional	Raw product processed when of good quality, typically 3–7 days after harvest. Reduced need to maximize shelf life; about 7 days postprocessing life required. Good temperature management vital. Several deliveries weekly to end-users. Can better respond to short-term demands. Vacuum- and gas-flushing common; differentially permeable films.
Local	Raw product quality may vary greatly because processed 7–14 days after harvest. Relatively short postprocessing life required or expected. Good temperature management required but is often deficient. Small quantities processed and delivered. More labor intensive; discard large amounts of unusable product. Simpler and less costly packaging; less use of vacuum- or gas-flushing techniques.

Adapted from A. A. Kader (ed.), *Postharvest Technology of Horticultural Crops,* 3rd ed. (University of California, Division of Agriculture and Natural Resources publication 3311, 2002).

TABLE 8.41. PHYSIOLOGY AND STORAGE CHARACTERISTICS OF FRESH-CUT VEGETABLES (ALL PRODUCTS SHOULD BE STORED AT 32–41°F)

Vegetable	Fresh-cut Product	Respiration Rate in Air at 41°F ($mL\ CO_2 \cdot kg^{-1} \cdot h^{-1}$)	Common Quality Defects (other than microbial growth)	Beneficial Atmosphere	
				$\%O_2$	$\%CO_2$
Asparagus tips	Trimmed spears	40	Browning, softening	10–20	10–15
Beans, snap	Cut	15–18	Browning	2–5	3–12
Beet	Cubed	5	Leakage; color loss	5	5
Broccoli	Florets	20–35	Yellowing, off–odors	3–10	5–10
Cabbage	Shredded	13–20	Browning	3–7	5–15
Carrot	Sticks, shredded	7–10; 12–15	Surface drying (white blush), leakage	0.5–5	10
Cauliflower	Florets	—	Discoloration; off odors	5–10	<5
Celery	Sticks	2–3	Browning, surface drying	—	—
Cucumber	Sliced	5	Leakage	—	—
Garlic	Peeled clove	20	Sprout growth, discoloration	3	5–10
Jicama	Sticks	5–10	Browning; texture loss	3	10
Leek	Sliced	25	Discoloration	5	5
Lettuce, iceberg	Chopped, shredded	6; 10	Browning of cut edges	<0.5–3	10–15
Lettuce, other	Chopped	10–13	Browning of cut edges	1–3	5–10

TABLE 8.41. PHYSIOLOGY AND STORAGE CHARACTERISTICS OF FRESH-CUT VEGETABLES (ALL PRODUCTS SHOULD BE STORED AT 32–41°F) (*Continued*)

Vegetable	Fresh-cut Product	Respiration Rate in Air at 41°F ($mL\ CO_2 \cdot kg^{-1} \cdot h^{-1}$)	Common Quality Defects (other than microbial growth)	Beneficial Atmosphere	
				$\%O_2$	$\%CO_2$
Melons					
Cantaloupe	Cubed	5–8	Leakage; softening; glassiness (translucency)	3–5	5–15
Honeydew	Cubed	2–4	Leakage; softening; glassiness (translucency)	2–3	5–15
Watermelon	Cubed	2–4	Leakage; softening	3–5	5–15
Onion, bulb	Sliced, diced	8–12	Texture, juice loss, discoloration	2–5	10–15
Onion, green	Chopped	25–30	Discoloration, growth; leakage	—	—
Pepper	Sliced, diced	3; 6	Texture loss, browning	3	5–10
Potato	Sticks, peeled	4–8	Browning, drying	1–3	6–9
Rutabaga	Cubed	10	Discoloration, drying	5	5
Spinach	Cleaned, cut	6–12	Off-odors; rapid deterioration of small pieces	1–3	8–10
Squash, summer	Cubed, sliced	12–24	Browning; leakage	1	—
Strawberry	Sliced; topped	12; 6	Loss of texture, juice, color	1–2	5–10
Tomato	Sliced	3	Leakage	3	3

Adapted from A. A. Kader (ed.), *Postharvest Technology of Horticultural Crops,* 3rd ed. (University of California, Division of Agriculture and Natural Resources Publication 3311, 2002).

11
CONTAINERS FOR VEGETABLES

TABLE 8.42. SHIPPING CONTAINERS FOR FRESH VEGETABLES

Vegetable	Container[1]	Approximate Net Weight (lb)[2, 3]
Artichoke	Carton by count or loose pack	23
Asparagus	Pyramid cartons or crates	30
	Cartons or crates, bunched	28
	Lugs or cartons, loose	25
	Cartons, 16 1½-lb bunches	24–25
	Lugs or cartons, loose	21
	Pyramid carton or crate, ½	20
	Carton, bunched	20
	Pyramid carton or crate, ½	15–17
	Carton, bunched	14
	Carton, loose	15
	Carton, ½	12
	Carton or crate, ⅓	12–13
	Carton or crate	11
Bean	Wirebound crate or hamper, bushel	26–31
	Carton or crate	25–30
Green	Carton	20–22
	Presnipped bags, retail	12 oz[3]
	Presnipped bags, foodservice	10
Yellow	Wirebound crate or hamper	30
	Carton	25–30
	Carton	15
Beet, Bunched	Wirebound crate or carton, 12 bunches	45
	Carton or crate, 24 bunches	38
Topped	Mesh sack	50
	Sack	25
	Carton or crate, 12 bunches	20
Belgian endive	Carton	10
Bitter melon	Crate	40
	Carton or crate	30

TABLE 8.42. SHIPPING CONTAINERS FOR FRESH VEGETABLES (*Continued*)

Vegetable	Container[1]	Approximate Net Weight (lb)[2, 3]
	Carton	20
	Carton	10
	Carton	5
Boniato	Carton or sack	50
	Carton	10
Brussels sprouts	Cartons, loose	25
	Flats or cartons, 16 12-oz cello bags	10
	8 1-lb clamshells	8
	24 1-lb Vexar bags	25
Cabbage	Bulk bin	2,000
	Bulk bin	1,000
	Flat crate	50–60
	Carton or mesh sack	50
	Crate, 1¾ bushel	50
	Carton	45
	Carton (savoy)	20
Calabaza	Bin	800
	Carton or sack	50
Carrot	Table carton	50
	48 1-lb film bags	48
	Table poly bags	25
	24 2-lb poly bags	48
	12 2-lb poly bags	24
	5 10-lb poly bags	50
	16 3-lb poly bags	48
	10 5-lb poly bags	50
Bunched	Carton	26
Baby, peeled	Carton 20 1-lb bags	20
	24 1-lb bags	24
	40 1-lb bags	40
	10 2-lb bags	20
	12 2-lb bags	24
	20 2-lb bags	40
	4 5-lb bags	20

TABLE 8.42. SHIPPING CONTAINERS FOR FRESH VEGETABLES (*Continued*)

Vegetable	Container[1]	Approximate Net Weight (lb)[2, 3]
	8 5-lb bags	40
	73 3-oz bags	14
Foodservice	Poly jumbo	25
	Poly jumbo	50
Cauliflower	Long Island wirebound crate	60
	Catskill carton	50
	Carton, 12- and 16-count film-wrapped trimmed heads	25–30
Celeriac	Crate, 1⅑ bushel	35
	Crate	20
	Carton, 12 count	
Celery	Carton or crate	50–60
Hearts	Carton	28
	Carton	18
Chayote	Crate	50
	Crate	40
	Carton	30
	1-layer flat, 24 count	
	Carton	20
Corn	Carton or crate	50
	Carton, crate, or sack	42
	Wirebound crate	42
	Sack	37
	Carton, 48 count	
	12 × 4 packaged tray pack	
	12 × 3 packaged tray pack	
Cucumber	Carton or crate, 1⅑ bushel	55
	Carton, 3.56 decaliter	55
	Carton, 48 count	30
	Carton or crate, ⅝ bushel	28
	Carton, 36–42 count	28
	Carton, 36–42 count (CA)	24
	Carton, 24 count	22
Greenhouse	Carton, film wrapped	16
	Carton or flat, film wrapped	12

TABLE 8.42. SHIPPING CONTAINERS FOR FRESH VEGETABLES (*Continued*)

Vegetable	Container[1]	Approximate Net Weight (lb)[2, 3]
Daikon	Carton or crate	50
	Carton or crate	45
	Carton or crate, 1½ bushel	40
	Box, crate, or lug	20
	Carton	10
	Carton	5
Eggplant	Carton, crate, or basket; bushel or 1⅑ bushel	33
	Carton, 3.56 decaliter	33
	Carton, crate, or lug	25
	L.A. lug or carton, 18–24 count	
	Lug, ½ and ⅝ bushel	17
Chinese	Lug	26
	Carton	25
	Carton or crate, ½ and ⅝ bushel	15
Italian	Lug	26
	Carton or crate, ½ and ⅝ bushel	15
Japanese	Carton or crate, ½ and ⅝ bushel	15
Endive/Escarole	Carton, 24 count	34
	Crate, 3-wire celery, 24 count	30–40
	Crate, 1⅝ bushel	Various
	Crate, ⅝ bushel	Various
Garlic	Carton	5
	Carton	10
	Carton	15
	Carton	22
	Carton	30
	Bag	3
	Bag	3 16-oz[3]
	Cello bag or tray; 2, 3, 4 count	
Ginger	Carton	30
	Carton	20
	Carton or film bag	5

TABLE 8.42. SHIPPING CONTAINERS FOR FRESH VEGETABLES (*Continued*)

Vegetable	Container[1]	Approximate Net Weight (lb)[2, 3]
Greens (collards and dandelion)	Crate, 1⅖, 1⅗ bushel	30–35
	Basket, crate or carton; bushel	20–25
	Crate or carton, 12–24 bunch count	
Haricot vert	Tray	11
	Tray	10
	Tray	5
Jerusalem artichoke	Carton	25
	Carton	20
	Carton	10
	Carton	5
Jicama	Crate, 1⅑ bushel	45
	Wirebound crate	40
	Carton or crate	20
	Carton	10
Leek	Carton, 12 bunches	30
	Carton, 24 bunches	24–30
	Carton or crate, ⅘ bushel	20
	Carton, 10 1-lb film bag	10
Lettuce, iceberg	Carton; 18, 24, 30 count	50
	Carton	30
	Carton; 15, 16 count	20
Boston	Crate, 1⅑ bushel	22
	Carton or crate, 24 count	20
	Flat, carton, or crate	10
	Basket or carton, 12 q	5
Bibb	Flat, carton, or crate	10
	Basket or carton, 12 q	5
	Basket, greenhouse	5
Leaf	Carton or crate, 24 count	25
	Crate, ⅘ bushel	20
	Crate, 1⅔ bushel	14
	Basket or carton, 24 q	10
	Carton	3

TABLE 8.42. SHIPPING CONTAINERS FOR FRESH VEGETABLES (*Continued*)

Vegetable	Container[1]	Approximate Net Weight (lb)[2, 3]
	Carton	2
Processed iceberg	Carton, chopped	20
	Carton, chopped or cleaned/cored	30
	Bins	1,000
Romaine	Carton; ⅔, 24 count (West)	40
	Carton	40
	Carton, 1.3 bushel	28
	Carton or crate, 1⅑ bushel	22
	Carton, 24 count (East)	22
	Carton, 12 count	18
Lo Bok	Crate	45
	Crate	40
	Carton, crate, or lug	25
Long bean	Carton	40
	Crate	30
	Carton	10
	Carton	5
Melon		
Cantaloupe	Bin	1,000
	Jumbo crate	80
	1¾ bushel cartons or crates	60
	Carton or crate, ⅓	54
	Carton or crate, ½	40
	Carton or crate, 1⅑ bushel	40
	Bushel basket	40
	Single-layer pack	18–21
Honeydew	Flat crate	35
	Carton, ⅔, various count	30
	Carton	30
Mixed	Flat crate	35
	Carton, various count	30
Mushroom	Carton, 12 1-lb trays	12
	Carton	10
	Carton, 18 8-oz or 8 1-lb trays	8
	Carton, 12 8-oz trays	6

TABLE 8.42. **SHIPPING CONTAINERS FOR FRESH VEGETABLES (*Continued*)**

Vegetable	Container[1]	Approximate Net Weight (lb)[2, 3]
	Carton	5
	Basket, 4 q	3
Napa	WGA crate	70
	Crate, celery	50
	Carton	50
	Crate, 1.3 bushel	45
	Carton	45
	Carton or crate, 1⅑ bushel	40
	Carton	30
Okra	Basket, crate, hamper; bushel	30
	Hamper, ¾ bushel	23
	Crate or flat, 5/9 bushel	18
	Basket, crate or lug	15
Onion		
Bulb	Carton, crate, sack	50
	Master container, 10 5-lb bags	50
	Master container, 16 3-lb bags	48
	Master container, 24 2-lb sacks	48
	Master container, 15 3-lb sacks	45
	Master container, 20 2-lb sacks	40
	Carton	40
	Master container, 12 3-lb sacks	36
	Master container, 16 2-lb sacks	32
	Sack, reds, boilers	25
	Carton or bag	25
	Master container, 12 2-lb sacks	24
	Carton, sack, or bag	10
	Bag or carton	5
Green	Carton, bunched bulb-type	28
	Carton or crate, bunched 24 count	20
	Carton, bunched 48 count	13
	Carton, bunched 36 count	11
Pak choi	WGA Crate	70
	Crate	60

TABLE 8.42. SHIPPING CONTAINERS FOR FRESH VEGETABLES (*Continued*)

Vegetable	Container[1]	Approximate Net Weight (lb)[2, 3]
	Carton or crate	50
	Carton or crate	40
	Carton or crate	35
	Carton or crate	30
Parsley	Carton or wirebound crate, 60 count	
	Carton or wirebound crate, 30 count	
	Carton or crate, 1⅑ bushel, bunched	21
	Carton, crate, or basket, bunched	11
Parsnip	Carton or crate, ½ bushel	25
	Film sack	20
	Carton, 12 1-lb bags	12
Pea		
Green	Basket, crate, or hamper; 1 bushel	30
	Crate, 1⅑ bushel	30
Edible pod	Carton	10
Southern	Hamper, 1 bushel	25
Pepino	Carton, 1 layer	10
	Carton	8
Pepper		
Bell	Carton, 1⅑ bushel	35
	Carton or crate (Mexico)	30
	Carton or crate, bushel, and 1⅑ bushel	28
	Carton, 3.56 decaliter	28
	Carton	25
	Carton, ½ bushel	14–15
	Flat carton (Netherlands)	11
Chiles: jalapeno,	Crate or carton, ½ bushel	
yellow wax,	Crate or carton, ⅝ bushel	
others	Bin	500

TABLE 8.42. SHIPPING CONTAINERS FOR FRESH VEGETABLES (*Continued*)

Vegetable	Container[1]	Approximate Net Weight (lb)[2,3]
	Crate or carton 1⅑ bushel	
	Cases, bulk	10
Potato	Sack	100
	Carton or sack	50
	Baled, 5 10-lb bags	50
	Baled, 10 5-lb bags	50
	Carton, 60–100 count	
Prickly pear	Carton	18
	Carton, 35 count	10
Pumpkin	Bin	1,000
	Carton, crate, or sack	50
	Carton or crate, ½ bushel	25
Radicchio	Carton or lug	7
Radish		
Topped	Sack or bag, loose	40
	Bag	25
	Bag, resealable or conventional	14
	Basket or carton	12
	Bag, 30 6-oz or 24 8-oz	12
	Bag, 4 5-lb	20
Bunched	Carton or crate, 48 count	35
	Carton or lug, ⅘ bushel	30
	Carton, 24 count	25
	Carton or crate, 24 count	20
	Carton or crate, 24 count	15
Rhubarb	Carton or lug	20
	Carton	15
Rutabaga	Carton or bag, 1 bushel	50
	Carton or bag, ½ bushel	25
Salad mix	Carton, 4 5-lb bags	20
	Carton, 2 10-lb bags	20
	Carton, 3 24 count (retail)	
	Carton, mesclun	3

TABLE 8.42. SHIPPING CONTAINERS FOR FRESH VEGETABLES (*Continued*)

Vegetable	Container[1]	Approximate Net Weight (lb)[2, 3]
Salsify	Carton	22
	Carton	10
	Bag	4
Spinach	Carton or crate, 1½ bushel	32
	Container, bushel	25
	Carton, 24-bunch count	20
	Carton, 12-bunch count	20
	Bag, 12 10-lb	120
	Bag, 24-q	10
	Carton, 12 10-oz bags	8
Sprouts		
Alfalfa	Carton or flat	5
	Bag	5
	Bag	1
Bean	Carton or film bag	10
	Carton	6
	Film bag	5
Radish	Various containers	50
	Carton or crate, 1⅑ bushel	40
	Various containers	30
	Various containers	25
	Carton	20
	Various containers	20
	Carton	10
	Various containers	10
	Various containers	8
	Carton	5
	Various containers	4
Squash		
Summer	Container, bushel and 1⅑ bushel	42
	Carton or crate	35
	Carton or crate, ¾ bushel	30
	Carton or lug (California, Mexico)	26

TABLE 8.42. SHIPPING CONTAINERS FOR FRESH VEGETABLES (*Continued*)

Vegetable	Container[1]	Approximate Net Weight (lb)[2, 3]
	Container, ½ or ⅝ bushel	21
	Basket or carton, 8-q	10
Winter	Carton or crate, 1⅑ bushel	50
	Carton or crate	40
	Carton or crate	35
Strawberry	Flat, 12 1-pt baskets	12
	Flat, 6 1-qt baskets	12
	Flat, 12 10-oz clamshells	7.5
	Flat, 12 8.8-oz clamshells	6.6
	Crate, 8 16-oz clamshells	9
	Tray, ½	5
	Flat, 4 2-lb clamshells	8
Sweet potato	Carton	40
	Carton	20
	Carton	10
	Poly bag	5
	Poly bag	3
Taro	Carton, crate, or sack	50
	Carton	10
Tomatillo	Carton	40
	Carton	30
	Carton	10
Tomato		
Round	Carton, loose	25
	Carton, flats	20
	Lug, 3-layer	
	Lug, 2-layer	
Cherry	Flat, 12 1-pt baskets	
Roma	Carton, loose 25	
Greenhouse	Flat, 1 layer	15
Grape	Clamshells: 8, 12, 24-oz	
	Containers, 12 1-pt	
	Containers, bulk	20

TABLE 8.42. SHIPPING CONTAINERS FOR FRESH VEGETABLES (*Continued*)

Vegetable	Container[1]	Approximate Net Weight (lb)[2, 3]
Turnip	Basket or sack, bushel	50
	Carton, bunched	40
	Basket, carton, crate, bag; ½ bushel	25
	Carton, 12-count bunch	20
Watermelon	Bulk	45,000
	Bin	1,050
	Carton, various count	85
	Carton, seedless	65
	Carton, icebox	35
	Bins: 24, 30, and 36 in.	
Winter melon	Bin	800
	Crate	70
	Carton, crate, or sack	50
	Various containers	50
	Various containers	40
	Various containers	30
	Various containers	20
	Various containers	10
Yucca	Carton, crate, or sack	50
	Various containers	50
	Various containers	40
	Various containers	30
	Various containers	20
	Various containers	10
	Cartons	10

Adapted from *The Packer Sourcebook,* Vance Publishing Corp., 10901 W. 84th Terr., Lenexa, KS 66214-1632 (2004). Reprinted by permission from *The Packer. The Packer* does not review or endorse products, services, or opinions.

[1] Other containers are being developed and used in the marketplace. The requirements of each market should be determined.

[2] Actual weights larger and smaller than those shown may be found. The midpoint of the range should be used if a single value is desired.

[3] Other weight as shown.

TABLE 8.43. STANDARDIZED SHIPPING CONTAINER DIMENSIONS DESIGNED FOR A 40 × 48-INCH PALLET

Outside Base Dimensions (in.)	Containers per Layer	Layers may be Cross-stacked
15¾ × 11¾	10	Yes
19¾ × 11¾	8	No
19¾ × 15¾	6	No
23¾ × 15¾	5	Yes

Adapted from S. A. Sargent, M. A. Ritenour and J. K. Brecht, " Handling, Cooling, and Sanitation Techniques for Maintaining Postharvest Quality" in *Vegetable Production Handbook* (University of Florida, 2005–2006).

TABLE 8.45. TRANSPORT EQUIPMENT INSPECTION

Most carriers check their transport equipment before presenting it to the shipper for loading. The condition of the equipment is critical to maintaining the quality of the products. Therefore, the shipper also should check the equipment to ensure it is in good working order and meets the needs of the product. Carriers provide guidance on checking and operating the refrigeration systems.

All transportation equipment should be checked for:

- Cleanliness—the load compartment should be regularly steam-cleaned.
- Damage—walls, floors, doors, ceilings should be in good condition.
- Temperature control—refrigerated units should be recently calibrated and supply continuous air circulation for uniform product temperatures.

Shippers should insist on clean equipment. A load of products can be ruined by:

- Odors from previous shipments
- Toxic chemical residues
- Insects nesting in the equipment
- Decaying remains of agricultural products
- Debris blocking drain openings or air circulation along the floor

Shipper should insist on well-maintained equipment and check for the following:

- Damage to walls, ceilings, or floors that can let in the outside heat, cold, moisture, dirt, and insects
- Operation and condition of doors, ventilation openings, and seals
- Provisions for load locking and bracing

For refrigerated trailers and van containers, the following additional checks are important:

- With the doors closed, have someone inside the cargo area check for light—the door gaskets must seal. A smoke generator also can be used to detect leaks.

- The refrigeration unit should cycle from high to low speed when the desired temperature is reached and then back to high speed.
- Determine the location of the sensing element that controls the discharge air temperature. If it measures return air temperature, the thermostat may have to be set higher to avoid chilling injury or freezing injury of the products.
- A solid return air bulkhead should be installed at the front of the trailer.
- A heating device should be available for transportation in areas with extreme cold weather.
- Equipment with a top air delivery system must have a fabric air chute or metal ceiling duct in good condition.

Adapted from B. M. McGregor, *Tropical Products Handbook,* USDA Agr. Handbook 668 (1987).

12
VEGETABLE MARKETING

TABLE 8.46. CHARACTERISTICS OF DIRECT MARKETING ALTERNATIVES FOR FRESH VEGETABLES

Grower Characteristics	Pick-Your-Own	Roadside Market	Farmer's Market
Harvesting cost	Customer assumes the cost.	Usual cost.	Usual cost.
Transportation cost	Customer assumes the cost.	Usually minimal for produce.	Depends on grower's distance to market.
Selling cost	Field attendant is needed. Harvesting instructions should be provided. Advertising.	Checkout attendant is needed. Advertising.	Checkout attendant is needed.
Grower liability	Liable for accidents. Absorbs damages to property and crop.	Liable for accidents at market.	Owner of market is responsible.
Market investment	Containers. Locational signs. Available parking.	Building or stand. Available parking. Containers.	Usually parking or building space is rented. Containers.

Volume of produce desired	Enough for customer traffic demands.	Enough to visibly attract customers to stop. Variety is helpful.	Enough to justify transportation and other costs.
Prices received for produce	Often lower than other alternatives because transportation and harvesting cost is assumed by the customer. Producer sets the price.	Producer sets the price given perceived demand competitive conditions.	Producer sets the price. There may be competition from other sellers.
Quality	Can sell whatever the customers will pick.	Can classify produce and sell more than one grade.	Ability to sell may depend on the competing qualities available from other growers.
Other	Balance between number of pickers and amount needing to be harvested sometimes is difficult to achieve.	Sometimes other items besides produce are sold to supplement income. Produce spoilage can be minimized if adequate cooling facilities are used.	Sometimes other items besides produce are sold to supplement income. Bulk sales are sometimes recommended.

Adapted from *Cucurbit Production and Pest Management*, Oklahoma Cooperative Extension Circular E-853 (1986).

TABLE 8.47. CHARACTERISTICS OF SOME WHOLESALE MARKETING ALTERNATIVES FOR FRESH VEGETABLES

Grower Consideration	Terminal Market	Cooperative and Private Packing Facilities	Peddling to Grocer or Restaurant	Wholesale/Broker
Harvesting cost	Usual cost.	Sometimes harvesting equipment is provided.	Usual cost.	Usual cost.
Transportation cost	Depends on distance to market.	Sometimes transportation is provided.	Depends on distance traveled.	Depends on prior arrangements for delivery or pickup.
Prices received for produce	Grower is usually the price taker.	Prices received by growers depend on market prices, costs, and revenues.	Buyer and grower may compromise on price, or grower fixes price.	Grower is usually the price taker.
Required volume	Usually large quantities are needed.	Depends on the products to be sold.	Depends on the size of outlets and route.	Usually large quantities are needed.

Market investment	Truck or some transportation arrangements. Specialized containers are required.	Relatively low on a per-unit basis.	Truck. Containers.	Depends on arrangements. Usually minimal costs to grower. Specialized containers are required.
Quality	Must meet buyer's standards or U.S. grades.	Must meet buyer's standards or U.S. grades.	High quality is needed.	Must meet standards or U.S. grades so produce can be handled in bulk.
Other	Good source of market information. Can move very large quantities at one time. Many buyers are located at terminal markets.	May provide technical assistance to growers. Firms help in planning of growing and selling. Equipment may be shared by growers.	Long-term outlet for consistent quality. Good price for quality produce. Difficult to enter market and develop customers.	Good wholesaler/ broker can sell produce quickly at good prices. A long-term buyer/ seller relationship is desirable. Broker does not necessarily take title of produce.

Adapted from *Cucurbit Production and Pest Management,* Oklahoma Cooperative Extension Circular E-853 (1986).

ADDITIONAL SOURCES OF POSTHARVEST INFORMATION

J. A. Bartz and J. K. Brecht, *Postharvest Physiology and Pathology of Vegetables,* 2nd ed. (New York: Marcel Dekker, 2003).

A. A. Kader (ed.), *Postharvest Technology of Horticultural Crops* (University of California Agriculture and Natural Resource Publication 3211, 2002).

S. J. Kays and R. E. Powell, *Postharvest Biology* (Athens, Ga.: Exon, 2004).

A. L. Snowdon, *A Color Atlas of Post-Harvest Disease and Disorders of Fruits and Vegetables,* vol. 1, *General Introduction and Fruits* (Boca Raton, Fla.: CRC Press, 1990).

A. L. Snowdon, *A Color Atlas of Post-Harvest Disease and Disorders of Fruits and Vegetables,* vol. 2, *Vegetables* (Boca Raton, Fla.: CRC Press, 1992).

PART 9

VEGETABLE SEEDS

01
SEED LABELS

LABELING VEGETABLE SEEDS

Seeds entering into interstate commerce must meet the requirements of the Federal Seed Act. Most state seed laws conform to federal standards. However, the laws of the individual states vary considerably with respect to the kinds and tolerances for noxious weeds. The noxious weed seed regulations and tolerances, if any, may be obtained from the State Seed Laboratory of any state.

Vegetable seed in packets or in larger containers must be labeled in any form that is clearly legible with the following required information:

- *Kind, variety, and hybrid.* The name of the kind and variety and hybrid, if appropriate, must be on the label. Words or terms that create a misleading impression as to the history or characteristics of kind or variety may not be used.
- *Name of shipper or consignee.* The full name and address of either the shipper or consignee must appear on the label.
- *Germination.* Vegetable seeds in containers of 1 lb or less with germination equal to or more than the standards need not be labeled to show the percentage germination or date of test. Vegetable seeds in containers of more than 1 lb must be labeled to show the percentage of germination, the month and year of test, and the percentage of hard seed, if any.
- *Lot number.* The lot number or other lot identification of vegetable seed in containers of more than 1 lb must be shown on the label and must be the same as that used in the records pertaining to the same lot of seed.
- *Seed treatment.* Any vegetable seed that has been treated must be labeled in no smaller than 8-point type to indicate that the seed has been treated and to show the name of any substance used in such treatment.

Adapted from Federal Seed Act Regulations, http://www.ams.usda.gov/lsg/seed.htm.

02
SEED GERMINATION TESTS

TABLE 9.1. REQUIREMENTS FOR VEGETABLE SEED GERMINATION TESTS

					Additional Directions	
Seed	Substrata[1]	Temperature[2] (°F)	First Count (Days)	Final Count (Days)	Specific Requirements	Fresh and Dormant Seed
Artichoke	B, T	68–86	7	21		
Asparagus	B, T, S	68–86	7	21		
Asparagus bean	B, T, S	68–86	5	8[3]		
Bean						
Garden	B, T, S, TC	68–86; 77	none	8		Use 0.3–0.6% $Ca(NO_3)_2$ to moisten substratum for retesting if hypocotyl collar rot is observed in initial test.
Lima	B, T, C, S	68–86	5	9[3]		
Runner	B, T, S	68–86	5	9[3]		
Beet	B, T, S	68–86	3	14	Presoak seeds in water for 2 hrs.	

TABLE 9.1. REQUIREMENTS FOR VEGETABLE SEED GERMINATION TESTS (*Continued*)

Seed	Substrata[1]	Temperature[2] (°F)	First Count (Days)	Final Count (Days)	Additional Directions: Specific Requirements	Additional Directions: Fresh and Dormant Seed
Broad bean	S, C	64	4	14[3]		Prechill at 50°F for 3 days.
Broccoli	B, P, T	68–86	3	10		Prechill at 41° or 50°F for 3 days; KNO_3 and light.
Brussels sprouts	B, P, T	68–86	3	10		Prechill at 41° or 50°F for 3 days; KNO_3 and light.
Burdock, great	B, T	68–86	7	14		
Cabbage	B, P, T	68–86	3	10		Prechill at 41° or 50°F for 3 days; KNO_3 and light.
Cabbage, Chinese	B, T	68–86	3	7		
Cabbage, tronchuda	B, P	68–86	3	10		Prechill at 41° or 50°F for 3 days; KNO_3 and light.
Cantaloupe	B, T, S	68–86	4	10	Keep substratum on dry side; remove excess moisture.	

Cardoon	B, T	68–86	7	21		
Carrot	B, T	68–86	6	14		
Cauliflower	B, P, T	68–86	3	10		Prechill at 41° or 50°F for 3 days; KNO_3 and light.
Celeriac	P	59–77; 68	10	21	Light; 750–1,250 lux from cool-white fluorescent source.	
Celery	P	59–77; 68	10	21	Light; 750–1,250 lux from cool-white fluorescent source.	
Chard, Swiss	B, T, S	68–86	3	14	Presoak seed in water for 2 hrs.	
Chicory	P, TS	68–86	5	14	Light; KNO_3 or soil.	
Chives	B, T	68	6	14		
Citron	B, T	68–86	7	14	Soak seeds 6 hrs.	Test at 86°F.
Collards	B, P, T	68–86	3	10		Prechill at 41° or 50°F for 3 days; KNO_3 and light.
Corn, sweet	B, T, S, TC, TCS	68–86; 77	4	7		
Corn salad	B, T	59	7	28		Test at 50°F.
Cowpea	B, T, S	68–86	5	8[3]		
Cress						
Garden	B, P, T	59	4	10		Light.
Upland	P	68–86		7	Light; KNO_3.	Make first count when necessary or desirable.
Water	P	68–86	4	14	Light.	

TABLE 9.1. REQUIREMENTS FOR VEGETABLE SEED GERMINATION TESTS (*Continued*)

Seed	Substrata[1]	Temperature[2] (°F)	First Count (Days)	Final Count (Days)	Additional Directions	
					Specific Requirements	Fresh and Dormant Seed
Cucumber	B, T, S	68–86	3	7	Keep substratum on dry side; remove excess moisture.	
Dandelion	P, TB	68–86	7	21	Light, 750–1,250 lux.	
Dill	B, T	68–86	7	21		
Eggplant	P, TB, RB, T	68–86	7	14		Light; KNO_3.
Endive	P, TS	68–86	5	14	Light, KNO_3 or soil.	
Fennel	B, T	68–86	6	14		
Kale	B, P, T	68–86	3	10		Prechill at 41° or 50°F for 3 days; KNO_3 and light.
Kale, Chinese	B, P, T	68–86	3	10		Prechill at 41° or 50°F for 3 days; KNO_3 and light.
Kale, Siberian	B, P, T	68–86; 68	3	7		
Kohlrabi	B, P, T	68–86	3	10		Prechill at 41° or 50°F for 3 days; KNO_3 and light.
Leek	B, T	68	6	14		

Lettuce	P	68	none	7	Light.	Prechill at 50°F for 3 days or test at 59°F.
Mustard, India	P	68–86	3	7	Light.	Prechill at 50°F for 7 days and test for 5 additional days.
Mustard, spinach	B, T	68–86	3	7		
Okra	B, T	68–86	4	14[3]		
Onion	B, T	68	6	10		
Alternate method	S	68	6	12		
Onion, Welsh	B, T	68	6	10		
Pak choi	B, T	68–86	3	7		
Parsley	B, T, TS	68–86	11	28		
Parsnip	B, T, TS	68–86	6	28		
Pea	B, T, S	68	5	8[3]		
Pepper	TB, RB, T, B, P	68–86	6	14		Light and KNO_3.
Pumpkin	B, T, S	68–86	4	7	Keep substratum on dry side; remove excess moisture.	
Radish	B, T	68	4	6		
Rhubarb	TB, TS	68–86	7	21	Light.	
Rutabaga	B, T	68–86	3	14		
Sage	B, T, S	68–86	5	14		
Salsify	B, T	59	5	10		Prechill at 50°F for 3 days.
Savory, summer	B, T	68–86	5	21		
Sorrel	P, TB, TS	68–86	3	14	Light.	Test at 59°F.
Soybean	B, T, S, TC, TCS	68–86; 77	5	8[3]		

TABLE 9.1. REQUIREMENTS FOR VEGETABLE SEED GERMINATION TESTS (*Continued*)

Seed	Substrata[1]	Temperature[2] (°F)	First Count (Days)	Final Count (Days)	Additional Directions: Specific Requirements	Additional Directions: Fresh and Dormant Seed
Spinach	TB, T	59, 50	7	21	Keep substratum on dry side; remove excess moisture.	
Spinach, New Zealand	T	59, 68	5	21	Soak fruits overnight (16 hrs), air-dry 7 hrs; plant in very wet towels; do not rewater unless later counts exhibit drying out.	On 21st day, scrape fruits and test for 7 additional days.
Squash	B, T, S	68–86	4	7	Keep substratum on dry side; remove excess moisture.	
Tomato	B, P, RB, T	68–86	5	14		Light; KNO_3.
Tomato, husk	P, TB	68–86	7	28	Light; KNO_3.	
Turnip	B, T	68–86	3	7		
Watermelon	B, T, S	68–86; 77	4	14	Keep substratum on dry side; remove excess moisture.	Test at 86°F.

Adapted from Association of Official Seed Analysts, *Rules for Testing Seeds, Germination Tests, Methods of Testing for Laboratory Germination* (2004), http://www.aosaseed.com.

[1] B = between blotters
TB = top of blotters
T = paper toweling, used either as folded towel tests or as roll towel tests in horizontal or vertical position
S = sand or soil
TS = top of sand or soil
P = covered petri dishes: with 2 layers of blotters; with 1 layer of absorbent cotton; with 5 layers of paper toweling; with 3 thicknesses of filter paper; or with sand or soil
C = creped cellulose paper wadding (0.3-in. thick Kimpak or equivalent) covered with a single thickness of blotter through which holes are punched for the seed that are pressed for about one-half their thickness into the paper wadding.
TC = on top of creped cellulose paper without a blotter
RB = blotters with raised covers, prepared by folding up the edges of the blotter to form a good support for the upper fold which serves as a cover, preventing the top from making direct contact with the seeds

[2] Temperature. A single number indicates a constant temperature. Two numerals separated by a dash indicate an alternation of temperature; the test is to be held at the first temperature for approximately 16 hrs and at the second temperature for approximately 8 hrs per day.

[3] Hard seeds. Seeds that remain hard at the end of the prescribed test because they have not absorbed water, due to an impermeable seed coat, are to be counted as *hard seed.* If at the end of the germination period provided for legume and okra swollen seeds or seeds of these kinds that have just started to germinate are still present, all seeds or seedlings except the above-stated shall be removed and the test continued for 5 additional days and the normal seedlings included in the percentage of germination.

TABLE 9.2. GERMINATION STANDARDS FOR VEGETABLE SEEDS IN INTERSTATE COMMERCE

The following germination standards for vegetable seeds in interstate commerce, which are construed to include hard seed, are determined and established under the Federal Seed Act.

Seed	%	Seed	%
Artichoke	60	Cress, garden	75
Asparagus	70	Cress, upland	60
Bean, asparagus	75	Cress, water	40
Bean, broad	75	Cucumber	80
Bean, garden	70	Dandelion	60
Bean, lima	70	Dill	60
Bean, runner	75	Eggplant	60
Beet	65	Endive	70
Broccoli	75	Kale	75
Brussels sprouts	70	Kale, Chinese	75
Burdock, great	60	Kale, Siberian	75
Cabbage	75	Kohlrabi	75
Cabbage, tronchuda	70	Leek	60
Cantaloupe	75	Lettuce	80
Cardoon	60	Mustard, India	75
Carrot	55	Mustard, spinach	75
Cauliflower	75	Okra	50
Celeriac	55	Onion	70
Celery	55	Onion, Welsh	70
Chard, Swiss	65	Pak choi	75
Chicory	65	Parsley	60
Chinese cabbage	75	Parsnip	60
Chives	50	Pea	80
Citron	65	Pepper	55
Collards	80	Pumpkin	75
Corn, sweet	75	Radish	75
Corn salad	70	Rhubarb	60
Cowpea (southern pea)	75	Rutabaga	75

TABLE 9.2. GERMINATION STANDARDS FOR VEGETABLE SEEDS IN INTERSTATE COMMERCE (*Continued*)

Seed	%	Seed	%
Sage	60	Spinach, New Zealand	40
Salsify	75	Squash	75
Savory, summer	55	Tomato	75
Sorrel	65	Tomato, husk	50
Soybean	75	Turnip	80
Spinach	60	Watermelon	70

Adapted from Federal Seed Act Regulations, http://www.ams.usda.gov/lsg/seed.htm (2005).

TABLE 9.3. ISOLATION DISTANCES BETWEEN PLANTINGS OF VEGETABLES FOR OPEN-POLLINATED SEED PRODUCTION

Self-pollinated Vegetables

Self-pollinated crops have little outcrossing. Consequently, the only isolation necessary is to have plantings spaced far enough apart to prevent mechanical mixture at planting or harvest. A tall-growing crop is often planted between different varieties.

Bean	Bean, lima	Chicory	Endive
Lettuce	Pea	Tomato	

Cross-pollinated Vegetables

Cross-pollination of vegetables may occur by wind or insect activity. Therefore, plantings of different varieties of the same crop or different crops in the same family that can cross with each other must be isolated. Some general isolation guidelines are provided; however, the seed grower should follow the recommendations of the seed company for whom the seed is being grown.

Wind-pollinated Vegetables	Distance (miles)
Beet	½–2 5 from sugar beet or Swiss chard
Sweet corn	1
Spinach	¼–3
Swiss chard	¾–5 5 for sugar beet or beet

TABLE 9.3. ISOLATION DISTANCES BETWEEN PLANTINGS OF VEGETABLES FOR OPEN-POLLINATED SEED PRODUCTION (*Continued*)

Insect-pollinated Vegetables	Distance (miles)
Asparagus	¼
Broccoli	½–3
Brussels sprouts	½–3
Cabbage	½–3
Cauliflower	½–3
Collards	¾–3 5 from other cole crops
Kale	¾–2 5 from other cole crops
Kohlrabi	½–3
Carrot	½–3
Celeriac	1
Celery	1
Chinese cabbage	1
Cucumber	1½ for varieties ¼ from other cucurbits
Eggplant	¼
Gherkin	¼
Leek	1
Melons	1½–2 for varieties ¼ from other cucurbits
Mustard	1
Onion	1–3
Parsley	½–1
Pepper	½
Pumpkin	1½–2 for varieties ¼ from other cucurbits
Radish	¼–2
Rutabaga	¼–2
Spinach	¼–2
Squash	1½–2 for varieties ¼ from other cucurbits
Turnip	¼–2

Adapted in part from *Seed Production in the Pacific Northwest*, Pacific Northwest Extension Publications (1985).

TABLE 9.4. CONDITIONS FOR CLASSES OF CERTIFIED VEGETABLE SEED

	Foundation				Registered				Certified			
Vegetable	Land[1]	Isolation[2]	Field[3]	Seed[4]	Land[1]	Isolation[2]	Field[3]	Seed[4]	Land[1]	Isolation[2]	Field[3]	Seed[4]
Bean	1	0	2,000	0.05	1	0	1,000	0.1	1	0	400	0.2
Bean, broad	1	0	2,000	0.05	1	0	1,000	0.1	1	0	500	0.2
Bean, mung	1	0	1,000	0.1	1	0	500	0.2	1	0	200	0.5
Corn, sweet	—	—	—	—	—	—	—	—	0	660	1,000	0.5
Okra	1	1,320	0	0	1	1,320	2,500	0.5	1	825	1,250	1.0
Onion	1	5,280	200	0	1	2,640	200	0.5	1	1,320	200	1.0
Pepper	1	200	0	0	1	100	300	0.5	1	30	150	1.0
Southern pea	1	0	2,000	0.1	1	0	1,000	0.2	1	0	500	0.5
Tomato	1	200	0	0	1	100	300	0.5	1	30	150	1.0
Watermelon	1	2,640	0	0	1	2,640	0	0.5	1	1,320	500	1.0

Adapted from Federal Seed Act Regulations, USDA, AMS. Certified Seed. Minimum Land, Isolation, Field, and Seed Standards (2004), http://www.ams.usda.gov/lsg/seed.htm.

[1] Years that must elapse after destruction of a previous crop of the same kind.

[2] Distance in feet from any contaminating source, but sufficient to prevent mechanical mixture.

[3] Minimum number of plants in which one off-type plant is permitted.

[4] Maximum percentage of off-type seeds permitted in cleaned seed.

VEGETABLE SEED PRODUCTION WEBSITES

Vegetable Seed Production, http://www.ag.ohio-state.edu/~seedsci/vsp01.html

Vegetable Seed Production—Dry Seeds, http://www.ag.ohio-state.edu/~seedsci/vsp02.html

Vegetable Seed Production—Wet Seeds, http://www.ag.ohio-state.edu/~seedsci/vsp03.html

Onion Seed Production in California, http://www.anrcatalog.ucdavis.edu/pdf8008.pdf

Cucurbit Seed Production in California, http://www.anrcatalog.ucdavis.edu/pdf7229.pdf

Carrot Seed Production, http://www.ars.usda.gov/research/docs.htm?docid=5235

Crop Profile for Table Beet Seed in Washington, http://www.ipmcenters.org/cropprofiles/docs/wabeetseed.html

Crop Profile for Cabbage Seed in Washington, http://www.ipmcenters.org/cropprofiles/docs/wacabbageseed.html

Crop Profile for Spinach Seed in Washington, http://www.ipmcenters.org/cropprofiles/docs/waspinachseed.html

Seed Production and Seed Sources of Organic Vegetables, http://edis.ifas.ufl.edu/hs227

Investigation of Organic Seed Treatments for Spinach Disease Control, http://vric.ucdavis.edu/scrp/sum-koike.html

TABLE 9.5. VEGETABLE SEED YIELDS[1]

Vegetable	Average Yield (lb/acre)	Range (lb/acre)
Asparagus		
o.p.[2]	925	380–2,800
F_1	500	
F_2	750	
Bean, snap	1,800	1,400–2,800
Bean, lima	2,220	1,500–3,000
Beet		
o.p.	1,950	1,800–2,500
F_1	1,150	900–1,400
Broccoli		
o.p.	725	350–1,000
F_1	375	250–500
Brussels sprouts		
o.p.	900	800–1,000
F_1	425	250–600
Cabbage		
o.p.	740	500–1,000
F_1	440	300–600
Cantaloupe		
o.p.	420	350–500
F_1	225	175–300
Carrot		
o.p.	840	500–1,000
F_1	450	200–800
Cauliflower		
o.p.	540	350–1,000
F_1	175	100–250
Celeriac	1,200	800–2,000
Celery	835	500–1,200
Chard, Swiss	1,600	1,000–2,000
Chicory	500	400–600
Chinese cabbage		
o.p.	900	800–1,000
F_1	400	300–500

TABLE 9.5. VEGETABLE SEED YIELDS[1] (*Continued*)

Vegetable	Average Yield (lb/acre)	Range
Cilantro	2,000	1,500–2,500
Corn, sweet		
su	1,940	1,500–2,500
sh_2	1,100	400–1,700
Cucumber		
beit alpha	450	350–550
pickle	650	450–850
slicer		
o.p.	500	275–600
F_1	290	200–550
Eggplant		
o.p.	640	500–775
F_1	500	400–625
Endive	735	650–800
Florence fennel		
o.p.	1,500	1,000–2,000
F_1	700	600–800
Kale		
o.p.	1,100	1,000–1,200
F_1	650	600–700
Kohlrabi		
o.p.	875	850–900
F_1	450	400–500
Leek		
o.p.	625	500–850
F_1	300	200–400
Lettuce	600	450–800
Mustard	1,325	1,300–1,350
New Zealand spinach	1,750	1,500–2,000
Okra		
o.p.	1,600	1,200–2,000
F_1	650	600–700
Onion		
o.p.	690	575–900
F_1	450	350–550
Parsley	900	600–1,200
Parsnip	975	600–1,300
Pea	2,085	1,000–3,000

TABLE 9.5. VEGETABLE SEED YIELDS[1] (*Continued*)

Vegetable	Average Yield (lb/acre)	Range
Pepper		
o.p.	170	100–300
F_1	125	100–150
Pumpkin		
o.p.	575	300–850
F_1	300	235–400
Radish		
o.p.	1,200	600–2,000
F_1	525	200–1,000
Rutabaga	2,200	1,800–2,500
Salisfy	800	600–1,000
Southern pea	1,350	1,200–1,500
Spinach		
o.p.	1,915	1,000–2,500
F_1	1,100	1,000–1,200
Squash, summer		
o.p.	760	400–1,200
F_1	360	250–425
Squash, winter		
o.p.	620	300–1,200
F_1	310	200–400
Tomatillo	600	500–700
Tomato		
F_1	125	75–170
Turnip		
o.p.	2,300	2,000–3,000
F_1	1,350	1,200–1,500
Watermelon		
o.p.	405	350–600
F_1	220	190–245
Triploid (seedless)	40	20–70

[1] Yields are from information provided by representations of several major seed companies. Yields of some hybrids may be very much lower because of difficulties of seed production.

[2] o.p. = open pollinated

F_1 = first-generation hybrid

F_2 = second-generation hybrid

06
SEED STORAGE

STORAGE OF VEGETABLE SEEDS

High moisture and temperature cause rapid deterioration in vegetable seeds. The control of moisture and temperature becomes more important the longer seeds are held. Low moisture in the seeds means longer life, especially if they must be held at warm temperatures. Kinds of seeds vary in their response to humidity.

The moisture content of seeds can be lowered by drying them in moving air at 120°F. This may be injurious to seeds with an initial moisture content of 25–40%. With such seeds, 110°F is preferred. It may require less than 1 hr to reduce the moisture content of small seeds or up to 3 hr for large seeds. The difference depends on the depth of the layer of seeds, the volume of air, dryness of air, and original moisture content of seed. When seeds cannot be dried in this way, seal them in airtight containers over, but not touching, some calcium chloride. Use enough calcium chloride so that the moisture absorbed from the seeds produces no visible change in the chemical. Dried silica gel can be used in place of the calcium chloride.

Bean and okra may develop hard seeds if their moisture content is lowered to 7% or below. White-seeded beans are likely to become hard if the moisture content is reduced to about 10%. Dark-colored beans can be dried to less than 10% moisture before they become hard. Hard seeds do not germinate satisfactorily.

The moisture content of seed reaches an equilibrium with the atmosphere after a period of about 3 weeks for small seeds and 3–6 weeks for large seeds.

Storage temperatures near 32°F are not necessary. Between 40 and 50°F is quite satisfactory when the moisture content of the seed is low.

If the moisture content is reduced to 4–5% and the seeds put in sealed containers, a storage temperature of about 70°F will be satisfactory for more than 1 year.

The 5-month limitation on the date of test does not apply when the following conditions are met:

a. The seed was packaged within 9 months after harvest.
b. The container does not allow water vapor penetration through any wall or seal greater than 0.05 g water per 100 sq in. of surface at 100°F with a relative humidity on one side of 90% and on the other side of 0%.

c. The seed in the container does not exceed the percentage of moisture, on a wet weight basis, as listed in the table.

d. The container is conspicuously labeled in not less than 8-point type with the information that the container is hermetically sealed, that the seed is preconditioned as to moisture content, and the calendar month and year in which the germination test was completed.

e. The percentage pf germination of vegetable seed at the time of packaging was equal or above Federal Standards for Germination (see pages 512–513).

TABLE 9.6. STORAGE OF VEGETABLE SEEDS IN HERMETICALLY SEALED CONTAINERS

Vegetable	Moisture (%)	Vegetable	Moisture (%)
Bean, garden	7.0	Lettuce	5.5
Bean, lima	7.0	Melon	6.0
Beet	7.5	Mustard, India	5.0
Broccoli	5.0	Onion	6.5
Brussels sprouts	5.0	Onion, Welsh	6.5
Cabbage	5.0	Parsley	6.5
Carrot	7.0	Parsnip	6.0
Cauliflower	5.0	Pea	7.0
Celeriac	7.0	Pepper	4.5
Celery	7.0	Pumpkin	6.0
Chard, Swiss	7.5	Radish	5.0
Chinese cabbage	5.0	Rutabaga	5.0
Chives	6.5	Spinach	8.0
Collards	5.0	Squash	6.0
Cucumber	6.0	Sweet corn	8.0
Eggplant	6.0	Tomato	5.5
Kale	5.0	Turnip	5.0
Kohlrabi	5.0	Watermelon	6.5
Leek	6.5	All others	6.0

Adapted from Federal Seed Act Regulations, http://www.ams.usda.gov/lsg/seed/seed_pub.htm (2005).

07
VEGETABLE VARIETIES

NAMING AND LABELING OF VEGETABLE VARIETIES

Every year, many new varieties of vegetable seed reach the U.S. marketplace. New varieties, when added to those already on the market, provide growers with a wide selection of seed. But, in order for them to buy intelligently, seed must be correctly named and labeled.

Marketing a product by its correct name might seem the most likely way to do business. However, USDA seed officials have found that seed, unfortunately, is sometimes named, labeled, or advertised improperly as it passes through marketing channels.

Marketing seed under the wrong name is misrepresentation. It can lead to financial loss for several participants in the seed marketing chain.

The grower, for example, buys seed to achieve specific objectives such as increased yield, competitiveness in a specialized market, or adaptability to growing conditions of a specific region. If seed is misrepresented and the grower buys seed other than he or she intended, the harvest may be less valuable than anticipated—or, worse yet, there may not even be a market for the crop.

In one case, a grower bought seed to grow cabbage to be marketed for processing into sauerkraut. As the cabbage matured, he found his crop was not suitable for processing and, even worse, that there was no market for the cabbage in his fields. In this case, improper variety labeling brought about financial hardship.

Seed companies and plant breeders also suffer in a market where problems with variety names exist. For instance, if the name of a newly released variety is misleading or confusing to the potential buyer, the variety may not attract the sales that it might otherwise.

This section outlines requirements for naming vegetable seed. It is based on the Federal Seed Act, a truth-in-labeling law intended to protect growers and home gardeners who purchase seed. Exceptions to the basic rules and the do's and don'ts of seed variety labeling and advertising also are explained.

WHO NAMES NEW VARIETIES

The originator or discoverer of a new variety may give that variety a name. If the originator or discoverer can't or chooses not to name a variety, someone else may give that variety a name for marketing purposes. In such a case, the name first used when the seed is introduced into commerce is the name of the variety.

It is illegal to change a variety name once the name is legally assigned. In other words, a buyer may not purchase seed labeled as variety *X* and resell it as variety *Y*. An exception to this rule occurs when the original name is determined to be illegal. In such an instance, the variety must be renamed according to the rules mentioned above. Another exception applies to a number of varieties that were already being marketed under several names before 1956. (See section on synonyms.)

WHAT'S IN A NAME

To fully understand what goes into naming a variety, you need to know the difference between a *kind* of seed and a *variety. Kind* is the term used for the seed of one or more related plants known by a common name, such as carrot, radish, tomato, or watermelon.

Variety is a subdivision of *kind.* A variety has different characteristics from another variety of the same kind of seed—for example, "Oxheart" carrot and "Danvers 126" carrot or "Charleston Gray" and "Mickylee" watermelon.

The rules for naming plants relate to both kinds and varieties of seed:

1. A variety must be given a name unique to the kind of seed to which it belongs. For instance, there can only be one variety of squash called "Dividend."
2. Varieties of two or more kinds of seed may have the same name if the kinds are not closely related. For example, there could be a "Dividend" squash and a "Dividend" tomato because squash and tomato are kinds of seed not closely related. On the other hand, it would not be permissible to have a "Dividend" squash and a "Dividend" pumpkin because the two kinds of seed are closely related.
3. Once assigned to a variety, the name remains exclusive. Even if "Dividend" squash has not been marketed for many years, a newly developed and different squash variety can't be given the name "Dividend" unless the original owner agrees to withdraw "Dividend" squash.
4. A company name may be used in a variety name as long as it is part of the original, legally assigned name. Once part of a legal variety name, the company name must be used by everyone, including another company that might market the seed.

When a company name is not a part of the variety name, it should not be used in any way that gives the idea that it *is* part of the variety name. For example, Don's Seed Company can't label or advertise "Dividend" squash variety as "Don's Dividend" because "Don's" may be mistaken to be part of the variety name.

The simplest way to avoid confusion is to separate the company and variety names in advertising or labeling.

5. Although the USDA discourages it, you may use descriptive terms in variety names as long as such terms are not misleading. "X3R," for instance, is accepted among pepper growers as meaning "Bacterial Leaf Spot, Race 1, 2, 3 resistant." It would be illegal to include "X3R" as part of a variety name if that variety were not "X3R" resistant. Similarly, if a cantaloupe variety is named "Burpee Hybrid PMT," the name would be illegal if the variety were not tolerant of powdery mildew.
6. A variety name should be clearly different in spelling and in sound. "Alan" cucumber would not be permissible if an "Allen" cucumber were already on the market.

HYBRIDS

Remember that a hybrid also is a variety. Hybrid designations, whether they are names or numbers, also are variety names. Every rule discussed here applies to hybrid seed as well as to nonhybrid seed.

In the case of hybrids, however, the situation is potentially more complex because more than one seed producer or company might use identical parent lines in producing a hybrid variety. One company could then produce a hybrid that was the same as one already introduced by another firm.

When this happens, both firms must use the same name because they are marketing the same variety.

If the people who developed the parent lines give the hybrid variety a name, that is the legal name. Otherwise, the proper name is the one given by the company that first introduced the hybrid seed into commerce.

USDA seed regulatory officials believe the following situation occurs far too often:

1. State University releases hybrid corn parent lines A and B.
2. John Doe Seed Company obtains seed lines of A and B, crosses the two lines, and is the first company to introduce the resulting hybrid into commerce under a variety name. John Doe Seed Company names this hybrid "JD 5259."

3. La Marque Seeds, Inc., obtains lines A and B, makes the same cross, and names the resulting hybrid variety "SML 25." There has been no change in the A and B lines that would result in a different variety. La Marque ships the hybrid seed, labeled "SML 25," in interstate commerce, and violates the Federal Seed Act because the seed should have been labeled "JD 5259."

SYNONYMS—VARIETIES WITH SEVERAL NAMES

As noted earlier, the name originally assigned to a variety is the name that must be used forever. It can't be changed unless it is illegal.

This does not mean that all varieties must be marketed under a single name. In fact, some old varieties may be marketed legally under more than one name. If several names for a single variety of a vegetable seed were in broad general use before July 28, 1956, those names still may be used.

Here are some examples:

The names "Acorn," "Table Queen," and "Des Moines" have been known for many years to represent a single squash variety. They were in broad general use before July 28, 1956, so seed dealers may continue to use these names interchangeably.

With the exception of old varieties with allowable synonym names, all vegetable and agricultural varieties may have *only one* legally recognized name, and that name must be used by anyone who represents the variety name in labeling and advertising. This includes interstate seed shipments and seed advertisements sent in the mail or in interstate or foreign commerce.

IMPORTED SEED

Seed imported into the United States can't be renamed if the original name of the seed is in the Roman alphabet.

For example, cabbage seed labeled "Fredrikshavn" and shipped to the United States from Denmark can't be given a different variety name such as "Bold Blue."

Seed increased from imported seed also can't be renamed. If "Fredrikshavn" were increased in the United States, the resulting crop still couldn't be named "Bold Blue."

Seed with a name that is not in the Roman alphabet must be given a new name. In such a case, the rules for naming the variety are the same as stated previously.

BRAND NAMES

USDA officials have found evidence of confusion over the use of variety names and brand or trademark names. This includes names registered with the Trademark Division of the U.S. Patent Office.

Guidelines:

1. The brand or trademark name must be clearly identified as being other than part of the variety name.
2. A brand name must never take the place of a variety name.
3. If a brand or trademark name is part of a variety's name, that trademark loses status. Anyone marketing the variety under its name is required to use the exact, legal variety name, including brand or trademark.

SUMMARY

If the naming, labeling, and advertising of a seed variety is truthful, it is probably in compliance with the Federal Seed Act.

Keep these simple rules in mind to help eliminate violations and confusion in the marketing of seed:

- Research the proposed variety name before adopting it.
- Make sure the name cannot be confused with company names, brands, trademarks, or names of other varieties of the same kind of seed.
- Never change the variety name, whether marketing seed obtained from another source or from your own production—for example, hybrid seed that already has a legal name.

Adapted from Seed Regulatory and Testing Programs, http://www.ams.usda.gov/lsg/seed/facts.htm.

SELECTION OF VEGETABLE VARIETIES

Selection of the variety (technically, *cultivar*) to plant is one of the most important decisions the commercial vegetable grower must make each season. Each year, seed companies and experiment stations release dozens of new varieties to compete with those already available. Growers should evaluate some new varieties each year on a trial basis to observe performance on their own farms. A limited number of new varieties should be evaluated so that observations on plant performance and characteristics and yields can be noted and recorded. It is relatively easy to establish a trial but time-consuming to make all the observations necessary to make a decision on adoption of a new variety. Some factors to consider before adopting a variety follow:

Yield: The variety should have the potential to produce crops at least equivalent to those already grown. Harvested yield is usually much less than potential yield because of market restraints.

Disease resistance: The most economical and effective means of pest management is through the use of varieties with genetic resistance to disease. When all other factors are about equal, it is prudent to select a variety with the needed disease resistance.

Horticultural quality: Characteristics of the plant habit as related to climate and production practices and of the marketed plant product must be acceptable.

Adaptability: Successful varieties must perform well under the range of environmental conditions usually encountered on the individual farm.

Market acceptability: The harvested plant product must have characteristics desired by the packer, shipper, wholesaler, retailer, and consumer. Included among these qualities are packout, size, shape, color, flavor, and nutritional quality.

During the past few years there has been a decided shift to hybrid varieties in an effort by growers to achieve earliness, higher yields, better quality, and greater uniformity. Seed costs for hybrids are higher than for open-pollinated varieties because seed must be produced by controlled pollination of the parents of the hybrid.

Variety selection is a dynamic process. Some varieties retain favor for many years, whereas others are used only a few seasons if some special situation, such as plant disease or marketing change, develops. If a variety was released by the USDA or a university, many seed companies may carry it. Varieties developed by a seed company may be available only from that source, or may be distributed through many sources.

The Cooperative Extension Service in most states publishes annual or periodic lists of recommended varieties. These lists are usually available in county extension offices.

Adapted from D. N. Maynard, "Variety Selection," in Stephen M. Olson and Eric Simonne (eds.), *Vegetable Production Handbook for Florida*. (Gainesville, Fla.: Florida Cooperative Extension Service, 2004–2005), 17, http://edis.ifas.ufl.edu/CV102.

08
VEGETABLE SEED SOURCES

SOME SOURCES OF VEGETABLE SEEDS[1]

-A-

Abbott & Cobb
P.O. Box 307
Feasterville, PA 10953-0307
Ph (215) 245-6666
Fax (215) 245-9043
http://www.acseed.com

Abundant Life Seeds
P.O. Box 157
Saginaw, OR 97472
Ph (541) 767-9606
Fax (866) 514-7333
http://www.abundantlifeseeds.com

American Takii, Inc.
301 Natividad Road
Salinas, CA 93906
Ph (831) 443-4901
Fax (831) 443-3976
http://www.takii.com

Arkansas Valley Seed Solutions
4625 Colorado Boulevard
Denver, CO 80216
Ph (877) 957-3337
Fax (303) 320-7516
http://www.seedsolutions.com

-B-

Bakker Brothers USA
P.O. Box 519
Caldwell, ID 83606
Ph (208) 459-4420
Fax (208) 459-4457
http://www.bakkerbrothers.nl

Baker Creek Heirloom Seeds
2278 Baker Creek Road
Mansfield, MO 65704
Ph (417) 924-8917
Fax (417) 924-8887
http://www.rareseeds.com

Bejo Seeds, Inc.
1972 Silver Spur Place
Oceano, CA 93445
Ph (805) 473-2199
http://www.bejoseeds.com

BHN Seed
P.O. Box 3267
Immokalee, FL 34142
Ph (239) 352-1100
Fax (239) 352-1981
http://www.bhnseed.com

Bonanza Seeds International, Inc.
3818 Railroad Avenue
Yuba City, CA 95991
Ph (530) 673-7253
Fax (530) 673-7195
http://www.bonanzaseeds.com

Bountiful Gardens Seeds
18001 Shafer Ranch Road
Willits, CA 95490-9626
Ph (707) 459-6410
Fax (707) 459-1925
http://www.bountifulgardens.org

W. Brotherton Seed Co., Inc.
P.O. Box 1136
Moses Lake, WA 98837
Ph (509) 765-1816
Fax (509) 765-1817
http://www.vanwaveren.de/uk/mitte/brotherton.htm

Bunton Seed Co.
939 E. Jefferson Street
Louisville, KY 40206
Ph (800) 757-7179
Fax (502) 583-9040
http://www.buntonseed.com

Burgess Seed & Plant Co.
905 Four Seasons Road
Bloomington, IL 61701
Ph (309) 622-7761
http://www.eburgess.com

W. Atlee Burpee & Co.
300 Park Avenue
Warminster, PA 18974
Ph (800) 333-5808
Fax (800) 487-5530
http://www.burpee.com

-C-

California Asparagus Seed
2815 Anza Avenue
Davis, CA 95616
Ph (530) 753-2437
Fax (530) 753-1209
http://www.calif-asparagus-seed.com

Carolina Gourds and Seeds
259 Fletcher Avenue
Fuquay Varina, NC 27526
Ph (919) 577-5946
http://www.carolinagourdsandseeds.com

Champion Seed Co.
529 Mercury Lane
Brea, CA 92621-4894
Ph (714) 529-0702
Fax (714) 990-1280
http://www.championseed.com

Chesmore Seed Co.
P.O. Box 8368
St. Joseph, MO 64508-8368
Ph (816) 279-0865
Fax (816) 232-6134
http://www.chesmore.com

Alf Chrisianson Seed Co.
P.O. Box 98
Mount Vernon, WA 98273
Ph (360) 366-9727
Fax (360) 419-3035
http://www.chriseed.com

Clifton Seed Company
P.O. Box 206
Faison, NC 28341
Ph (800) 231-9359
Fax (910) 267-2692
http://www.cliftonseed.com

Comstock, Ferre & Co.
263 Main Street
Wethersfield, CT 06109
Ph (860) 571-6590
Fax (860) 571-6595
http://www.comstockferre.com

The Cook's Garden
PO Box 6530
Warminster, PA 18974
Ph (800) 457-9703
http://www.cooksgarden.com

Corona Seeds Worldwide
590-F Constitution Avenue
Camarillo, CA 93012
Ph (805) 388-2555
Fax (805) 445-8344
http://www.coronaseeds.com

Crookham Co.
P.O. Box 520
Caldwell, ID 83606-0520
Ph (208) 459-7451
Fax (208) 454-2108
http://www.crookham.com

Crop King Inc.
5050 Greenwich Road
Seville, OH 44273-9413
Ph (330) 769-2002
Fax (330) 769-2616
http://www.cropking.com

Cutter Asparagus Seed
516 Young Avenue
Arbuckle, CA 95912
Ph (530) 475-3647
Fax (953) 476-2422
http://www.asparagusseed.com

-D-

DeRuiter Seeds, Inc.
13949 W. Colfax Avenue
Building No. 1, Suite 220
Lakewood, CO 80401
Ph (303) 274-5511
Fax (303) 274-5514
http://www.deruiterusa.com

Dominion Seed House
P.O. Box 2500
Georgetown, ONT
Canada L7G 4A2
Ph (905) 873-3037
Fax (800) 282-5746
http://www.dominion-seed-house.com

-E-

Elsoms Seeds Ltd.
Pinchbeck Road
Spalding
Lincolnshire PE11 1QG
England, UK
Ph 0 1775 715000
Fax 0 1775 715001
http://www.elsoms.com

Enza Zaden
7 Harris Place
Salinas, CA 93901
Ph (831) 751-0937
Fax (831) 751-6103
http://www.enzazaden.nl/site/uk/

-F-

Farmer Seed & Nursery Co.
Division of Plantron, Inc.
818 NW Fourth Street
Faribault, MN 52201
Ph (507) 334-1623
http://www.farmerseed.com

Henry Field Seed & Nursery Co.
P.O. Box 397
Aurora, IN 47001-0397
Ph (513) 354-1494
Fax (513) 354-1496
http://www.henryfields.com

-G-

Germania Seed Co.
P.O. Box 31787
Chicago, IL 60631
Ph (800) 380-4721
Fax (800) 410-4721
http://www.germaniaseed.com

Fred C. Gloeckner & Co., Inc.
600 Mamaroneck Avenue
Harrison, NY 10528-1631
Ph (800) 345-3787
Fax (914) 698-2857
http://www.fredgloeckner.com

Golden Valley Seed
P.O. Box 1600
El Centro, CA 92243
Ph (760) 337-3100
Fax (760) 337-3135
http://www.goldenvalleyseed.com

Gurney's Seed & Nursery Co.
P. O. Box 4178
Greendale, IN 47025-4178
Ph (513) 354-1492
Fax (513) 354-1493
http://www.gurneys.com

-H-

Harris Moran Seed Co.
P.O. Box 4938
Modesto, CA 95352
Ph (209) 579-7333
Fax (209) 527-8684
http://www.harrismoran.com

Harris Seeds
P.O. Box 24966
Rochester, NY 14624-0966
Ph (800) 544-7938
Fax (877) 892-9197
http://www.harrisseeds.com

The Chas. C. Hart Seed Co.
P.O. Box 9169
Wethersfield, CT 06109
Ph (860) 529-2537
Fax (860) 563-7221
http://www.hartseed.com

Hazera Genetics Ltd.
2255 Glades Road, Suite 123A
Boca Raton, FL 33431
Ph (561) 988-1315
Fax (561) 988-1319
http://www.hazera.co.il/

Heirloom Seeds
P.O. Box 245
West Elizabeth, PA 15088-0245
Ph (412) 384-0852
http://www.heirloomseeds.com

Hollar and Company, Inc.
P.O. Box 106
Rocky Ford, CO 81067
Ph (719) 254-7411
Fax (719) 254-3539
http://www.hollarseeds.com

Hydro-Gardens, Inc.
P.O. Box 25845
Colorado Springs, CO 80932
Ph (800) 936-5845
Fax (888) 693-0578
http://www.hydro-gardens.com

-I-

Illinois Foundation Seeds, Inc.
P.O. Box 722
Champaign, IL 61824-0722
Ph (217) 485-6260
Fax (217) 485-3687
http://www.ifsi.com

-J-

Jersey Asparagus Farms, Inc.
105 Porchtown Road
Pittsgrove, NJ 08318
Ph (856) 358-2548
Fax (856) 358-6127
http://www.jerseyasparagus.com

Johnny's Selected Seeds
955 Benton Avenue
Winslow, ME 04901
Ph (866) 838-1073
http://www.johnnyseeds.com

Jordan Seeds, Inc.
6400 Upper Aston Road
Woodbury, MN 55125
Ph (651) 738-3422
Fax (651) 731-7690
http://www.jordanseeds.com

J. W. Jung Seed Co.
335 S. High Street
Randolph, WI 53957-0001
Ph (800) 247-5864
Fax (800) 692-5864
http://www.jungseed.com

-K-

Keithly-Williams Seeds
P.O. Box 177
Holtville, CA 92250
Ph (800) 533-3465
Fax (760) 356-2409
http://www.keithlywilliams.com

Known-You Seed Co., Ltd.
26, Chung Cheng 2nd Road
Kaohsiung, Taiwan
Republic of China
http://www.knownyou.com

-L-

Livingston Seed Co.
880 Kinnear Road
Columbus, OH 43212
Ph (614) 488-1163
Fax (614) 488-4857
http://www.livingstonseed.com

-M-

Earl May Seed & Nursery Co.
208 N. Elm Street
Shenandoah, IA 51603
Ph (712) 246-1020
Fax (712) 246-1760
http://www.earlmay.com

Mesa Maize Co.
60936 Falcon Road
Olathe, CO 81425
http://www.mesamaize.com

McFayden Seed Co., Ltd.
30 Ninth Street
Brandon, Manitoba
Canada, R7A 6A6
Ph (800) 205-7111
http://www.mcfayden.com

Henry F. Michell Co.
P.O. Box 60160
King of Prussia, PA 19406-0160
Ph (800) 422-4678
http://www.michells.com

Monsanto Company
800 N. Lindbergh Boulevard
St. Louis, MO 63167
Ph (314) 694-1000
http://www.monsanto.com

Mushroompeople
P.O. Box 220
Summertown, TN 38483-0220
Ph (800) 692-6329
Fax (800) 386-4496
http://www.mushroompeople.com

-N-

Native Seeds/SEARCH
526 N. Fourth Avenue
Tucson, AZ 85705-8450
Ph (520) 622-5561
Fax (520) 622-5591
http://www.nativeseeds.org

New England Seed Co.
3580 Main Street
Hartford, CT 06120
Ph (800) 825-5477
Fax (877) 229-8487
http://www.neseed.com

Nichol's Garden Nursery
1190 Old Salem Road NE
Albany, OR 97321-4580
Ph (800) 422-3985
Fax (800) 231-5306
http://www.nicholsgardennursery.com

Nirit Seeds Ltd.
Moshav Hadar-Am
42935 Israel
Ph (972) 9 832 24 35
Fax (972) 9 832 24 38
http://www.niritseeds.com

NK Lawn & Garden Co.
P.O. Box 24028
Chattanooga, TN 37422-4028
Ph (800) 328-2402
Fax (423) 697-8001
http://www.nklawnandgarden.com

Nourse Farms, Inc.
41 River Road
South Deerfield, MA 01373
Ph (413) 665-2658
Fax (413) 665-7888
http://www.noursefarms.com

Nunhems Seed
P.O. Box 18
Lewisville, ID 83431
Ph (208) 754-8666
Fax (208) 754-8669
http://www.nunhems.com

-O-

OSC
Box 7
Waterloo, Ontario
Canada N2J 3Z9
Ph (519) 886-0557
Fax (519) 886-0605
http://www.oscseeds.com

Oriental Vegetable Seeds
Evergreen Y. H. Enterprises
P.O. Box 17538
Anaheim, CA 92817
Ph (714) 637-5769
http://www.evergreenseeds.com

Ornamental Edibles
5723 Trowbidge Way
San Jose, CA 95138
Ph (408) 528-7333
Fax (408) 532-1499
http://www.ornamentaledibles.com

Orsetti Seed Co., Inc.
2301 Technology Parkway
P.O. Box 2350
Hollister, CA 95023
Ph (831) 636-4822
Fax (831) 636-4814
http://www.orsettiseed.com

Outstanding Seed Company
354 Center Grange Road
Monaca, PA 15061
Ph (800) 385-9254
http://www.outstandingseed.com

-P-

D. Palmer Seed Co., Inc.
8269 S. Highway 95
Yuma, AZ 85365
Ph (928) 341-8494
Fax (928) 341-8496
http://www.dpalmerseed.com

Paramount Seeds, Inc.
P.O. Box 1866
Palm City, FL 34991
Ph (772) 221-0653
Fax (772) 221-0102
http://www.paramount-seeds.com

Park Seed Co.
1 Parkton Avenue
Greenwood, SC 29647
Ph (800) 213-0076
http://www.parkseed.com

Penn State Seed Co.
Box 390, Route 309
Dallas, PA 18612-9781
Ph (800) 847-7333
Fax (570) 675-6562
http://www.pennstateseed.com

Pepper Gal
P.O. Box 23006
Ft. Lauderdale, FL 33307-3007
Ph (954) 537-5540
Fax (954) 566-2208
http://www.peppergal.com

Pinetree Garden Seeds
P.O. Box 300
New Gloucester, ME 04260
Ph (207) 926-3400
Fax (888) 527-3337
http://www.superseeds.com

-R-

Redwood City Seed Co.
P.O. Box 361
Redwood City, CA 94064
Ph (650) 325-7333
Fax (650) 325-4056
http://www.ecoseeds.com

Renee's Garden Seeds
7389 W. Zayante Road
Felton, CA 95018
Ph (888) 880-7228
Fax (831) 335-7227
http://www.reneesgarden.com

Rijk Zwaan
2274 Portola Drive
Salinas, CA 93908
Ph (831) 484-9486
Fax (831) 484-9486
http://www.rijkzwaan.com

Rupp Seeds, Inc.
17919 County Road B
Wauseon, OH 43567
Ph (419) 337-1841
Fax (419) 337-5491
http://www.ruppseeds.com

Rispens Seeds, Inc.
P.O. Box 310
Beecher, IL 60401
Ph (888) 874-0241
Fax (708) 746-6115
http://www.rispenseeds.com

-S-

Sakata Seed America, Inc.
P.O. Box 880
Morgan Hill, CA 95038-0880
Ph (408) 778-7758
Fax (408) 778-7751
http://www.sakata.com

Seeds of Change
621 Old Santa Fe Trail #10
Santa Fe, NM 87501
Ph (5888 762-7333
Fax (505) 438-7052
http://www.seedsofchange.com

Seedway, Inc.
1225 Zeager Road
Elizabethtown, PA 17022
Ph (800) 952-7333
Fax (800) 645-2574
http://www.seedway.com

Seminis Inc.
2700 Camino del Sol
Oxnard, CA 93030-7967
Ph (805) 647-1572
http://www.seminis.com

Shamrock Seed Co.
3 Harris Place
Salinas, CA 93901-4856
Ph (831) 771-1500
Fax (831) 771-1517
http://www.shamrockseed.com

R.H. Shumway
334 W. Stroud Street
Randolph, WI 53956-1274
Ph (800) 342-9461
Fax (888) 437-2773
http://www.rhshumway.com

Siegers Seed Co.
13031 Reflections Drive
Holland, MI 49424
Ph (616) 786-4999
Fax (616) 994-0333
http://www.siegers.com

Snow Seed Co.
12855 Rosehart Way
Salinas, CA 93908
Ph (831) 758-9869
Fax (831) 757-4550
http://www.snowseedco.com

Southern Exposure Seed Exchange
P.O. Box 460
Mineral, VA 23117
Ph (540) 894-9480
Fax (540) 894-9481
http://www.southernexposure.com

Southwestern Seed Co.
P.O. Box 11449
Casa Grande, AZ 85230
Ph (520) 836-7595
Fax (520) 836-0117
http://www.southwesternseed.com

Stokes Seeds, Inc.
P.O. Box 548
Buffalo, NY 14240-0548
Ph (716) 695-6980
http://www.stokesseeds.com

Sugar Creek Seed, Inc.
P.O. Box 508
Hinton, OK 73047
Ph (405) 542-3920
Fax (405) 542-3921
http://www.sugarcreekseed.com

Sutter Seeds
P.O. Box 1357
Colusa, CA 95932
Ph (530) 458-2566
Fax (530) 458-2721
http://www.sutterseed.com

Syngenta Seeds, Inc.
Rogers Brand Vegetable Seeds
P.O. Box 4188
Boise, ID 83711-4188
Ph (208) 322-7272
Fax (208) 378-6625
http://www.rogersadvantage.com

-T-

The Territorial Seed Co.
P.O. Box 158
Cottage Grove, OR 97424-0061
Ph (541) 942-9547
Fax (888) 657-3131
http://www.territorial-seed.com

Thompson & Morgan, Inc.
P.O. Box 1308
Jackson, NJ 08527-0308
Ph (800) 274-7333
Fax (888) 466-4769
http://www.wholesale.thompson-morgan.com/us/

Tokita Seed Co., Ltd.
1069 Nakagawa
Omiya-shi
Saitama-ken 300-8532
Japan
http://www.tokitaseed.co.jp

Tomato Grower's Supply Co.
P.O. Box 2237
Ft. Myers, FL 33902
Ph (888) 478-7333
http://www.tomatogrowers.com

Otis S. Twilley Seed Co., Inc.
121 Gary Road
Hodges, SC 29653
Fax (864) 227-5108
http://www.twilleyseed.com

-U-

United Genetics
800 Fairview Road
Hollister, CA 95023
Ph (831) 636-4882
Fax (831) 636-4883
http://www.unitedgenetics.com

U S Seedless, LLC
P.O. Box 3006
Falls Church, VA 22043
Ph (703) 903-9190
Fax (703) 903-9456
http://www.usseedless.com

-V-

Vermont Bean Seed Co., Inc.
334 W. Stroud Street
Randolph, WI 53956-1274
Ph (800) 349-1071
Fax (888) 500-7333
http://www.vermontbean.com

Vesey's Seeds Ltd.
P.O. Box 9000
Charlottetown, PEI
Canada C1A 8K6
Ph (902) 368-7333
Fax (800) 686-0329
http://www.veseys.com

Victory Seed Co.
P.O. Box 192
Molalla, OR 97038
Ph & Fax (503) 829-3126
http://www.victoryseeds.com

Vilmorin Inc.
2551 N. Dragoon
Tucson, AZ 85175
Ph (520) 884-0011
Fax (520) 884-5102
http://www.vilmorin.com

Virtual Seeds
92934 Coyote Drive
Astoria, OR 97103
Ph (503) 458-0919
http://www.virtualseeds.com

-W-

West Coast Seeds
3925 Sixty-fourth Street
Delta, BC
Canada V4K 3N2
Ph (604) 952-8820
Fax (8770 482-8822
http://www.westcoastseeds.com

Willhite Seed Co.
Box 23
Poolville, TX 76487-0023
Ph (800) 828-1840
Fax (817) 599-5843
http://www.willhiteseed.com

Wyatt-Quarles Seed Co.
P.O. Box 739
Garner, NC 27529
Ph (919) 772-4243
http://www.wqseeds.com

-Y-

Arthur Yates & Co. Ltd.
P.O. Box. 6672
Silverwater BC
NSW 1811
Australia
http://www.yates.com.au

-Z-

Zeraim Gedera
P.O. Box 103
Gedera 70750
Israel
http://www.zeraimgedera.com

Information in this section was correct at the time of preparation. However, there are frequent changes in phone area codes, postal codes, addresses, and even company names. Suggest using your computer search engine if there is difficulty locating a specific concern.

PART 10

APPENDIX

01
SOURCES OF INFORMATION ON VEGETABLES

The Agricultural Experiment Station and Cooperative Extension Service in each state has been the principal source of information on vegetables in printed form. Most of this information is now on the Internet because of the high cost of producing, storing, and distributing printed information.

Websites providing information on vegetables from the state agricultural university in most states are listed below. If you are unable to access a site, try using the search engine on your computer.

State	Website	Address/URL
Alabama	Commercial Vegetable Production	http://www.aces.edu/dept/com_veg
Arizona	Vegetables	http://cals.arizona.edu/crops/vegetables/vegetables.html
Arkansas	Vegetables	http://aragriculture.org/horticulture/vegetables/default.asp
California	Vegetable & Research Information Center	http://vric.ucdavis.edu
	Vegetable Crops	http://www.anrcatalog.ucdavis.edu/InOrder/Shop/Shop.asp
	Postharvest Technology	http://postharvest.ucdavis.edu.
Colorado	Specialty Crops Program	http://www.specialtycrops.colostate.edu/SCP_about.htm
Connecticut	Vegetables	http://137.99.85.230/FMPro
	Agricultural Experiment Station	http://www.caes.state.ct.us/Publications/publications.htm
Delaware	Vegetable Program	http://www.rec.udel.edu/veggie/veggie2001.htm

State	Website	Address/URL
Florida	Vegetable Crops	http://edis.ifas.ufl.edu/TOPIC_Vegetables
	Watermelons	http://watermelons.ifas.ufl.edu
Georgia	Vegetables	http://www.uga.edu/~hort/comveg.htm
Hawaii	Vegetables	http://www.ctahr.hawaii.edu/fb/vege.htm
Illinois	Commercial Vegetable Production	http://www.nres.uiuc.edu/outreach/pubs.html
Indiana	Purdue Fruit and Vegetable Connection	http://www.hort.purdue.edu/fruitveg/vegmain.shtml
Iowa	Commercial Vegetables	http://www.public.iastate.edu/~taber/Extension/extension.html
Kansas	Horticulture Library	http://www.oznet.ksu.edu/library/hort2
Kentucky	Vegetable Information	http://www.uky.edu/Ag/Horticulture/comveggie.html
Louisiana	Commercial Vegetable Production Recommendations	http://www.louisianalawnandgarden.org/en/crops_livestock/crops/vegetables/
Maryland	Crops, Livestock & Nursery	http://www.agnr.umd.edu/MCE/Publications/Category.cfm?ID=C
Massachusetts	Vegetable Program	http://www.umassvegetable.org
Michigan	Michigan Vegetable Information Network	http://web4.msue.msu.edu/veginfo/index.cfm?doIntro=1
Minnesota	Fruits and Vegetables	http://horticulture.coafes.umn.edu/http://horticulture_coafes_umn_edu_fruitveg.html
Mississippi	Commercial Horticulture	http://msucares.com/crops/comhort/index.html
Missouri	Horticulture Publications	http://muextension.missouri.edu/explore/agguides/hort
Nebraska	Horticulture	http://ianrpubs.unl.edu/horticulture

State	Website	Address/URL
New Hampshire	Cooperative Extension's Vegetable Program	http://www.ceinfo.unh.edu/Agric/AGFVC/FVCVEG.htm
New Jersey	Vegetable and Herb Crops	http://www.rcre.rutgers.edu/pubs/subcategory.asp?cat=3&sub=24
New York	Commercial Fruits and Vegetables	http://www.hort.cornell.edu/extension/commercial/comfrveg.html
	Vegetable Research and Extension	http://www.hort.cornell.edu/department/faculty/rangarajan/Veggie/index.html
North Carolina	Vegetable Crops	http://www.cals.ncsu.edu/hort_sci/veg/vegmain.html
Ohio	Vegetable Crops	http://extension.osu.edu/crops_and.livestock/vegetable_crops.php
	The Extension Vegetable Lab	http://www.oardc.ohio-state.edu/Kleinhenz/Stuff/tm-wrkgrp1.htm
Oklahoma	Vegetable Trial Report	http://www.okstate.edu/ag/asnr/hortla/vegtrial/index.htm
Oregon	Vegetable Production Guides	http://oregonstate.edu/Dept/NWREC/vegindex.html
Pennsylvania	Vegetable Crop Resources	http://hortweb.cas.psu.edu/extension/vegcrp.html
South Carolina	Vegetable & Fruit Program	http://virtual.clemson.edu/groups/hort/vegprog.htm
Tennessee	Field and Commercial Crops	http://www.utextension.utk.edu/publications/fieldCrops/default.asp
Texas	Vegetable Web Sites	http://aggie-horticulture.tamu.edu/extension/infolinks.html
	Vegetable IPM	http://vegipm.tamu.edu
Vermont	Vermont Vegetable and Berry Page	http://www.uvm.edu/vtvegandberry
Virginia	Vegetable-Commercial Production	http://www.ext.vt.edu/cgi-bin/WebObjects/Docs.woa/wa/getcat?cat=ir-fv-vegc

State	Website	Address/URL
Washington	Vegetable Research and Extension	http://agsyst.wsu.edu/vegtble.htm
	Vegetables	http://cecommerce.uwex.edu/showcat.asp?id=18
Ontario Canada	Vegetable Production Information	http://www.gov.on.ca/omafra/english/crops/hort/vegetable.html

02
SOME PERIODICALS FOR VEGETABLE GROWERS

American Fruit Grower
American Vegetable Grower
Florida Grower
Greenhouse Grower
Productores de Hortalizas
Meister Media Worldwide
3377 Euclid Avenue
Willoughby, OH 44094-5992
Ph (440) 942-2000
http://www.meistermedia.com

Carrot Country
http://www.columbiapublications.com/carrotcountry

Onion World
http://www.columbiapublications.com/onionworld

Potato Country
http://www.columbiapublications.com/potatocountry

The Tomato Magazine
http://www.columbiapublications.com/tomatomagazine
Columbia Publishing
413-B N. Twentieth Avenue
Yakima, WA 98902

Citrus & Vegetable Magazine
Subscription Service Center
P.O. Box 83
Tifton, GA 31793
http://www.citrusandvegetable.com

The Grower
Subscription Service Center
400 Knightsbridge Parkway
Lincolnshire, IL 60069-3613
Fax (847) 634-4373
http://www.growermagazine.com

The Packer
Circulation Department
P.O. Box 2939
Shawnee Mission, KS 66201-1339
Ph (800) 621-2845
Fax (913) 438-0657
http://www.thepacker.com

The Produce News
482 Hudson Terrace
Englewood Cliffs, NJ 07632
Ph (800) 753-9110
http://www.producenews.com

Western Grower and Shipper Magazine
P.O. Box 2130
Newport Beach, CA 92614
http://www.wga.com

Vegetable Growers News
http://www.vegetablegrowersnews.com

Spudman
http://www.spudman.com

Fresh Cut
http://www.freshcut.com

Fruit Grower News
http://www.fruitgrowersnews.com

Great American Publishing
P.O. Box 128
Sparta, MI 49345
Ph (616) 887-9008
Fax (616) 887-2666

Growing for Market
P.O. Box 3747
Lawrence, KS 66046
Ph (785) 748-0605
Fax (785) 748-0609
http://www.growingformarket.com

Potato Grower Magazine
http://www.potatogrowers.com

03
U.S. UNITS OF MEASUREMENT

Length
1 foot = 12 inches
1 yard = 3 feet
1 yard = 36 inches
1 rod = 16.5 feet
1 mile = 5280 feet

Area
1 acre = 43,560 square feet
1 section = 640 acres
1 section = 1 square mile

Volume
1 liquid pint = 16 liquid ounces
1 liquid quart = 2 liquid pints
1 liquid quart = 32 liquid ounces
1 gallon = 8 liquid pints
1 gallon = 4 liquid quarts
1 gallon = 128 liquid ounces
1 peck = 16 pints (dry)
1 peck = 8 quarts (dry)
1 bushel = 4 pecks
1 bushel = 64 pints (dry)
1 bushel = 32 quarts (dry)

Mass or Weight
1 pound = 16 ounces
1 hundredweight = 100 pounds
1 ton = 20 hundredweight
1 ton = 2000 pounds
1 unit (fertilizer) = 1% ton = 20 pounds

04
CONVERSION FACTORS FOR U.S. UNITS

Multiply	By	To Obtain
Length		
feet	12.	inches
feet	0.33333	yards
inches	0.08333	feet
inches	0.02778	yards
miles	5,280.	feet
miles	63,360.	inches
miles	1,760.	yards
rods	16.5	feet
yards	3.	feet
yards	36.	inches
yards	0.000568	miles
Area		
acres	43,560.	square feet
acres	160.	square rods
acres	4,840.	square yards
square feet	144.	square inches
square feet	0.11111	square yards
square inches	0.00694	square feet
square miles	640.	acres
square miles	27,878,400.	square feet
square miles	3,097,600.	square yards
square yards	0.0002066	acres
square yards	9.	square feet
square yards	1,296.	square inches
Volume		
bushels	2,150.42	cubic inches
bushels	4.	pecks

Multiply	By	To Obtain
bushels	64.	pints
bushels	32.	quarts
cubic feet	1,728.	cubic inches
cubic feet	0.03704	cubic yards
cubic feet	7.4805	gallons
cubic feet	59.84	pints (liquid)
cubic feet	29.92	quarts (liquid)
cubic yards	27.	cubic feet
cubic yards	46,656.	cubic inches
cubic yards	202.	gallons
cubic yards	1,616.	pints (liquid)
cubic yards	807.9	quarts (liquid)
gallons	0.1337	cubic feet
gallons	231.	cubic inches
gallons	128.	ounces (liquid)
gallons	8.	pints (liquid)
gallons	4.	quarts (liquid)
gallons water	8.3453	pounds water
pecks	0.25	bushels
pecks	537.605	cubic inches
pecks	16.	pints (dry)
pecks	8.	quarts (dry)
pints (dry)	0.015625	bushels
pints (dry)	33.6003	cubic inches
pints (dry)	0.0625	pecks
pints (dry)	0.5	quarts (dry)
pints (liquid)	28.875	cubic inches
pints (liquid)	0.125	gallons
pints (liquid)	16.	ounces (liquid)
pints (liquid)	0.5	quarts (liquid)
quarts (dry)	0.03125	bushels
quarts (dry)	67.20	cubic inches
quarts (dry)	2.	pints (dry)
quarts (liquid)	57.75	cubic inches
quarts (liquid)	0.25	gallons
quarts (liquid)	32.	ounces (liquid)
quarts (liquid)	2.	pints (liquid)

Multiply	By	To Obtain
Mass or Weight		
ounces (dry)	0.0625	pounds
ounces (liquid)	1.805	cubic inches
ounces (liquid)	0.0078125	gallons
ounces (liquid)	0.0625	pints (liquid)
ounces (liquid)	0.03125	quarts (liquid)
pounds	16.	ounces
pounds	0.0005	tons
pounds of water	0.01602	cubic feet
pounds of water	27.68	cubic inches
pounds of water	0.1198	gallons
tons	32,000.	ounces
tons	20.	hundredweight
tons	2,000.	pounds
Rate		
feet per minute	0.01667	feet per second
feet per minute	0.01136	miles per hour
miles per hour	88.	feet per minute
miles per hour	1.467	feet per minute

05
METRIC UNITS OF MEASUREMENT

Length

1 millimeter = 1,000 microns
1 centimeter = 10 millimeters
1 meter = 100 centimeters
1 meter = 1,000 millimeters
1 kilometer = 1,000 meters

Area

1 hectare = 10,000 square meters

Volume

1 liter = 1,000 milliliters

Mass or Weight

1 gram = 1,000 milligrams
1 kilogram = 1,000 grams
1 quintal = 100 kilograms
1 metric ton = 1,000 kilograms
1 metric ton = 10 quintals

06
CONVERSION FACTORS FOR SI AND NON-SI UNITS

To Convert Column 1 into Column 2, Multiply by	Column 1 SI Unit	Column 2 non-SI Unit	To Convert Column 2 into Column 1, Multiply by
Length			
0.621	kilometer, km (10^3 m)	mile, mi	1.609
1.094	meter, m	yard, yd	0.914
3.28	meter, m	foot, ft	0.304
1.0	micrometer, μ (10^{-6} m)	micron, μ	1.0
3.94×10^{-2}	millimeter, mm (10^{-3} m)	inch, in	25.4
10	nanometer, nm (10^{-9} m)	Angstrom, Å	0.1
Area			
2.47	hectare, ha	acre	0.405
247	square kilometer, km^2 $(10^3 m)^2$	acre	4.05×10^{-3}
0.386	square kilometer, km^2 $(10^3 m)^2$	square mile, m^2	2.590
2.47×10^{-4}	square meter, m^2 $(10^3 m)^2$	acre	4.05×10^3
10.76	square meter, m^2 $(10^3 m)^2$	square foot, ft^2	9.29×10^{-2}
1.55×10^{-3}	square millimeter, m^2 $(10^{-6} m)^2$	square inch, in^2	645
Volume			
6.10×10^4	cubic meter, m^3	cubic inch, in^3	1.64×10^{-5}
2.84×10^{-2}	liter, L (10^{-3} m^3)	bushel, bu	35.24

1.057	liter, L (10^{-3} m^3)	quart (liquid), qt	0.946
3.53×10^{-2}	liter, L (10^{-3} m^3)	cubic foot, ft^3	28.3
0.265	liter, L (10^{-3} m^3)	gallon	3.78
33.78	liter, L (10^{-3} m^3)	ounce (fluid), oz	2.96×10^{-2}
2.11	liter, L (10^{-3} m^3)	pint (fluid), pt	0.473
1.06	liter, L (10^{-3} m^3)	quart (liquid), qt	0.946
9.73×10^{-3}	$meter^3$, m^3	acre-inch	102.8
35.7	$meter^3$, m^3	cubic foot, ft^3	2.80×10^{-2}

Mass

2.20×10^{-3}	gram, g (10^{-3} kg)	pound, lb	454
3.52×10^{-2}	gram, g	ounce (avdp), oz	28.4
2.205	kilogram, kg	pound, lb	0.454
10^2	kilogram, kg	quintal (metric), q	10^2
1.10×10^{-3}	kilogram, kg	ton (2,000 lb), ton	907
1.102	megagram, Mg (tonne)	ton (U.S.), ton	0.907

Yield and Rate

0.893	kilogram per hectare, kg ha^{-1}	pound per acre, lb $acre^{-1}$	1.12
7.77×10^{-2}	kilogram per cubic meter, kg m^{-3}	pound per bushel, lb bu^{-1}	12.87
1.49×10^{-2}	kilogram per hectare, kg ha^{-1}	bushel per acre, 60 lb	67.19
1.59×10^{-2}	kilogram per hectare, kg ha^{-1}	bushel per acre, 56 lb	62.71
1.86×10^{-2}	kilogram per hectare, kg ha^{-1}	bushel per acre, 48 lb	53.75
0.107	liter per hectare, L ha^{-1}	gallon per acre	9.35
893	megagram per hectare, Mg ha^{-1}	pound per acre, lb $acre^{-1}$	1.12×10^{-3}
0.446	megagram per hectare, Mg ha^{-1}	ton (2,000 lb) per acre, ton $acre^{-1}$	2.24
2.10	meter per second, m s^{-1}	mile per hour	0.477

To Convert Column 1 into Column 2, Multiply by	Column 1 SI Unit	Column 2 non-SI Unit	To Convert Column 2 into Column 1, Multiply by
Specific Surface			
10	square meter per kilogram, $m^2\ kg^{-1}$	square centimeter per gram, $cm^2\ g^{-1}$	0.1
10^3	square meter per kilogram, $m^2\ kg^{-1}$	square millimeter per gram, $mm^2\ g^{-1}$	10^{-3}
Pressure			
9.90	megapascal, MPa (10^6 Pa)	atmosphere	0.101
10	megapascal, MPa (10^6 Pa)	bar	0.1
1.00	megagram per cubic meter, $Mg\ m^{-3}$	gram per cubic centimeter, $g\ cm^{-3}$	1.00
2.09×10^{-2}	pascal, Pa	pound per square foot, $lb\ ft^{-2}$	47.9
1.45×10^{-4}	pascal, Pa	pound per square inch, $lb\ in^{-2}$	6.90×10^3
Temperature			
1.00 (°K – 273)	Kelvin, K	Celsius, °C	1.00 (°C + 273)
(9/5°C) + 32	Celsius, °C	Fahrenheit, °F	5/9 (°F – 32)

Energy, Work, Quantity of Heat

9.52×10^{-4}	joule, J	British thermal unit, Btu	1.05×10^{3}
0.239	joule, J	calorie, cal	4.19
10^{7}	joule, J	erg	10^{-7}
0.735	joule, J	foot-pound	1.36
2.387×10^{-5}	joule per square meter, J m^{-2}	calorie per square centimeter (langley)	4.19×10^{4}
10^{5}	newton, N	dyne	10^{-5}
1.43×10^{-3}	watt per square meter, W m^{-2}	calorie per square centimeter minute (irradiance), cal cm^{-2} min^{-1}	698

Transpiration and Photosynthesis

3.60×10^{-2}	milligram per square meter second, mg m^{-2} s^{-1}	gram per square decimeter hour, g dm^{-2} h^{-1}	27.8
5.56×10^{-3}	milligram (H_2O) per square meter second, mg m^{-2} s^{-1}	micromole (H_2O) per square centimeter second, μmol cm^{-2} s^{-1}	180
10^{-4}	milligram per square meter second, mg m^{-2} s^{-1}	milligram per square centimeter second, mg cm^{-2} s^{-1}	10^{4}
35.97	milligram per square meter second, mg m^{-2} s^{-1}	milligram per square decimeter hour, mg dm^{-2} h^{-1}	2.78×10^{-2}

Angle

5.73	radian, rad	degrees (angle),°	1.75×10^{-2}

To Convert Column 1 into Column 2, Multiply by	Column 1 SI Unit	Column 2 non-SI Unit	To Convert Column 2 into Column 1, Multiply by
Water Measurement			
9.73×10^{-3}	cubic meter, m^3	acre-inches, acre-in	102.8
9.81×10^{-3}	cubic meter per hour, $m^3\ h^{-1}$	cubic feet per second, $ft^3\ s^{-1}$	101.9
4.40	cubic meter per hour, $m^3\ h^{-1}$	U.S. gallons per minute, gal min^{-1}	0.227
8.11	hectare-meters, ha-m	acre-feet, acre-ft	0.123
97.28	hectare-meters, ha-m	acre-inches, acre-in	1.03×10^{-2}
8.1×10^{-2}	hectare-centimeters, ha-cm	acre-feet, acre-ft	12.33
		Concentrations	
1	centimole per kilogram, cmol kg^{-1} (ion exchange capacity)	milliequivalents per 100 grams, meq 100 g^{-1}	1
0.1	gram per kilogram, g kg^{-1}	percent, %	10
1	megagram per cubic meter, Mg m^{-3}	gram per cubic centimeter, g cm^{-3}	1
1	milligram per kilogram, mg kg^{-1}	parts per million, ppm	1

07
CONVERSIONS FOR RATES OF APPLICATION

1 ton per acre = 20.8 grams per square foot
1 ton per acre = 1 pound per 21.78 square feet
1 ton per acre furrow slice (6-inch depth) = 1 gram per 1,000 grams soil
1 gram per square foot = 96 pounds per acre
1 pound per acre = 0.0104 grams per square foot
1 pound per acre = 1.12 kilograms per hectare
100 pounds per acre = 0.2296 pounds per 100 square feet
grams per square foot × 96 = pounds per acre
kilograms per 48 square feet = tons per acre
pounds per square feet × 21.78 = tons per acre

08
WATER AND SOIL SOLUTION—CONVERSION FACTORS

Concentration

1 decisiemens per meter (dS/m) = 1 millimho per centimeter (mmho/cm)
1 decisiemens per meter (dS/m) = approximately 640 milligrams per liter salt
1 part per million (ppm) = 1/1,000,000
1 percent = 0.01 or 1/100
1 ppm × 10,000 = 1 percent
ppm × 0.00136 = tons per acre-foot of water
ppm = milligrams per liter
ppm = 17.12 × grains per gallon
grains per gallon = 0.0584 × ppm
ppm = 0.64 × micromhos per centimeter (in range of 100–5,000 micromhos per centimeter)
ppm = 640 × millimhos per centimeter (in range of 0.1–5.0 millimhos per centimeter)
ppm = grams per cubic meter
mho = reciprocal ohm
millimho = 1000 micromhos
millimho = approximately 10 milliequivalents per liter (meq/liter)
milliequivalents per liter = equivalents per million
millimhos per centimeter = EC × 10^3 (EC × 1,000) at 25°C (EC = electrical conductivity)
micromhos per centimeter = EC × 10^6 (EC × 1,000,000) at 25°C
millimhos per centimeter = 0.1 siemens per meter
millimhos per centimeter = (EC × 10^3) = decisiemens per meter (dS/m)
1,000 micromhos per centimeter = approximately 700 ppm
1,000 micromhos per centimeter = approximately 10 milliequivalents per liter
1,000 micromhos per centimeter = 1 ton of salt per acre-foot of water
milliequivalents per liter = 0.01 × (EC × 10^6) (in range of 100–5,000 micromhos per centimeter)
milliequivalents per liter = 10 × (EC × 10^3) (in range of 0.1–5.0 millimhos per centimeter)

Pressure and Head

1 atmosphere at sea level = 14.7 pounds per square inch
1 atmosphere at sea level = 29.9 inches of mercury
1 atmosphere at sea level = 33.9 feet of water

1 atmosphere = 0.101 megapascal (MPa)
1 bar = 0.10 megapascal (MPa)
1 foot of water = 0.8826 inch mercury
1 foot of water = 0.4335 pound per square inch
1 inch of mercury = 1.133 feet water
1 inch of mercury = 0.4912 pound per square inch
1 inch of water = 0.07355 inch mercury
1 inch of water = 0.03613 pound per square inch
1 pound per square inch = 2.307 feet water
1 pound per square inch = 2.036 inches mercury
1 pound per square foot = 47.9 pascals

Weight and Volume (U.S. Measurements)
1 acre-foot of soil = about 4,000,000 pounds
1 acre-foot of water = 43,560 cubic feet
1 acre-foot of water = 12 acre-inches
1 acre-foot of water = about 2,722,500 pounds
1 acre-foot of water = 325,851 gallons
1 cubic-foot of water = 7.4805 gallons
1 cubic foot of water at 59°F = 62.37 pounds
1 acre-inch of water = 27,154 gallons
1 gallon of water at 59°F = 8.337 pounds
1 gallon of water = 0.1337 cubic foot or 231 cubic inches

Flow (U.S. Measurements)
1 cubic foot per second = 448.8 gallons per minute
1 cubic foot per second = about 1 acre-inch per hour
1 cubic foot per second = 23.80 acre-inches per hour
1 cubic foot per second = 3600 cubic feet per hour
1 cubic foot per second = about 7½ gallons per second
1 gallon per minute = 0.00223 cubic feet per second
1 gallon per minute = 0.053 acre-inch per 24 hours
1 gallon per minute = 1 acre-inch in 4½ hours
1000 gallons per minute = 1 acre-inch in 27 minutes
1 acre-inch per 24 hours = 18.86 gallons per minute
1 acre-foot per 24 hours = 226.3 gallons per minute
1 acre-foot per 24 hours = 0.3259 million gallons per 24 hours

U.S.-Metric Equivalents
1 cubic meter = 35.314 cubic feet
1 cubic meter = 1.308 cubic yards

1 cubic meter = 1000 liters
1 liter = 0.0353 cubic feet
1 liter = 0.2642 U.S. gallon
1 liter = 0.2201 British or Imperial gallon
1 cubic centimeter = 0.061 cubic inch
1 cubic foot = 0.0283 cubic meter
1 cubic foot = 28.32 liters
1 cubic foot = 7.48 U.S. gallons
1 cubic foot = 6.23 British gallons
1 cubic inch = 16.39 cubic centimeters
1 cubic yard = 0.7645 cubic meter
1 U.S. gallon = 3.7854 liters
1 U.S. gallon = 0.833 British gallon
1 British gallon = 1.201 U.S. gallons
1 British gallon = 4.5436 liters
1 acre-foot = 43,560 cubic feet
1 acre-foot = 1,233.5 cubic meters
1 acre-inch = 3,630 cubic feet
1 acre-inch = 102.8 cubic meters
1 cubic meter per second = 35.314 cubic feet per second
1 cubic meter per hour = 0.278 liter per second
1 cubic meter per hour = 4.403 U.S. gallons per minute
1 cubic meter per hour = 3.668 British gallons per minute
1 liter per second = 0.0353 cubic feet per second
1 liter per second = 15.852 U.S. gallons per minute
1 liter per second = 13.206 British gallons per minute
1 liter per second = 3.6 cubic meters per hour
1 cubic foot per second = 0.0283 cubic meter per second
1 cubic foot per second = 28.32 liters per second
1 cubic foot per second = 448.8 U.S. gallons per minute
1 cubic foot per second = 373.8 British gallons per minute
1 cubic foot per second = 1 acre-inch per hour (approximately)
1 cubic foot per second = 2 acre-feet per day (approximately)
1 U.S. gallon per minute = 0.06309 liter per second
1 British gallon per minute = 0.07573 liter per second

Power and Energy

1 horsepower = 550 foot-pounds per second
1 horsepower = 33,000 foot-pounds per minute
1 horsepower = 0.7457 kilowatts
1 horsepower = 745.7 watts

1 horsepower-hour = 0.7457 kilowatt-hour
1 kilowatt = 1.341 horsepower
1 kilowatt-hour = 1.341 horsepower-hours
1 acre-foot of water lifted 1 foot = 1.372 horsepower-hours of work
1 acre-foot of water lifted 1 foot = 1.025 kilowatt-hours of work

09
HEAT AND ENERGY EQUIVALENTS AND DEFINITIONS

1 joule = 0.239 calorie
1 joule = Nm (m^2 kg s^{-2})
temperature of maximum density of water = 3.98°C (about 39°F)
1 British thermal unit (Btu) = heat needed to change 1 pound water at maximum density of 1°F
1 Btu = 1.05506 kilojoules (kJ)
1 Btu/lb = 2.326 kJ/kg
1 Btu per minute = 0.02356 horsepower
1 Btu per minute = 0.01757 kilowatts
1 Btu per minute = 17.57 watts
1 horsepower = 42.44 Btu per minute
1 horsepower-hour = 2547 Btu
1 kilowatt-hour = 3415 Btu
1 kilowatt = 56.92 Btu per minute
1 pound water at 32°F changed to solid ice requires removal of 144 Btu
1 pound ice in melting takes up to 144 Btu
1 ton ice in melting takes up to 288,000 Btu

USEFUL WEBSITES FOR UNITS AND CONVERSIONS

Weights, Measures, and Conversion Factors for Agricultural Commodities and Their Products (USDA Agricultural Handbook 697, 1992), http://www.ers.usda.gov/publications/ah697

Metric Conversions (2002), http://www.extension.iastate.edu/agdm/wholefarm/html/c6-80.html

A Dictionary of Units—Part 1 (2004), http://www.projects.ex.ac.uk/trol/dictunit/dictunit1.htm

A Dictionary of Units—Part 2 (2004), http://www.projects.ex.ac.uk/trol/dictunit/dictunit2.htm

U.S. Metric Association (2005), http://lamar.colostate.edu/~hillger

INDEX

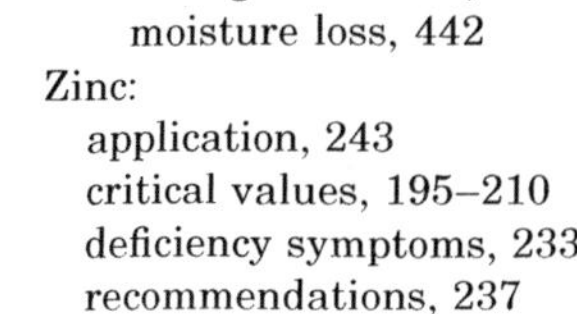

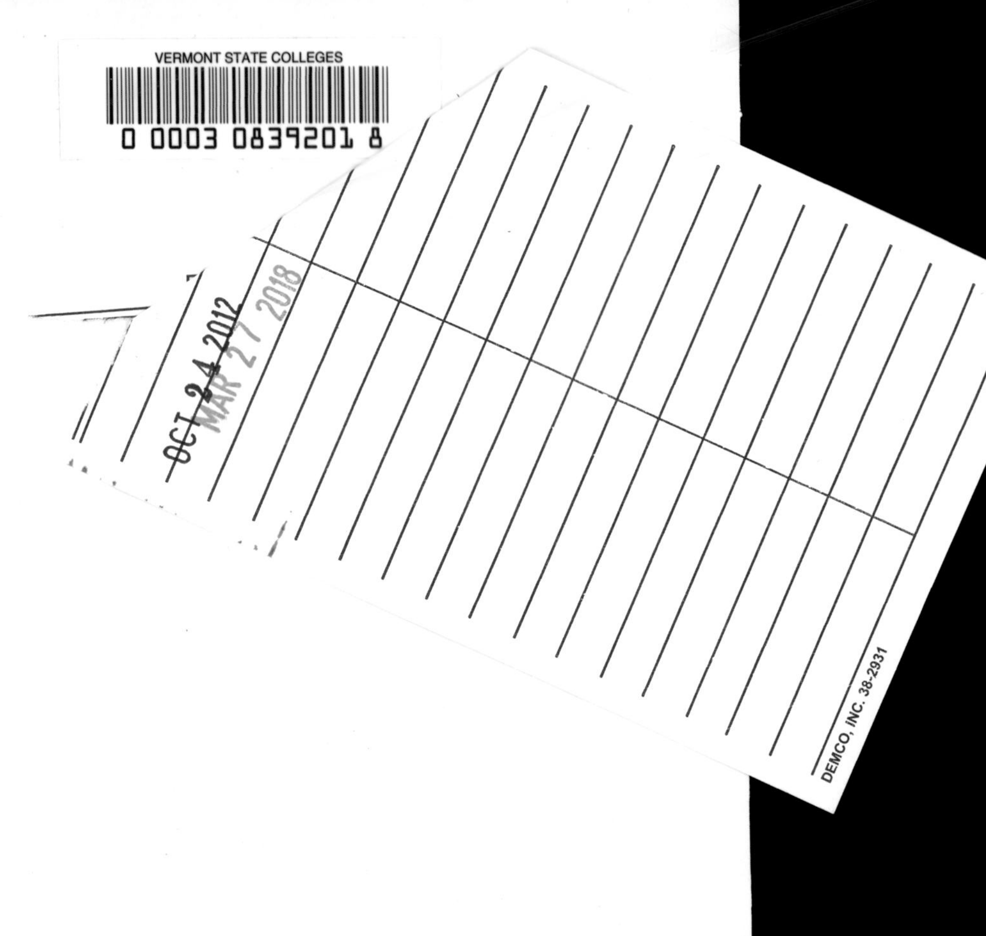
OCT 24 2012
MAR 27 2018
DEMCO, INC. 38-2931